Plant Functional Traits for Improving Productivity

Narendra Kumar • Hukum Singh
Editors

Plant Functional Traits for Improving Productivity

Springer

Editors
Narendra Kumar
College of Agriculture, Central Agriculture University (I)
Krydemkulai, Meghalaya, India

Hukum Singh
Plant Physiology Discipline,
Forest Genetics and Tree Improvement Division
Forest Research Institute
Dehradun, Uttarakhand, India

ISBN 978-981-97-1509-1 ISBN 978-981-97-1510-7 (eBook)
https://doi.org/10.1007/978-981-97-1510-7

This Springer imprint is published by the registered company Springer Nature Singapore Pte Ltd.
The registered company address is: 152 Beach Road, #21-01/04 Gateway East, Singapore 189721, Singapore

Preface

The plant functional traits-based approach offers a valuable framework for improving crop productivity in agricultural systems. This approach enables researchers to select and develop crop varieties with desirable traits that enhance productivity by focusing on the fundamental characteristics influencing a plant's performance, growth, and response to changing climatic and environmental conditions. Plant functional traits provide insights into how plants interact with their environment. By studying the relationships between specific traits and environmental factors such as temperature, water availability, or nutrient levels, researchers can better understand how plants respond and adapt to different conditions. This understanding allows for selecting crop varieties better suited to specific environments, improving productivity. Plant functional traits include leaf area, root architecture, photosynthetic capacity, and nutrient uptake efficiency. Researchers can develop varieties that exhibit improved growth, resource utilization, and stress tolerance by identifying and selecting traits positively correlated with high crop productivity. This targeted selection helps optimize crop performance and ultimately leads to increased yields. Moreover, climate change and environmental variability pose significant agricultural challenges. The plant functional traits-based approach can help address these challenges by selecting and breeding traits that enhance a plant's resilience and adaptability. Traits associated with drought tolerance, heat resistance, or pest and disease resistance can help crops thrive in changing conditions, ensuring more stable and productive agricultural systems. In addition, efficient utilization of water, nutrients, and light resources is crucial for sustainable agriculture. Plant functional traits that promote resource use efficiency, such as deep and extensive root systems or high photosynthetic rates, can improve crop productivity while minimizing resource inputs. This approach helps optimize resource allocation and reduces the environmental impact of agriculture. The plant functional traits-based approach aligns well with precision agriculture practices. Farmers can implement targeted management practices, such as precision irrigation or nutrient application, tailored to the needs of specific crop varieties by understanding the relationships between specific traits and environmental conditions. This precision approach optimizes resource use, minimizes wastage, and enhances overall crop productivity. Incorporating plant functional traits in breeding programs enhances the efficiency of genetic improvement efforts. Breeders can accelerate the development of crop varieties with desired characteristics by focusing on specific traits. Integrating genetic tools, such as marker-assisted or genomic selection, further enhances the precision and speed of trait selection and breeding. This book will provide a valuable framework for improving crop productivity, including agroforestry crops. The researchers and breeders can develop crop varieties that are better adapted to their environments, exhibit improved resource use efficiency, and display enhanced stress tolerance by understanding, selecting, and breeding for desirable traits. This approach contributes to sustainable agriculture, increased food production, and resilience in changing climatic and environmental conditions.

Krydemkulai, Meghalaya, India Narendra Kumar
Dehradun, Uttarakhand, India Hukum Singh

Contents

Editors and Contributors

About the Editors

Narendra Kumar earned his PhD in plant physiology from GB Pant University of Agriculture and Technology, Pantnagar, India. Later, he served as an assistant professor in various reputed academic institutes, especially Surajmal Agarwal (Pvt.) Kanya Mahavidyalaya, Kichha US Nagar, Dev Bhoomi Group of Institute, Dehradun, India; Uttaranchal University, Dehradun, India; Doon (PG) College of Agriculture Science and Technology, Dehradun, India. He also worked as a research associate at Forest Research Institute, Dehradun, India. His research areas focus primarily on plant physiology, ecophysiology, plant response to abiotic stresses, climate change adaptation and mitigation, functional traits, and climate change modeling. He has published numerous articles in national and international peer-reviewed journals. He is currently employed as Assistant Professor, College of Agrculture, Central Agriculture University (I), Kyrdemkulai, Meghalaya.

Hukum Singh is a scientist at Forest Genetics and Tree Improvement Division, Forest Research Institute, Dehradun, India, and has been actively involved in research and teaching. He is a member of various national and international scientific societies/organizations. His research areas focus primarily on eco-physiology, plant response to abiotic stresses, green technology, climate change adaptation and mitigation, functional traits, carbon-water-energy flux, and climate change modeling. He has published many articles in national and international peer-reviewed reputed journals.

Contributors

Anjana A. V Department of Genetics and Plant Breeding, C.P. College of Agriculture, S.D. Agricultural University, S.K. Nagar, Gujarat, India

Arushi Arora Department of Plant Breeding and Genetics, Punjab Agricultural University, Ludhiana, Punjab, India

Pardeep Bansal Department of Botany, K.L.P. College, Rewari, Haryana, India

Santan Barthwal Plant Physiology Discipline, Forest Genetics and Tree Improvement Division, Forest Research Institute, Dehradun, India

Rajneesh Bhardwaj School of Agriculture, Graphic Era Hill University Dehradun, Dehradun, Uttarakhand, India

Amit Bijlwan Department of Agrometeorology, G.B. Pant University of Agriculture and Technology, Pantnagar, Uttarakhand, India

Satish Chandra Department of Botany, Government Degree College Tyuni, Dehradun, Uttarakhand, India

Divya Chaudhary Department of Genetics and Plant Breeding, G.B. Pant University of Agriculture and Technology, Pantnagar, Uttarakhand, India

Ajay Chauhan Centre of Excellence on Sustainable Land Management, Indian Council of Forestry Research and Education, Dehradun, India

Anjali Chauhan School of Applied and Life Sciences (SALS), Uttaranchal University, Dehradun, Uttarakhand, India

Harsh K. Chauhan Department of Botany, D.S.B. Campus, Kumaun University, Nainital, Uttarakhand, India

Rahul Chhawri Department of Botany, Maharshi Dayanand University, Rohtak, Haryana, India

Akhil Donga Department of Genetics and Plant Breeding, C.P. College of Agriculture, S.D. Agricultural University, S.K. Nagar, Gujarat, India

Sonali Gaur Department of Genetics and Plant Breeding, G.B. Pant University of Agriculture and Technology, Pantnagar, India

A. R. S. S. H. Gupta Department of Genetics and Plant Breeding, C.P. College of Agriculture, S.D. Agricultural University, S.K. Nagar, Gujarat, India

Siddhant Gupta Department of Agrometeorology, G.B. Pant University of Agriculture and Technology, Pantnagar, Uttarakhand, India

Roomesh Kumar Jena Department of Agrometeorology, G.B. Pant University of Agriculture and Technology, Pantnagar, Uttarakhand, India

Ankita Jha ICAR-Indian Institute of Water Management, Bhubaneswar, Odisha, India

Amit Joshi Department of Biochemistry, Kalinga University, Naya-Raipur, Chhattisgarh, India

Hem C. Joshi Department of Botany, D.S.B. Campus, Kumaun University, Nainital, Uttarakhand, India

Jyoti Forest Biotechnology Division, Genetics and Tree Improvement Division, Forest Research Institute, Dehradun, India

Chandra Kanta School of Applied and Life Sciences (SALS), Uttaranchal University, Dehradun, Uttarakhand, India

Rowndel Khwairakpam School of Agriculture, Graphic Era Hill University Dehradun, Dehradun, Uttarakhand, India

Abhishek Kumar Forest Research Institute, Dehradun, India

Amit Kumar Department of Forestry, North Eastern Hill University, Tura Campus, Tura, Meghalaya, India

Garima Kumari Forest Ecology and Climate Change Division, Forest Research Institute, Dehradun, India

Kishan Kumar Plant Physiology Discipline, Forest Genetics and Tree Improvement Division, Forest Research Institute, Dehradun, India

Manoj Kumar GIS Center, Forest Research Institute, Dehradun, India

Mukesh Kumar Department of Vegetable Science, G.B. Pant University of Agriculture and Technology, Pantnagar, Uttarakhand, India

Narendra Kumar College of Agriculture, Central Agriculture University (I), Krydemkulai, Meghalaya, India

Jharna Maiti Department of Biochemistry, Kalinga University, Naya-Raipur, Chhattisgarh, India

Apurva Malik Forest Ecology and Climate Change Division, Forest Research Institute, Dehradun, India

K. C. Meena Department of Botany, Maharshi Dayanand University, Rohtak, Haryana, India

Haryana Forest Department, Panchkula, Haryana, India

Arvind Singh Negi School of Agriculture, Graphic Era Hill University Dehradun, Dehradun, Uttarakhand, India

Neeraj Pal Department of Molecular Biology and Genetic Engineering, G.B. Pant University of Agriculture and Technology, Pantnagar, Uttarakhand, India

Prabha Pant Department of Botany, D.S.B. Campus, Kumaun University, Nainital, Uttarakhand, India

Parth J. Patel Department of Genetics and Plant Breeding, C.P. College of Agriculture, S.D. Agricultural University, S.K. Nagar, Gujarat, India

Km. Prachi Patanjali Herbal Research Division, Patanjali Research Foundation, Haridwar, Uttarakhand, India

Bhamare Deepak Prashant School of Agricultural Biotechnology, Punjab Agricultural University, Ludhiana, Punjab, India

Rajeev Ranjan Department of Agrometeorology, G.B. Pant University of Agriculture and Technology, Pantnagar, Uttarakhand, India

Amrender Singh Rao Centre of Excellence on Sustainable Land Management, Indian Council of Forestry Research and Education, Dehradun, India

Department of Botany, Maharshi Dayanand University, Rohtak, Haryana, India

Balwant Rawat School of Agriculture, Graphic Era Hill University Dehradun, Dehradun, Uttarakhand, India

Tannu Ruhil Plant Physiology Discipline, Forest Genetics and Tree Improvement Division, Forest Research Institute, Dehradun, India

Ishwar Prakash Sharma Patanjali Herbal Research Division, Patanjali Research Foundation, Haridwar, Uttarakhand, India

Ritu Sharma Department of Genetics and Plant Breeding, C.P. College of Agriculture, S.D. Agricultural University, S.K. Nagar, Gujarat, India

Shreya Department of Genetics and Plant Breeding, C.P. College of Agriculture, S.D. Agricultural University, S.K. Nagar, Gujarat, India

Hukum Singh Plant Physiology Discipline, Forest Genetics and Tree Improvement Division, Forest Research Institute, Dehradun, India

Manish Singh Forest Ecology and Climate Change Division, Forest Research Institute, Dehradun, India

Jwalpa Tyagi GIS Center, Forest Research Institute, Dehradun, India

Sreya Venadan Department of Biotechnology, ICAR Directorate of Onion and Garlic Research, Pune, Maharashtra, India

Shulbhi Verma Department of Genetics and Plant Breeding, C.P. College of Agriculture, S.D. Agricultural University, S.K. Nagar, Gujarat, India

Surender Singh Yadav Department of Botany, Maharshi Dayanand University, Rohtak, Haryana, India

Santhoshi Yalla Department of Genetics and Plant Breeding, C.P. College of Agriculture, S.D. Agricultural University, S.K. Nagar, Gujarat, India

Abbreviations

2AP	2-Acetyl-1-pyrroline
ABA	Abscisic acid
ABMs	Agent-based models
AlaAT	Alanine aminotransferase
AN3	ANGUSTIFOLIA3
ANNs	Artificial neural networks
ANPP	Aboveground net primary productivity
Bt	*Bacillus thuringiensis*
Cas	CRISPR-associated protein
CER9	ECERIFERUM9
CESA3	Cellulose synthase A3
CH_4	Methane
CO_2	Carbon dioxide
CRISPR	Clustered regularly interspaced short palindromic repeats
CWM	Community-weighted mean
CWR	Crop wild relatives
Cyt b6f	Cytochrome b6f
DH	Doubled haploid
DRO1	Deeper rooting 1
EPF	Epidermal patterning factor
EPFL	Epidermal patterning factor-like
FokI	Flavobacterium okeanokoites I
GBS	Genotyping by sequencing
GE	Genetically engineered
GEBV	Genomic estimated breeding values
GEI	Genotype environment interactions
GET	Gene editing techniques
GHG	Greenhouse gas
GS	Genomic selection
GWAS	Genome-wide association studies
HSFs	Heat stress transcription factors
HSPs	Heat shock proteins
HTP	High-throughput phenotyping
IBMs	Individual-based models

ICPs	Insecticidal proteins
IoT	Integrating Internet of Things
IPCC	Intergovernmental Panel on Climate Change
LA	Leaf area
LAI	Leaf area index
LES	Leaf economic spectrum
LMA	Leaf mass per area
LNC	Leaf nitrogen content
LTR	Long-term response
MAB	Marker-assisted backcrossing
MAI	Marker-assisted introgression
MAS	Marker-assisted selection
MGIDI	Multi-trait genotype-ideotype distance index
MRI	Magnetic resonance imaging
N_2O	Nitrous oxide
NGS	Next-generation sequencing
NIP	Nodulin 26-like intrinsic proteins
NO^{3-}	Nitrate ions
NUE	Nutrient use efficiency
NUpE	Nitrogen uptake efficiency
NUtE	Nitrogen utilization efficiency
OsBADH2	Betaine aldehyde dehydrogenase
PAM	Protospacer adjacent motif
pegRNA	Prime editing guide RNA
PET	Positron emission tomography
PFTs	Plant functional traits
PIP	Plasma membrane intrinsic proteins
PIs	Protease inhibitors
PQ	Plastoquinone
PS I	Photosystem I
QTL	Quantitative trait loci
RGA	Rapid generation advance
RGR	Relative growth rate
RNAi	RNA interference
ROS	Reactive oxygen species
SAR	Systemic acquired resistance
SDD	Stomatal density and distribution
SDMs	Species distribution models
sgRNA	Single guide RNA
SLA	Specific leaf area
SM	Seed mass
SNP	Single nucleotide polymorphism
SRL	Specific root length
SRL	specific root lengths
SSD	Single seed descent

TALENs	Transcription activator-like effector nucleases
TE	Transpiration efficiency
TILLING	Targeting induced local lesions in genomes
TMM	Too many mouths
VInv	Vacuolar invertase
WGS	Whole genome sequencing
WUE	Water use efficiency
X-ray CT	X-ray computed tomography

Plant Functional Trait: Concept and Significance

1

Jwalpa Tyagi and Manoj Kumar

Abstract

Plant functional traits refer to physio-morpho-phenological qualities that indirectly impact fitness or performance by influencing the life history strategies of plants. Unfortunately, climate change increase stressors that are making plants less resilient. This changes the structure of forests and the ecosystem services they provide, which causes these traits to change over time. Functional traits have been widely used in two key areas of research; firstly, they are employed to describe how communities respond to environmental changes, and secondly, functional traits are used to analyze the impact of community shifts on processes of the ecosystem. The researchers can assess how these alterations influence the functioning of ecosystems by examining changes in traits within a community. Numerous methodologies have been elucidated in analyzing pre-determined trait sets in relation to the environment. This chapter explores the diverse aspects of plant functional traits, emphasizing their role in shaping ecosystems and addressing the pressing challenges of our time. Through an integrative approach that combines ecological research, data synthesis, and technological advancements, this chapter aims to deepen our understanding of plant functional traits and their significance or implications for the sustainable management of natural resources. Hence, measuring and understanding traits enhance our comprehension of ecological processes, contributing to advancing conservation and restoration efforts.

Keywords

Plant functional trait · Ecosystem services · Community shifts · Sustainable management · Climate change

J. Tyagi (✉) · M. Kumar
GIS Center, Forest Research Institute, Dehradun, India

N. Kumar, H. Singh (eds.), *Plant Functional Traits for Improving Productivity*,
https://doi.org/10.1007/978-981-97-1510-7_1

1.1 Introduction

Plants, as the primary producers in terrestrial ecosystems, play a crucial role in shaping the Earth's environment and supporting diverse life-forms. Understanding the functional traits of plants is essential for unraveling the intricacies of their ecological roles and responses to environmental changes. Plant functional traits encompass a wide range of characteristics that influence the fitness and performance of plants in their respective habitats. These traits include morphological, physiological, and phenological attributes that dictate how plants interact with their environments (Singh et al. 2014, 2021, 2024; Kumar et al. 2020a, b, c). The study of plant functional traits provides valuable insights into ecosystem dynamics, biodiversity patterns, and ecosystem services.

One of the key aspects of plant functional traits is their role in mediating ecological processes and interactions. For instance, leaf traits such as specific leaf area (SLA) and leaf nitrogen content are indicative of a plant's strategy for resource acquisition and utilization (Singh et al. 2010a, b; Kumar et al. 2021a, b). These traits influence photosynthetic rates, nutrient cycling, and ecosystem productivity, thereby shaping the structure and functioning of ecosystems (Wright et al. 2004). Root traits, such as root depth and root biomass, are critical for nutrient uptake, water absorption, and soil stabilization, affecting plant performance and ecosystem stability (Liu et al. 2010). Understanding how these traits vary among plant species and in response to environmental gradients is essential for predicting ecosystem responses to global change.

The significance of studying plant functional traits extends beyond ecological research and has implications for biodiversity conservation and ecosystem management. Plant traits are closely linked to species distribution patterns and community assembly processes. The concept of trait-based approaches allows researchers to predict how plant communities may respond to disturbances, land-use changes, and climate variability (Violle et al. 2007; Kumar et al. 2022). Furthermore, trait-based assessments contribute to the identification of keystone species and the development of effective conservation strategies (Díaz et al. 2013). Integrating functional trait information into biodiversity monitoring and management practices enhances our ability to preserve and restore ecosystems.

In recent years, advances in technology and analytical techniques have facilitated large-scale data collection and synthesis of plant functional trait information. Global initiatives, such as the TRY plant trait database (Kattge et al. 2011a), have compiled trait data from a vast array of plant species worldwide. These databases enable researchers to conduct meta-analyses, assess trait-environment relationships, and generate predictive models for future scenarios. The wealth of available trait data contributes to a more comprehensive understanding of plant functioning and its implications for ecosystem services. Essential elements of plant biology and plant functional traits affect community dynamics, ecosystem services, and ecological processes. This chapter aims to investigate the various facets of functional traits in plants, highlighting their significance in the formation of ecosystems and tackling the most important issues of our day. This chapter intends to expand our knowledge of plant functional traits and their significance for the sustainable management of natural resources through an integrative strategy that integrates ecological research, data synthesis, and technology improvements.

1.2 The Evolution of Plant Functional Traits

Since 2003, there has existed a growing focus on examining the development of plant functional traits via natural selection. Scientists have utilized different methods, including resurrection ecology (Franks et al. 2018), artificial selection (Delph et al. 2011), and experimental evolution, to monitor the alterations in functional traits throughout multiple generations. These methods have also been used to examine limitations on the development of plant functional trait modifications. Sheth and Angert (2016) utilized artificial selection to demonstrate that *Mimulus cardinalis* may be less able to adapt at its northern area boundary than at its southern area boundary if there is insufficient genetic variety during flowering time. While the frequency of these investigations is insufficient for drawing assumptions about the principles controlling the evolution of plant functional traits, they have produced diverse perspectives on the emergence of these traits in particular study systems.

Since 2003, a number of studies have been published that have examined the link between plant functional traits and species-level fitness in addition to individual-level research (Poorter et al. 2008; Visser et al. 2016). The thought behind many of these cross-species research is that changes in the life cycle and resource availability lead to the repeated development of functional trait combinations that represent distinct ecological approaches (trait spectrum) (Table 1.1). Trait spectra offer insights into the correlation between functional traits and fitness by establishing a direct connection between ecological processes and patterns of resource allocation in plants (Donovan et al. 2011). Research conducted across different species to investigate the expected correlation between these characteristics and reproductive success has yielded conflicting findings (Swenson et al. 2020). Some studies have suggested that alterations in functional features can serve as indicators for changes in growth rate or mortality, whereas other studies have not found such associations. There seems to be evidence that the number of researchers with training in both evolutionary biology and ecophysiology has increased since 2003. First, there is an unprecedented volume of studies discussing how functional traits in plants have evolved. More than half of the 11,300 publications published on the advancement of plant functional traits after 2003 were released after 2015 (Fig. 1.1). The coming decades have a great deal of promise for expanding our knowledge of how plant functional traits have evolved, given the increasing rate of publication.

As we go toward a more comprehensive integration between plant evolutionary and ecophysiology biology, the previously held assumptions regarding the fundamental steps and processes of functional trait evolution may become less precise. This observation aligns with the historical application of plant ecophysiology, in which the presumed ecological importance of certain features has not always been substantiated by actual investigation. Modeling and empirical investigations

Table 1.1 Possibilities for the investigation of the plant functional trait evolution

Studies giving future directions	References
Understanding how the size of the entire genome affects characteristics and the genetic composition of useful qualities other than the few well-researched traits, as well as how both genome size and architecture may limit the evolution of traits or allow for quick adaptation in response to selection	Guilherme Pereira and Des Marais (2020)
Determining if selection truly operates in a general way on traits and comprehending the proportional influence of abiotic and biotic perticular pressures on the development of these traits in plants	Caruso et al. (2020), Petipas et al. (2020)
By concentrating on individuals within a specific environmental setting, their functional traits can be more effectively linked to the vital rates and demographic changes that characterize fitness	Swenson et al. (2020)
Examining the intricacies of less studied belowground traits, such as how mutualisms influence functional trait variation and the root microbiome's function as an expanded phenotype	Maherali (2020), Petipas et al. (2020)
A deeper comprehension of functional trait plasticity, including how plasticity evolved, the way trait plasticity flows into abiotic and biotic relationships, and the creation of more effective techniques for integrating plasticity toward phylogenetic comparative approaches	Petipas et al. (2020)
Integrating processes of trait-trait correlations across dimensions, including intraspecific variation between and within populations, interspecific variance between species, and variation within individuals over place and time	Mason and De Bello (2013), Caruso et al. (2020)

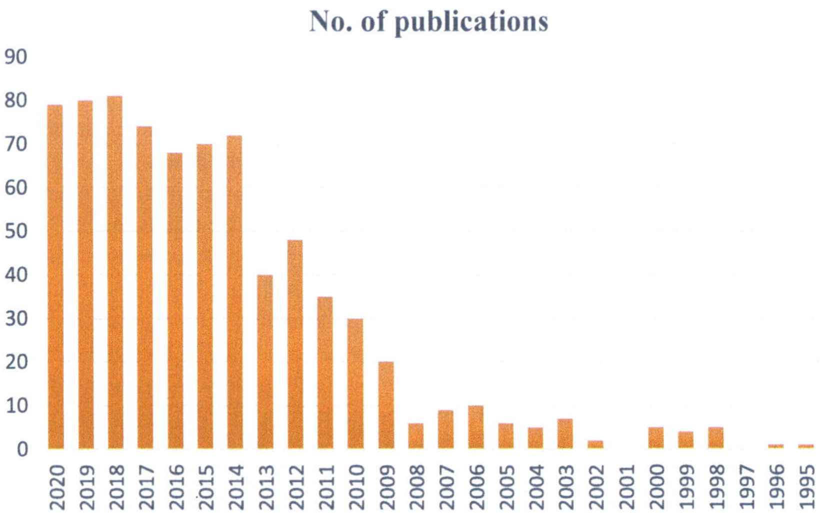

Fig. 1.1 Quantification of articles on the evolution of plant functional traits in the last 25 years

indicate that desert cactus spines most likely developed for thermoregulation, rather than for defense against parched mammals, as previously thought (Chave et al. 2009). Currently, there are two main types of extensive research that have significantly influenced the advancement of theories related to the evolution of functional traits in plants. The first type involves models that examine the growth of functional traits in phylogenies containing thousands of species, as demonstrated by studies conducted by Moray et al. (2016). The second type of research focuses on illustrating the variation of functional traits in different biomes or along environmental gradients, as shown by studies conducted by Moles et al. (2014) and Sterck et al. (2014). When developing hypotheses about the evolutionary process, conducting comprehensive investigations such as this one is an ideal starting point (Weber and Agrawal 2012; Shipley et al. 2016). Nevertheless, it is necessary to do smaller-scale empirical or observational research in order to authenticate these concepts (DeMalach et al. 2022).

1.3 The Concept of Plant Functional Traits

Functional ecology is the study of species' physiological and anatomical features with the objective of better understanding of ecological patterns and the processes leading up to community assembly (Keddy 1992). Since then, traits have been added to evaluate functional diversity and link community structure to ecosystem-level processes (Nalley et al. 2014). Plant functional traits are defined and viewed from a variety of angles (McGill et al. 2006; Violle et al. 2007). "A well-defined, measurable assets of organisms, usually measured at the individual level and used comparatively across species," is how a trait is commonly defined. According to McGill et al. (2006), functional traits are those that have a significant impact on an organism's performance. Plant functional traits are characteristics that may be observed and measured that relate a plant's physiological, morphological, and phenological features to its functions (Schellberg and Pontes 2012). Plant functional traits include stem and belowground traits such as specific root length or bark thickness; whole plant traits such as plant height or life form; leaf traits such as leaf phenology, specific leaf area (SLA), or leaf size; and lastly regenerative traits such as seed mass, mode of dispersal, or capacity for resprouting (Kumar et al. 2021c, d) (Fig. 1.2). Therefore, a trait may indicate a species' capacity to proliferate, spread, or endure in a certain habitat. Combining morphological and physiological characteristics with climate variables offers a practical means of classifying different plant species.

The concept of plant functional traits arose in 1806, but was recognized in 1997. Traits are classified as "soft traits" and "hard traits." Diaz et al. (2004) stated that soft traits are used more because they are relatively easy to quantify. Soft traits do not directly drive ecosystem processes but are often good correlates of hard traits, which in turn can drive ecosystem productivity. Specific leaf area (SLA), for example, is a soft trait as it does not directly drive ecosystem processes but shows high correlation for certain variables like net photosynthetic activity, leaf longevity, and relative growth rate which is closely related to ecosystem function (Lundholm et al. 2015). The distinction of traits like life cycle (annual or perennial), phenology (evergreen or deciduous), leaf shape (broadleaf or needle-leaf), and habit (woody or

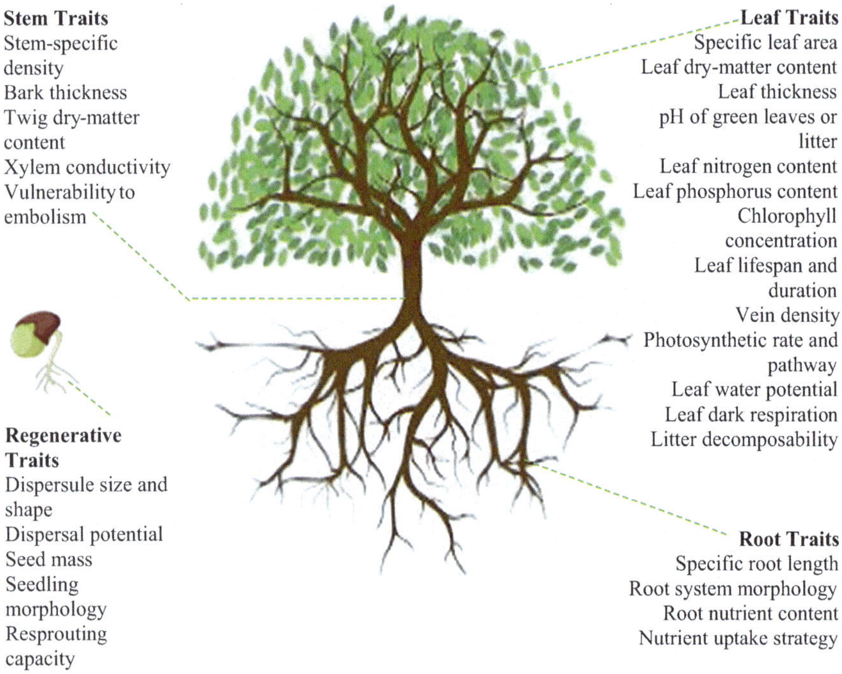

Fig. 1.2 Major plant functional traits

nonwoody) are particularly useful offsets as these characteristics can be observed through remote sensing techniques which in turn helps to determine stomatal conductance, photosynthetic activity, and allocation factor of the photosynthate (Bonan et al. 2002).

Despite the challenges this presents for incredibly varied communities, hard and soft traits are necessary for assessing or collecting trait values on a wide range of species (Keddy 1992). Quickly and easily measurable "soft traits" are frequently investigated in order to help with this (Hodgson et al. 1999; Liu et al. 2010). Although soft traits are not as closely linked to certain functions as "hard" traits are, they can still be used as substitutes as evaluating them does not require as much effort. Hard characteristics take longer to measure and are therefore more difficult to quantify for large numbers of species in various global regions (Walker and Langridge 2002; Lavorel and Garnier 2002). The distinction between "soft" and "hard" traits is based on how strongly a trait is linked to a species' reaction or effect; hard characteristics are indicative of trustworthy, mechanistic markers of a species' function that are responsible for community or ecosystem effects. The size of the body is arguably the most commonly used soft characteristic in aquatic environments. In plankton, it is a good indicator of many hard qualities like growth rates and nutrient intake. Similar to this, in terrestrial plants, seed morphology and size are readily quantifiable soft characteristics that facilitate the interpretation of the harder-to-observe hard trait of dispersal patterns.

Functional traits can be defined as various characteristics of organisms, such as their biochemistry, morphology, physiology, phenology, structure, or behaviour, which affect their response to the environment and/or their impact on ecosystem properties. This definition also encompasses the influence of traits on the functioning of ecosystem. This description highlights not only the specific characteristics that ecologists assess but also the two primary purposes of trait-based strategies: to quantify the impacts of traits on ecosystems and to examine the links between traits and environmental responses. Functional traits are commonly used to describe various aspects of an organism's physiology, behaviour, or morphology. These traits include measures such as basal metabolic rate, frost tolerance, nutrient accumulation, and photosynthetic rate. They also encompass morphological characteristics like seed size, leaf mass per area (LMA), body mass, and wood density. Additionally, behavioral traits such as predator evasion strategies, feeding strategy, and trophic level are considered. These concepts have been extensively studied and discussed by researchers such as Weiher et al. (1999), Poff et al. (2006), and Litchman and Klausmeier (2008).

Further impact qualities that define an organism are those that dictate its impact on the environment and, in turn, the advantages or disadvantages that human civilizations obtained from them. A species' ability to successfully establish and thrive in a certain habitat, as well as its capacity to adapt and survive in the face of changing environmental conditions, are largely determined by response traits. Moreover, several functional traits that are often evaluated have been studied as factors influencing or outcomes influenced by other factors. Let us take leaf mass area (LMA) as an example. The nutrient cycle will be slower in combinations of plants with high LMA compared to mixtures with low LMA. Leaves with higher bulk per unit area decay more slowly.

1.4 Plant Functional Trait Data Availability

For effective classification of forests into plant functional types, the necessary data for the relevant plant functional traits need to be retrieved. The manual collection and measurement of plant functional traits is cumbersome. Remote sensing provides an alternative method to retrieve data across multiple forest communities over a large geographic region. The use of remote sensing to retrieve plant functional traits is a relatively new approach, so the methods are not well defined. There is also a scarcity of research on the subject. None of the procedures till date appears to be designed specifically for mapping these traits over broad mountainous areas which are found to be highly relevant to identify and analyze the natural ecosystem responses (Cai et al. 2014). From airborne hyperspectral, LIDAR data and models like PROSAIL, various traits such as leaf area index (LAI), water content in the leaf, chlorophyll content in the leaf, specific leaf area (SLA), equivalent water thickness, LMA, and carbon and nitrogen concentrations have been retrieved (Cleland et al. 2007; Huber et al. 2016). Indices are being developed for the retrieval of functional trait data through the use of multispectral images. Aside from defining and grouping traits, determining which plant traits to measure using remote sensing and how to measure them is also difficult.

Research networks and organizations have launched a number of international functional databases. For example, Global Plant Trait Network Database (GLOPNET), a multi-investigator group, researches global features of plants. Their collection measures a global "leaf economic spectrum" and contains significant physiological, chemical, and structural leaf parameters for 2548 plant species distributed over 175 areas or sites (Reich et al. 2007; Wright et al. 2005). Scientists can access and utilize GLOPNET. Under the direction of DIVERSITAS, and IGBP, TRY is a network of scientists studying vegetation. It was established in 2007, and over 7000 species' values of traits are represented by the TRY database, which globally incorporates 93 plant trait repositories (Kattge et al. 2011b). Public and private data can both be found in the TRY database. Downloading the public data is free of charge. The bulk of the data, however, are private and cannot be accessed without the relevant administrators' consent. The lengthy process needed to access the majority of TRY's information limits its utility to some degree. It is expected that this obstruction will be lifted in the near future.

1.5 Significance of Plant Functional Traits

1.5.1 Plant Functional Traits as Measures of Ecosystem Services

The species and functional makeup of an ecosystem, with a focus on plant communities, have an impact on how well it functions (Hooper et al. 2005; Suding and Goldstein 2008). The features of plants that control how they react to and interact with their surroundings, trophic levels, and other species are known as plant functional traits. Studying the consequences of diverse disturbances and large-scale land and climate change has shown this method to be accurate and successful. Functional characteristics provide accurate markers of diverse ecosystem responses and activities, smoothly transitioning from functions at the ecosystem level to plants, which are widely acknowledged as the basis of an ecosystem (Kremen 2005). The reason for this is that changes in a plant's traits directly and often decisively affect the ecosystem functioning (Suding et al. 2008). This is significant because these functional characteristic responses and assessments have been conceptually and quantitatively linked to ecosystem services, which are the components of ecosystems that are either indirectly or directly profitable to humanity (Bardgett and Van Der Putten 2014).

Functional traits specific to plants are essential for forming ecosystems and for the provision of ecosystem services. The advantages that ecosystems offer to humans are known as ecosystem services, and functional characteristics of plants play a major role in the provisioning, control, support, and cultural aspects of these services. Plant functional traits represent one way to accurately and thoroughly explain ecosystem services, given the growing body of evidence suggesting that species composition and functional effects, rather than species diversity or richness, more accurately forecast ecosystem processes and, consequently, ecosystem services (Fig. 1.3) (Díaz and Cabido 1997). Plant functional traits influence important mechanisms that determine ecosystem services. For instance, traits associated with water-use efficiency, such as leaf shape and stomatal conductance, impact water regulatory services. Plants having characteristics that maximize water consumption contribute to improved water infiltration, enhanced groundwater recharge, and reduced runoff (McIntyre et al. 2015). Plant functional features are intimately linked to delivering ecosystem services, including fuel, food, and timber. Traits such as growth rate, seed size, and nutrient-use efficiency determine plant productivity (Violle et al. 2007).

Agricultural systems that contain a range of plant traits can promote resilience to shifts in environmental conditions, leading to steady provisioning services. Root features, including root architecture and symbiotic connections with mycorrhizal fungi, impact the uptake of nutrients and soil stability (Comas et al. 2014). Biodiverse plant communities with distinctive traits give chances for pleasure, education, and cultural activities. Ecosystems with a range of plant functional traits display increased adaptability and resilience to environmental changes, such as climatic variability and land-use changes. Functional diversity guarantees that some species within a population exhibit features appropriate to existing conditions, contributing to the stability and durability of ecosystems (de Bello et al. 2010).

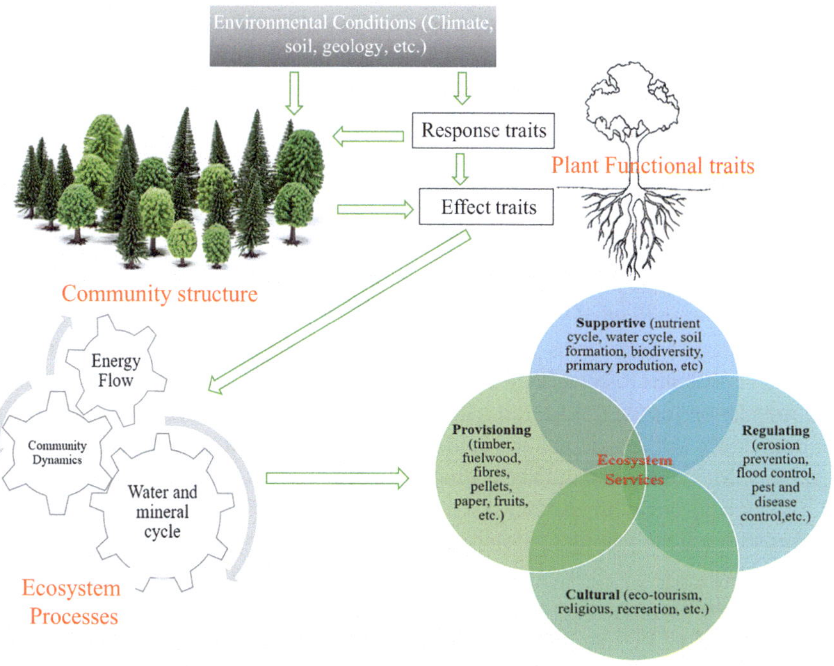

Fig. 1.3 Plant functional trait determining the ecosystem functioning

In conclusion, functional characteristics of plants are crucial to the supply of ecosystem services, influencing the functioning, resilience, and structure of ecosystems. The significance of these features rests in their potential to drive processes that enable the supply of important services to human communities. Understanding the relation between plant functional features and ecosystem services is vital for sustainable land management, conservation practices, and the formulation of policies that promote the health and operation of ecosystems.

1.5.2 Functional Traits Forecasting Plant Community Response to Changes Globally

Functional trait-based approaches to global change focus on establishing a predictive ability based on the many effects of changes in the environment on plant functional methods, rather than only defining an ecological response (Gornish and Prather 2014; Violle et al. 2014). A mechanistic knowledge of how changes in species composition may affect ecological processes is made easier by focusing on features, which also provides a foundation for encouraging tolerance to worldwide changes (Garnier et al. 2004; Laliberté and Legendre 2010). Understanding ecological feedback also becomes possible by it (Bardgett and Wardle 2010). Determining which traits regulate adaptation to worldwide variables allows for broader generalizations, providing predictions for how animals with identical qualities may respond in different environments (Gornish and Prather 2014).

Along with abiotic variables like climate and resource availability, a range of biotic factors, including competition and facilitation, also influence a species' capacity within a community (Grubb 1994; Vellend et al. 2013). Functional qualities may be a poor predictor of global change in some situations, but they may be a good one in others. One way to address this issue is to look into how the same species responds to similar changes brought about by global change when cultivated in different biotic environments (Gornish and Prather 2014). By examining two complementary community-level metrics, functional diversity (Mason and De Bello 2013) and community-weighted mean (CWM) trait values (Garnier et al. 2004), one can scale from individual species traits through the community to ecosystem functioning. Grime's mass ratio hypothesis (Grime 1998), which postulates that each species contributes to ecosystem function in accordance with its biomass, serves as the foundation for the CWM trait values. For ecosystem processes, the whole range of trait values within a population could be more important than the average of those qualities. Many functional diversity methods can be used to quantify trait variation (Laliberté and Legendre 2010; Mason and De Bello 2013). For understanding how communities are adjusting to global change, functional diversity and CWM are also helpful indicators (D'Antonio et al. 2016; de Tombeur et al. 2021; Klumpp and Soussana 2009).

It is projected that in most semiarid regions, including the southwestern United States, global climate change would result in more intense precipitation events, severe droughts, and a spike in the frequency of wildfires (Hufnagel and Garamvölgyi

2014; Syphard et al. 2007). Recent trends in precipitation have already affected the makeup of plant communities in many locations, frequently in parallel with species-specific mortality rates (Kelly and Goulden 2008). It has also been shown that an increase in the frequency of fires modifies the composition of plant communities (Díaz-Delgado et al. 2002). Plant functional features ought to be associated with these changes, even though it might not always be possible to forecast these linkages in reality. Increased drought conditions may favor species with features that help them live in drier soils, such as high water-use efficiency (Dudley 1996; Kimball et al. 2016). Conversely, species that are drought-resistant and have features like quick growth and early flowering may become more prevalent in drier areas (Franks 2011). A full knowledge of the conditions under which each approach performs best, including its context of reliance, is important for generalization to different systems.

1.5.2.1 Response of Leaf Traits to Changing Climatic Conditions

A leaf's persistence and productivity time determine how much carbon it fixes overall in its lifetime. Species with long leaf lives and high LMA generally have lesser leaf nitrogen content per leaf mass in terms of LMA and leaf productivity; on the other hand, species with short leaf lives and low LMA generally have higher leaf nitrogen contents per leaf mass. The association between characters is more stable than that of a single individual (Singh and Verma 2013a, b; Kumar et al. 2015, 2017). The alterations in the relationships between leaf traits showed that different leaf traits were affected by environmental changes in different ways and that plants responded to the altered environment to varying degrees based on these changes in different features. There was a direct correlation between leaf characteristics and environmental limitations.

1.5.2.2 Response of Root Traits to Changing Climatic Conditions

Plants absorb nutrients and water through their fine roots. Fine root features can convey information about the surrounding environment and serve as a reliable indication of plant growth and expansion. Fine roots are essential to maintaining plant growth and adapting to climate change. However, there is currently very little research on the characteristics of plant roots. Although there is little chance of a global climate shift, there are extensive and well-coordinated biological responses. Global climate change causes complex ecosystem communities to continually reorganize as a result. It is unclear how long-term environmental changes and short-term disturbances are related, as well as how population expansion affects the survival and stability of communities. Plant functional characteristics are significantly impacted by global changes. Currently, the primary areas of study in this discipline include species division, phenology, plant physiology, and community composition (Singh et al. 2013a, b; Kumar et al. 2016, 2019, 2021e). Furthermore, investigations on physiological and ecological aspects are typically conducted in short-term experiments, whereas functional characteristics research still lacks long-term, large-scale experiments.

1.5.3 Plant Functional Traits Determining Trade-Off

Plant functional traits are essential in determining the ecological strategies and performance of plants in different environments. The importance of plant functional features frequently entails trade-offs, wherein certain traits may confer advantages under particular circumstances but are accompanied by liabilities or limits. Plant functional traits are linked to trade-offs in the allocation of resources. As the leaf economic spectrum, which emphasizes a trade-off between traits such as specific leaf area (SLA) and leaf longevity, species exhibiting a high SLA prioritize quick acquisition of resources but typically have leaves with shorter life spans. Conversely, species with a low SLA allocate more resources toward prolonging the life span of their leaves (Wright et al. 2004).

Reproductive approaches can include making trade-offs. There exists a clearly established trade-off between the size and number of seeds, as extensively reported. Plants encounter a predicament—whether to produce numerous little seeds to maximize quantity or produce fewer giant seeds to enhance quality. The allocation of resources to reproduction involves managing this trade-off in order to maximize fitness in various circumstances. Plants frequently face a compromise between their ability to withstand drought and their rates of growth. Species adapted to arid regions invest in features that assist water saving, such as deep roots and tiny leaves. Nevertheless, these changes might restrict the capacity for fast expansion, demonstrating the compromise between drought resistance and growth tactics.

Plants deploy diverse defense systems against herbivores, and they are subject to trade-offs. For instance, a trade-off arises between induced and constitutive defenses. Providing resources to constitutive defenses, such as chemical compounds or thorns, can incur losses when weighed against inducible defenses that are developed in response to herbivore attacks. Root traits also reveal trade-offs like a trade-off may occur between root diameter and mycorrhizal connections. Species with thicker roots may engage less in mycorrhizal connections, altering nutrient acquisition tactics.

Understanding these trade-offs in plant functional traits is critical for anticipating plant responses to environmental changes and competition dynamics. While each plant species may reflect a unique combination of features, the underlying patterns of trade-offs provide insights into the strategies that plants take to negotiate ecological obstacles. These trade-offs are crucial to forming plant communities and influencing ecosystem dynamics.

1.5.4 Influence of Plant Functional Traits on Life History Strategies

Plant functional traits play an important role in shaping the life history strategies of plants, influencing their growth, reproduction, and survival in diverse environments. Life history strategies refer to the suite of characteristics and trade-offs that organisms adopt to allocate resources throughout their life cycle. Plant functional traits have a profound impact on reproductive strategies as the traits related to flowering phenology, seed mass, and seed dispersal mechanisms influence the timing of reproduction, investment in individual offspring, and the spatial distribution of seeds (Leishman et al. 2000). Species with different trait combinations may exhibit strategies ranging from rapid reproduction with numerous small seeds to slower reproduction with fewer, larger seeds. Functional traits such as SLA and growth rate play a key role in determining a plant's growth strategy. Fast-growing species often exhibit traits associated with high photosynthetic rates and resource acquisition, allowing them to exploit favorable conditions and outcompete slower-growing species (Westoby et al. 2002). This influences the successional dynamics and competitive interactions within plant communities.

These traits also contribute to strategies for survival and stress tolerance. Traits like leaf toughness, drought tolerance, and root depth influence a plant's ability to withstand environmental stresses (Chapin et al. 1993). Species with stress-tolerant traits may allocate resources to defense mechanisms and exhibit slower growth but enhanced resilience to adverse conditions. Longevity and resource allocation are critical components of life history strategies. Traits related to leaf life span and allocation of resources to different plant parts influence the overall longevity of a plant and its ability to persist in a given environment (Reich et al. 1992). Deciduous species with shorter-lived leaves may invest more in rapid growth and reproduction, while evergreen species with longer-lived leaves may prioritize resource conservation.

Trade-offs between different functional traits are central to the concept of life history strategies. For instance, there may be a trade-off between growth rate and stress tolerance, where fast-growing species may be less stress-tolerant and vice versa. Understanding these trade-offs allows researchers to identify adaptive strategies that optimize fitness under specific environmental conditions. Plant functional traits are fundamental to the development and expression of life history strategies in plants. These traits represent adaptations that have evolved in response to environmental pressures, shaping the ecological roles and responses of plant species. The significance of plant functional traits in life history strategies provides a framework for understanding plant diversity, community dynamics, and ecosystem functioning.

1.5.5 The Functioning of Invasive Species in an Evolving Ecosystem Is Best Explained by Plant Functional Traits

One of the biggest threats to natural ecosystems on a global scale is the biological invasion of alien plant species (Bellard et al. 2016). Invasive species that are alien, are one of the biggest threats to the declination of biodiversity in the world, and they have a significant negative impact on forestry, aquaculture, and agriculture. According to Mooney and Drake (1987) and Siemann and Rogers (2003), introduced species really pose a bigger harm to native biodiversity as pollution, harvesting, and disease put together. They can result in large financial losses and have a detrimental effect on species composition, biodiversity, and ecosystem structure

(Bajwa et al. 2016; Normile 2010). The effects of invading species, however, rely on the context, meaning that they are not only determined by the functional traits of the invader but also by the recipient community's resistance to the invasion (Kueffer et al. 2013).

It is believed that functional variations among species correspond with invasiveness and impact the capacity to exploit available resources(Gallagher et al. 2015; Naeem et al. 2000). Most of the time, comparing trait values between alien invasive plant and native species allows for the deduction of functional differences between them (Leishman et al. 2007; Ordonez and Olff 2013). But if only some difference is known to accelerate invasion, distinctions are worthless. Because trait values are flexible, assessing whether an invasion will happen may depend on the setting of the contrast between invasive and native species and differences in trait values in relation to other factors that support invasion (Hulme 2008). Trait-based analysis is being used more often in invasion research to show how the fitness and niche needs of native and invasive species differ (Funk et al. 2018).

The position of alien species throughout the introduction-naturalization-invasion continuum is the primary issue that needs to be addressed in trait-based approaches (Blackburn et al. 2011). Invasive species, or alien species, which propagate over specific distances, may differ in certain aspects from non-invasive species, or foreign species that reproduce in the wild but do not spread quickly in the introduced region (Richardson and Pyšek 2006). The habitat context of invasions is a crucial topic that needs to be covered in the trait-based research of invasiveness. Different plant species invade different environments (Chytrý et al. 2008). On the other hand, varied subsets of gene that have adjusted to a variety of habitats make up alien species pools (Kalusová et al. 2014). It is possible that species traits which allow it to thrive in one plant community and establish itself there would not be beneficial in another.

The species traits that are essential to invasion ecology and that predict invasion success form the basis of the research of functional diversity (Jeschke and Strayer 2006). These traits, in particular, enable plants to maximize yield (Mahaut et al. 2020). The functional traits of the local flora, where certain organisms might be more saturated than others, lead to an invasion's success. Accelerated phenotypic plasticity in specific traits has been the subject of previous attempts and broad analysis (Richardson and Pyšek 2006). These studies suggest that high phenotypic plasticity in present traits may play a role in promoting invasion success by enabling the expression of advantageous phenotypes in a variety of ecosystems (Matesanz et al. 2015).

When comparing invasive and native species, the relative growth rate fluctuates. In general, invasiveness is linked to functional features, morphological leaf traits such particular leaf area, and physiological plant attributes (Godoy et al. 2012). Invasive woody species have been found to have high values (Grotkopp and Rejmánek 2007). Invasive species appear to respond more plastically to light levels than noninvasive species do (Yamashita et al. 2002). This response is important for plant survival and growth (do Nascimento et al. 2015). Furthermore, a number of present research have shown a strong correlation between biological invasion and reproductive dispersion parameters, including seed production, dry mass, manner of dispersal, and plant maximum height (Moravcová et al. 2015). While it is helpful in identifying characteristics linked to invasion success, all characteristics are significant for all processes influencing invasion when they are taken into account in a single dissimilarity index. Yet, different characteristic combinations may influence one's capacity for competition and tolerance to the environment (Krause et al. 2018).

1.5.6 Response of Functional Traits to Forest Management

Forest management is a complex and dynamic field that involves the sustainable utilization of forest resources while preserving ecosystem integrity. The importance of plant functional traits in forest management has gained prominence as researchers and practitioners recognize their role in shaping forest structure, composition, and ecological processes. Understanding and incorporating plant functional traits into forest management strategies can enhance the resilience of forests in the face of environmental changes, promote biodiversity conservation, and optimize the delivery of ecosystem services.

One key aspect of the importance of plant functional traits in forest management lies in their ability to serve as indicators of ecosystem health and functionality. Traits such as leaf area, SLA, and wood density can provide valuable information about a species' resource-use strategy, growth rate, and competitive ability (Chave et al. 2009). By assessing these traits of dominant tree species in a forest ecosystem, managers can gain insights into the potential responses of the forest to disturbances, climate change, or management interventions. Furthermore, plant functional traits play a major role in shaping forest regeneration and successional dynamics. Traits related to seed dispersal, germination, and early growth can influence the establishment and growth of tree seedlings (Moles et al. 2007). Understanding these traits allows forest managers to predict regeneration patterns and guide interventions such as selective logging or controlled burns to favor species with desirable functional traits for ecosystem restoration.

In sustainable forest management, the concept of functional diversity is increasingly recognized as essential for maintaining ecosystem stability and resilience (Lavorel and Garnier 2002). Functional diversity, based on a variety of plant functional traits, contributes to the ability of forests to adapt to changing environmental conditions and resist disturbances. Forest management practices that promote functional diversity, such as maintaining a mix of species with different growth forms and life histories, can enhance ecosystem stability and reduce vulnerability to pests, diseases, and climate extremes.

The application of plant functional traits in silvicultural practices is another area where their significance becomes evident. Silviculture involves the manipulation of forest stands to achieve specific objectives, such as timber production, wildlife habitat enhancement, or carbon sequestration. Incorporating knowledge of functional traits allows for the selection of tree species and varieties that are better suited to particular site conditions and management goals (Messier et al. 2013). For example, choosing species with traits associated with fast growth and high wood density may optimize timber production, while selecting species with traits promoting leaf litter decomposition can enhance nutrient cycling. The importance of plant functional traits in forest management is multifaceted, encompassing their role as indicators of ecosystem health, drivers of successional dynamics, contributors to functional diversity, and guides for silvicultural decision-making. By integrating the study of plant functional traits into forest management practices, managers can make more informed decisions that balance ecological, economic, and social objectives, contributing to the long-term sustainability of forest ecosystems.

1.6 Conclusion

Plant functional traits are fundamental characteristics that influence the performance and ecological role of plants in their respective environments. These traits provide valuable insights into the functional strategies employed by plants due to its multiple significance leading to cope with their surroundings, affecting various ecological processes, and contributing to the overall dynamics of ecosystem. Understanding the concept and significance of plant functional traits is crucial for unraveling the intricacies of plant ecology, biodiversity, and ecosystem functioning. These traits are observable and measurable features that reflect the adaptive strategies plants have evolved in response to environmental pressures.

References

Bajwa AA, Chauhan BS, Farooq M, Shabbir A, Adkins SW (2016) What do we really know about alien plant invasion? A review of the invasion mechanism of one of the world's worst weeds. Planta 244:39–57

Bardgett RD, Van Der Putten WH (2014) Belowground biodiversity and ecosystem functioning. Nature 515(7528):505–511

Bardgett RD, Wardle DA (2010) Aboveground-belowground linkages: biotic interactions, ecosystem processes, and global change. Oxford University Press

Bellard C, Cassey P, Blackburn TM (2016) Alien species as a driver of recent extinctions. Biol Lett 12(2):20150623

Blackburn TM, Pyšek P, Bacher S, Carlton JT, Duncan RP, Jarošík V, Wilson JRU, Richardson DM (2011) A proposed unified framework for biological invasions. Trends Ecol Evol 26(7):333–339

Bonan GB, Levis S, Kergoat L, Oleson KW (2002) Landscapes as patches of plant functional types: an integrating concept for climate and ecosystem models. Glob Biogeochem Cycles 16(2):5-1–5-23. https://doi.org/10.1029/2000gb001360

Cai D, Xiao Y, Yang W, Ye W, Wang B, Younas M, Wu J, Liu K (2014) Association mapping of six yield-related traits in rapeseed (Brassica napus L.). Theor Appl Genet 127:85–96

Caruso CM, Mason CM, Medeiros JS (2020) The evolution of functional traits in plants: is the giant still sleeping? Int J Plant Sci 181(1):1–8

Chapin FS III, Autumn K, Pugnaire F (1993) Evolution of suites of traits in response to environmental stress. Am Nat 142:S78–S92

Chave J, Coomes D, Jansen S, Lewis SL, Swenson NG, Zanne AE (2009) Towards a worldwide wood economics spectrum. Ecol Lett 12(4):351–366

Chytrý M, Jarošík V, Pyšek P, Hájek O, Knollová I, Tichý L, Danihelka J (2008) Separating habitat invasibility by alien plants from the actual level of invasion. Ecology 89(6):1541–1553

Cleland EE, Chuine I, Menzel A, Mooney HA, Schwartz MD (2007) Shifting plant phenology in response to global change. Trends Ecol Evol 22(7):357–365. https://doi.org/10.1016/j.tree.2007.04.003

Comas J, Domínguez C, Salas-Vázquez DI, Parera J, Díez S, Bayona JM (2014) Input and leaching potential of copper, zinc, and selenium in agricultural soil from swine slurry. Arch Environ Contam Toxicol 66:277–286

D'Antonio CM, August-Schmidt E, Fernandez-Going B (2016) Invasive species and restoration challenges. In: Foundations of restoration ecology. Island Press, pp 216–244

de Bello F, Lavorel S, Díaz S, Harrington R, Cornelissen JHC, Bardgett RD, Berg MP, Cipriotti P, Feld CK, Hering D (2010) Towards an assessment of multiple ecosystem processes and services via functional traits. Biodivers Conserv 19:2873–2893

de Tombeur F, Laliberté E, Lambers H, Faucon M, Zemunik G, Turner BL, Cornelis J, Mahy G (2021) A shift from phenol to silica-based leaf defences during long-term soil and ecosystem development. Ecol Lett 24(5):984–995

Delph LF, Steven JC, Anderson IA, Herlihy CR, Brodie ED III (2011) Elimination of a genetic correlation between the sexes via artificial correlational selection. Evolution 65(10):2872–2880

DeMalach N, Ke P, Fukami T (2022) The effects of ecological selection on species diversity and trait distribution: predictions and an empirical test. Ecology 103(1):e03567

Díaz S, Cabido M (1997) Plant functional types and ecosystem function in relation to global change. J Veg Sci 8(4):463–474

Diaz S, Hodgson JG, Thompson K, Cabido M, Cornelissen JHC, Jalili A, Montserrat-Martí G, Grime JP, Zarrinkamar F, Asri Y (2004) The plant traits that drive ecosystems: evidence from three continents. J Veg Sci 15(3):295–304

Díaz S, Purvis A, Cornelissen JHC, Mace GM, Donoghue MJ, Ewers RM, Jordano P, Pearse WD (2013) Functional traits, the phylogeny of function, and ecosystem service vulnerability. Ecol Evol 3(9):2958–2975

Díaz-Delgado R, Lloret F, Pons X, Terradas J (2002) Satellite evidence of decreasing resilience in Mediterranean plant communities after recurrent wildfires. Ecology 83(8):2293–2303

do Nascimento JCF, de Almeida Ferreira S, de Albuquerque Vasconcelos JL, da Silva-Filho JLQ, Barbosa BT, Bezerra MF, Rocha CRC, Beltrão EIC (2015) Fut3 role in breast invasive ductal carcinoma: investigating its gene promoter and protein expression. Exp Mol Pathol 99(3):409–415

Donovan LA, Maherali H, Caruso CM, Huber H, de Kroon H (2011) The evolution of the worldwide leaf economics spectrum. Trends Ecol Evol 26(2):88–95

Dudley SA (1996) Differing selection on plant physiological traits in response to environmental water availability: a test of adaptive hypotheses. Evolution 50(1):92–102

Franks SJ (2011) Plasticity and evolution in drought avoidance and escape in the annual plant Brassica rapa. New Phytol 190(1):249–257

Franks SJ, Hamann E, Weis AE (2018) Using the resurrection approach to understand contemporary evolution in changing environments. Evol Appl 11(1):17–28

Funk C, Davenport F, Harrison L, Magadzire T, Galu G, Artan GA, Shukla S, Korecha D, Indeje M, Pomposi C (2018) 18. Anthropogenic enhancement of moderate-to-strong El Niño events likely contributed to drought and poor harvests in southern Africa during 2016. Bull Am Meteorol Soc 99:S91–S96

Gallagher RV, Randall RP, Leishman M (2015) Trait differences between naturalized and invasive plant species independent of residence time and phylogeny. Conserv Biol 29(2):360–369

Garnier E, Cortez J, Billès G, Navas M-L, Roumet C, Debussche M, Laurent G, Blanchard A, Aubry D, Bellmann A (2004) Plant functional markers capture ecosystem properties during secondary succession. Ecology 85(9):2630–2637

Godoy O, Valladares F, Castro-Díez P (2012) The relative importance for plant invasiveness of trait means, and their plasticity and integration in a multivariate framework. New Phytol 195(4):912–922

Gornish ES, Prather CM (2014) Foliar functional traits that predict plant biomass response to warming. J Veg Sci 25(4):919–927
Grime JP (1998) Benefits of plant diversity to ecosystems: immediate, filter and founder effects. J Ecol 86(6):902–910
Grotkopp E, Rejmánek M (2007) High seedling relative growth rate and specific leaf area are traits of invasive species: phylogenetically independent contrasts of woody angiosperms. Am J Bot 94(4):526–532
Grubb M (1994) Global policies for global problems: the case of climate change. In: Economic policies for sustainable development. Springer, pp 283–308
Guilherme Pereira C, Des Marais DL (2020) The genetic basis of plant functional traits and the evolution of plant-environment interactions. Int J Plant Sci 181(1):56–74
Hodgson JG, Wilson PJ, Hunt R, Grime JP, Thompson K (1999) Allocating CSR plant functional types: a soft approach to a hard problem. Oikos 85:282–294
Hooper DU, Chapin FS III, Ewel JJ, Hector A, Inchausti P, Lavorel S, Lawton JH, Lodge DM, Loreau M, Naeem S (2005) Effects of biodiversity on ecosystem functioning: a consensus of current knowledge. Ecol Monogr 75(1):3–35
Huber N, Kienast F, Ginzler C, Pasinelli G (2016) Using remote-sensing data to assess habitat selection of a declining passerine at two spatial scales. Landsc Ecol 31:1919–1937
Hufnagel L, Garamvölgyi Á (2014) Impacts of climate change on vegetation distribution no. 2-climate change induced vegetation shifts in the new world. Appl Ecol Environ Res 12(2):355–422
Hulme PE (2008) Phenotypic plasticity and plant invasions: is it all jack? Funct Ecol 22(1):3–7
Jeschke JM, Strayer DL (2006) Determinants of vertebrate invasion success in Europe and North America. Glob Chang Biol 12(9):1608–1619
Kalusová V, Chytrý M, Peet RK, Wentworth TR (2014) Alien species pool influences the level of habitat invasion in intercontinental exchange of alien plants. Glob Ecol Biogeogr 23(12):1366–1375
Kattge J, Diaz S, Lavorel S, Prentice IC, Leadley P, Bönisch G, Garnier E, Westoby M, Reich PB, Wright IJ (2011a) TRY–a global database of plant traits. Glob Chang Biol 17(9):2905–2935
Kattge J, Ogle K, Bönisch G, Díaz S, Lavorel S, Madin J, Nadrowski K, Nöllert S, Sartor K, Wirth C (2011b) A generic structure for plant trait databases. Methods Ecol Evol 2(2):202–213
Keddy PA (1992) A pragmatic approach to functional ecology. Funct Ecol 6(6):621–626
Kelly AE, Goulden ML (2008) Rapid shifts in plant distribution with recent climate change. Proc Natl Acad Sci 105(33):11823–11826
Kimball S, Funk JL, Spasojevic MJ, Suding KN, Parker S, Goulden ML (2016) Can functional traits predict plant community response to global change? Ecosphere 7(12):e01602
Klumpp K, Soussana J (2009) Using functional traits to predict grassland ecosystem change: a mathematical test of the response-and-effect trait approach. Glob Chang Biol 15(12):2921–2934
Krause A, Pugh TAM, Bayer AD, Li W, Leung F, Bondeau A, Doelman JC, Humpenöder F, Anthoni P, Bodirsky BL (2018) Large uncertainty in carbon uptake potential of land-based climate-change mitigation efforts. Glob Chang Biol 24(7):3025–3038
Kremen C (2005) Managing ecosystem services: what do we need to know about their ecology? Ecol Lett 8(5):468–479
Kueffer C, Pyšek P, Richardson DM (2013) Integrative invasion science: model systems, multi-site studies, focused meta-analysis and invasion syndromes. *New Phytol* 200(3):615–633
Kumar N, Kumar N, Shukla A, Shankhdhar SC, Shankhdhar D (2015) Impact of terminal heat stress on pollen viability and yield attributes of rice (Oryza sativa L.). Cereal Res Commun 43(4):616–626
Kumar N, Shankhdhar SC, Shankhdhar D (2016) Impact of elevated temperature on antioxidant activity and membrane stability in different genotypes of rice (Oryza sativa L.). Indian J Plant Physiol (Now Plant Physiol Rep) 21(1):37–43
Kumar N, Suyal DC, Sharma IP, Verma A, Singh H (2017) Elucidating stress proteins in rice (Oryza sativa L.) genotype under elevated temperature: a proteomic approach to understand heat stress response. 3 Biotech 7:205

Kumar N, Jeena N, Singh H (2019) Elevated temperature modulates rice pollen structure: a study from foothill Himalayan agro-ecosystem in India. 3 Biotech 9:175

Kumar A, Dwivedi GK, Tewari S, Paul J, Anand R, Kumar N, Kumar P, Singh H, Kaushal R (2020a) Carbon mineralization and inorganic nitrogen pools under Terminalia chebula Retz.-based agroforestry system in Himalayan foothills, India. For Sci 66(5):634–643

Kumar A, Kumar P, Singh H, Kumar N (2020b) Adaptation and mitigation potential of roadside trees with bio-extraction of heavy metals under vehicular emissions and their impact on physiological traits during seasonal regimes. Urban For Urban Green 58:126900

Kumar A, Dwivedi GK, Tewari S, Paul J, Sah VK, Singh H, Kumar P, Kumar N, Kaushal R (2020c) Soil organic carbon pools under Terminalia chebula Retz. based agroforestry system in Himalayan foothills, India. Curr Sci 118(7):1098–1103

Kumar A, Kumar P, Singh H, Bisht S, Kumar N (2021a) Relationship of physiological plant functional traits with soil carbon stock in temperate forest of Garhwal Himalaya. Curr Sci 120(8):1368–1373

Kumar A, Kumar P, Singh H, Kumar N (2021b) Modulation of plant functional traits under essential plant nutrients during seasonal regime in natural forests of Garhwal Himalayas. Plant Soil 465:197–212

Kumar A, Tewari S, Singh H, Kumar P, Kumar N, Bisht S, Kushwaha S, Tamta N, Kaushal R (2021c) Biomass accumulation and carbon stocks in different agroforestry system prevalent in Himalayan foothills, India. Curr Sci 120(6):1083–1088

Kumar A, Kumar P, Singh H, Kumar N (2021d) Impact of plant functional traits on near saturated hydraulic conductivity of soil under different forests of Kempty watershed in Garhwal Himalaya. Indian J Soil Conserv 49(1):38–44. NAAS rating: 5.28. ISSN: 2455-7145

Kumar N, Jeena N, Kumar A, Khairakpam R, Singh H (2021e) Comparative response of rice cultivars to elevated air temperature in Bhabar region of Indian Himalaya: status on yield attributes. Heliyon 7:e07474

Kumar A, Singh H, Kumari G, Bisht S, Malik A, Kumar N, Singh M, Raturi A, Barthwal S, Thakur A, Kaushal R (2022) Adaptive resilience of roadside trees to vehicular emissions via leaf enzymatic, physiological, and anatomical trait modulations. Environ Pollut 313:120191

Laliberté E, Legendre P (2010) A distance-based framework for measuring functional diversity from multiple traits. Ecology 91(1):299–305

Lavorel S, Garnier E (2002) Predicting changes in community composition and ecosystem functioning from plant traits: revisiting the holy grail. Funct Ecol 16(5):545–556

Leishman MR, Wright IJ, Moles AT, Westoby M (2000) The evolutionary ecology of seed size. In: Seeds: the ecology of regeneration in plant communities. CABI Publishing, Wallingford, pp 31–57

Leishman MR, Haslehurst T, Ares A, Baruch Z (2007) Leaf trait relationships of native and invasive plants: community-and global-scale comparisons. New Phytol 176(3):635–643

Litchman E, Klausmeier CA (2008) Trait-based community ecology of phytoplankton. Annu Rev Ecol Evol Syst 39:615–639

Liu G, Freschet GT, Pan X, Cornelissen JHC, Li Y, Dong M (2010) Coordinated variation in leaf and root traits across multiple spatial scales in Chinese semi-arid and arid ecosystems. New Phytol 188(2):543–553

Lundholm J, Tran S, Gebert L (2015) Plant functional traits predict green roof ecosystem services. Environ Sci Technol 49(4):2366–2374. https://doi.org/10.1021/es505426z

Mahaut L, Cheptou P-O, Fried G, Munoz F, Storkey J, Vasseur F, Violle C, Bretagnolle F (2020) Weeds: against the rules? Trends Plant Sci 25(11):1107–1116

Maherali H (2020) Mutualism as a plant functional trait: linking variation in the mycorrhizal symbiosis to climatic tolerance, geographic range, and population dynamics. Int J Plant Sci 181(1):9–19

Mason NWH, De Bello F (2013) Functional diversity: a tool for answering challenging ecological questions. J Veg Sci 24(5):777–780

Matesanz S, Horgan-Kobelski T, Sultan SE (2015) Evidence for rapid ecological range expansion in a newly invasive plant. AoB Plants 7:plv038

McGill BJ, Enquist BJ, Weiher E, Westoby M (2006) Rebuilding community ecology from functional traits. Trends Ecol Evol 21(4):178–185

McIntyre JK, Davis JW, Hinman C, Macneale KH, Anulacion BF, Scholz NL, Stark JD (2015) Soil bioretention protects juvenile salmon and their prey from the toxic impacts of urban stormwater runoff. Chemosphere 132:213–219

Messier C, Puettmann KJ, Coates KD (2013) Managing forests as complex adaptive systems: building resilience to the challenge of global change. Routledge

Moles AT, Ackerly DD, Tweddle JC, Dickie JB, Smith R, Leishman MR, Mayfield MM, Pitman A, Wood JT, Westoby M (2007) Global patterns in seed size. Glob Ecol Biogeogr 16(1):109–116

Moles AT, Perkins SE, Laffan SW, Flores-Moreno H, Awasthy M, Tindall ML, Sack L, Pitman A, Kattge J, Aarssen LW (2014) Which is a better predictor of plant traits: temperature or precipitation? J Veg Sci 25(5):1167–1180

Mooney HA, Drake JA (1987) The ecology of biological invasions. Environ Sci Policy Sustain Dev 29(5):10–37

Moravcová L, Pyšek P, Jarošík V, Pergl J (2015) Getting the right traits: reproductive and dispersal characteristics predict the invasiveness of herbaceous plant species. PLoS One 10(4):e0123634

Moray C, Goolsby EW, Bromham L (2016) The phylogenetic association between salt tolerance and heavy metal hyperaccumulation in angiosperms. Evol Biol 43:119–130

Naeem S, Knops JMH, Tilman D, Howe KM, Kennedy T, Gale S (2000) Plant diversity increases resistance to invasion in the absence of covarying extrinsic factors. Oikos 91(1):97–108

Nalley JO, Stockenreiter M, Litchman E (2014) Community ecology of algal biofuels: complementarity and trait-based approaches. Ind Biotechnol 10(3):191–201

Normile D (2010) Saving forests to save biodiversity. American Association for the Advancement of Science

Ordonez A, Olff H (2013) Do alien plant species profit more from high resource supply than natives? A trait-based analysis. In: Global ecology and biogeography, vol 22(6). Wiley Online Library, pp 648–658

Petipas RH, Bowsher AW, Bekkering CS, Jack CN, McLachlan EE, White RA III, Younginger BS, Tiemann LK, Evans SE, Friesen ML (2020) Interactive effects of microbes and nitrogen on Panicum virgatum root functional traits and patterns of phenotypic selection. Int J Plant Sci 181(1):20–32

Poff NL, Olden JD, Vieira NKM, Finn DS, Simmons MP, Kondratieff BC (2006) Functional trait niches of north American lotic insects: traits-based ecological applications in light of phylogenetic relationships. J N Am Benthol Soc 25(4):730–755

Poorter L, Wright SJ, Paz H, Ackerly DD, Condit R, Ibarra-Manríquez G, Harms KE, Licona JC, Martinez-Ramos M, Mazer SJ (2008) Are functional traits good predictors of demographic rates? Evidence from five neotropical forests. Ecology 89(7):1908–1920

Reich PB, Walters MB, Ellsworth DS (1992) Leaf life-span in relation to leaf, plant, and stand characteristics among diverse ecosystems. Ecol Monogr 62(3):365–392

Reich PB, Wright IJ, Lusk CH (2007) Predicting leaf physiology from simple plant and climate attributes: a global GLOPNET analysis. Ecol Appl 17(7):1982–1988

Richardson DM, Pyšek P (2006) Plant invasions: merging the concepts of species invasiveness and community invasibility. Prog Phys Geogr 30(3):409–431

Schellberg J, Pontes LDS (2012) Plant functional traits and nutrient gradients on grassland. Grass Forage Sci 67(3):305–319

Sheth SN, Angert AL (2016) Artificial selection reveals high genetic variation in phenology at the trailing edge of a species range. Am Nat 187(2):182–193

Shipley B, De Bello F, Cornelissen JHC, Laliberté E, Laughlin DC, Reich PB (2016) Reinforcing loose foundation stones in trait-based plant ecology. Oecologia 180:923–931

Siemann E, Rogers WE (2003) Reduced resistance of invasive varieties of the alien tree Sapium sebiferum to a generalist herbivore. Oecologia 135:451–457

Singh H, Verma A (2013a) Physiological responses of rice cultivars to various nitrogen levels. Int J Agric Environ Biotechnol 6(3):383–388

Singh H, Verma A (2013b) Characterization and screening of high nitrogen efficient rice genotype to curtail environmental pollution. Int J Agric Environ Biotechnol 6(1):23–30

Singh H, Verma A, Shukla A (2010a) Comparative study of yield and yield components of hybrid and inbred genotypes of rice (Oryza Sativa L.). Int J Agric Environ Biotechnol 3:355–360

Singh H, Verma A, Krishnamoorthy M, Shukla A (2010b) Consequence of diverse nitrogen levels on leaf pigments in five rice genotypes under field emergent circumstance. Int J Bioresour Stress Manag 1:189–193

Singh H, Verma A, Rai SK (2013a) Biochemical evaluation of different rice genotypes grown at various nitrogen levels using SDS-PAGE. Curr Adv Agric Sci 5(1):144–146

Singh H, Verma A, Shukla A (2013b) Guttation fluid as a physiological marker for selection of nitrogen efficient rice (Oryza sativa L.) genotypes. Afr J Biotechnol 12(44):6276–6281

Singh H, Verma A, Ansari MW, Shukla A (2014) Physiological response of rice (Oryza sativa L.) genotypes to elevated nitrogen applied under field conditions. Plant Signal Behav 9:e29015. https://doi.org/10.4161/psb.29015

Singh S, Singh H, Sharma SK, Nautiyal R (2021) Seasonal variation in biochemical responses of bamboo clones in the sub-tropical climate of Indian Himalayan foothills. Heliyon 7(4):e06859. https://doi.org/10.1016/j.heliyon.2021.e06859

Singh M, Singh H, Kumar A, Kumar M, Barthwal S, Thakur A (2024) Soil nitrogen availability determines the CO2 fertilization effect on tree species (Neolamarckia cadamba): growth and physiological evidence. Environ Sustain. https://doi.org/10.1007/s42398-023-00300-w

Sterck F, Markesteijn L, Toledo M, Schieving F, Poorter L (2014) Sapling performance along resource gradients drives tree species distributions within and across tropical forests. Ecology 95(9):2514–2525

Suding KN, Goldstein LJ (2008) Testing the holy grail framework: using functional traits to predict ecosystem change. New Phytol 180:559–562

Suding KN, Lavorel S, Chapin FS III, Cornelissen JHC, DIAz S, Garnier E, Goldberg D, Hooper DU, Jackson ST, Navas M (2008) Scaling environmental change through the community-level: a trait-based response-and-effect framework for plants. Glob Chang Biol 14(5):1125–1140

Swenson NG, Worthy SJ, Eubanks D, Iida Y, Monks L, Petprakob K, Rubio VE, Staiger K, Zambrano J (2020) A reframing of trait–demographic rate analyses for ecology and evolutionary biology. Int J Plant Sci 181(1):33–43

Syphard AD, Radeloff VC, Keeley JE, Hawbaker TJ, Clayton MK, Stewart SI, Hammer RB (2007) Human influence on California fire regimes. Ecol Appl 17(5):1388–1402

Vellend M, Brown CD, Kharouba HM, McCune JL, Myers-Smith IH (2013) Historical ecology: using unconventional data sources to test for effects of global environmental change. Am J Bot 100(7):1294–1305

Violle C, Navas M, Vile D, Kazakou E, Fortunel C, Hummel I, Garnier E (2007) Let the concept of trait be functional! Oikos 116(5):882–892

Violle C, Reich PB, Pacala SW, Enquist BJ, Kattge J (2014) The emergence and promise of functional biogeography. Proc Natl Acad Sci 111(38):13690–13696

Visser MD, Bruijning M, Wright SJ, Muller-Landau HC, Jongejans E, Comita LS, De Kroon H (2016) Functional traits as predictors of vital rates across the life cycle of tropical trees. Funct Ecol 30(2):168–180

Walker BH, Langridge JL (2002) Measuring functional diversity in plant communities with mixed life forms: a problem of hard and soft attributes. Ecosystems 5:529–538

Weber MG, Agrawal AA (2012) Phylogeny, ecology, and the coupling of comparative and experimental approaches. Trends Ecol Evol 27(7):394–403

Weiher E, Van Der Werf A, Thompson K, Roderick M, Garnier E, Eriksson O (1999) Challenging theophrastus: a common core list of plant traits for functional ecology. J Veg Sci 10(5):609–620

Westoby M, Falster DS, Moles AT, Vesk PA, Wright IJ (2002) Plant ecological strategies: some leading dimensions of variation between species. Annu Rev Ecol Syst 33(1):125–159

Wright IJ, Reich PB, Westoby M, Ackerly DD, Baruch Z, Bongers F, Cavender-Bares J, Chapin T, Cornelissen JHC, Diemer M (2004) The worldwide leaf economics spectrum. Nature 428(6985):821–827

Wright IJ, Reich PB, Cornelissen JHC, Falster DS, Garnier E, Hikosaka K, Lamont BB, Lee W, Oleksyn J, Osada N (2005) Assessing the generality of global leaf trait relationships. New Phytol 166(2):485–496

Yamashita N, Koike N, Ishida A (2002) Leaf ontogenetic dependence of light acclimation in invasive and native subtropical trees of different successional status. Plant Cell Environ 25(10):1341–1356

Implication of Plant Functional Traits in Crop Improvement

2

Rajneesh Bhardwaj, Rowndel Khwairakpam, Arvind Singh Negi, and Balwant Rawat

Abstract

Plant functional traits are pivotal in revolutionizing crop improvement strategies, offering a robust framework to address agricultural sector challenges. In the face of climate change, resource limitations, and the imperative to sustainably feed a growing global population, functional traits encompassing morphological, physiological, and biochemical attributes provide valuable insights into plant adaptation, resource utilization efficiency, and responses to environmental stresses. Leveraging these traits holds great promise for enhancing agricultural productivity and sustainability. Despite their potential, realizing the full benefits of plant functional traits requires overcoming significant challenges. Complex trait interactions, genetic variability, and the need for high-throughput phenotyping present formidable obstacles. Trade-offs between traits and ensuring consumer acceptance of modified crop characteristics add further complexity. However, advancements in genomics and biotechnology enable precise trait manipulation, allowing breeders to target specific attributes with exceptional accuracy. Combining multiple functional traits is crucial for developing adaptable crops in diverse environments. In a rapidly changing climate, these traits facilitate the development of crops resilient to new challenges such as heat stress and water scarcity. With the help of optimizing resource-use efficiency, enhancing pest and disease resistance, and improving nutritional quality, crops developed through functional trait-based approaches can reduce agriculture's environmental footprint while meeting the nutritional demands of a burgeoning world population. Additionally, sustainability, market differentiation, and adaptive crop management present potential benefits, further underscoring the significance of plant functional traits in shaping the future of agriculture.

R. Bhardwaj (✉) · R. Khwairakpam · A. S. Negi · B. Rawat
School of Agriculture, Graphic Era Hill University Dehradun, Dehradun, Uttarakhand, India

N. Kumar, H. Singh (eds.), *Plant Functional Traits for Improving Productivity*,
https://doi.org/10.1007/978-981-97-1510-7_2

2.1 Introduction

Plant functional traits refer to the inherent characteristics or attributes of plants that exert influence on their ecological roles and interactions within the environment (Violle et al. 2007). These traits play a pivotal role in comprehending the adaptive mechanisms of plants to diverse ecological niches, their competitive dynamics with other species, and their responses to environmental fluctuations. Researchers leverage plant functional traits for the categorization and analysis of plant species, evaluation of ecosystem functions, and prognostication of ecosystem reactions to disturbances or shifts in climate. In the realm of crop enhancement, these traits prove highly advantageous. The prioritization of traits such as drought tolerance, disease resistance, and nutrient utilization efficiency can culminate in heightened crop yields and enhanced quality. Additionally, traits like heat tolerance and resistance to salinity facilitate the survival of crops in environments undergoing substantial changes. Features associated with water and nutrient efficiency contribute to resource conservation, fostering sustainability in agriculture. The promotion of traits supporting positive interactions with pollinators and other species contributes significantly to biodiversity conservation. Despite the potential rewards of utilizing functional traits, their incorporation into crop improvement is laden with formidable challenges. The intricate interactions among plant traits make predicting the outcomes of trait combinations exceptionally difficult. Limited access to a diverse pool of genetic resources containing desirable traits poses a hindrance to seamless trait integration. The selection and breeding process for functional traits may prolong breeding cycles, resulting in delays in the release of improved crop varieties. Moreover, certain genetically modified traits encounter regulatory impediments and public acceptance issues.

Crop improvement stands as a pivotal component in ensuring food security amid a rapidly expanding global population. While traditional breeding methods have successfully boosted crop yields, the quest for sustainable and efficient strategies has spurred the exploration of innovative approaches (Caligari 2001). One such approach gaining prominence involves the integration of plant functional traits into crop improvement programs. Encompassing a spectrum of characteristics, from morphological and physiological features to biochemical and molecular attributes, plant functional traits significantly impact a plant's performance in its environment (Singh et al. 2013a, b) (Fig. 2.1).

The comprehension and manipulation of plant functional traits present distinctive opportunities for augmenting crop performance. These traits dictate a plant's responses to environmental stimuli, influencing its growth, development, and overall productivity. Unleashing the potential of plant functional traits can contribute to the development of robust crops capable of thriving amidst changing climatic conditions, resisting pests and diseases, and yielding higher outputs (Martin and Isaac 2015).

The primary aim of this chapter is to acquaint readers with the significance of plant functional traits in the enhancement of crops. The chapter predominantly underscores the challenges and opportunities inherent in this approach. The

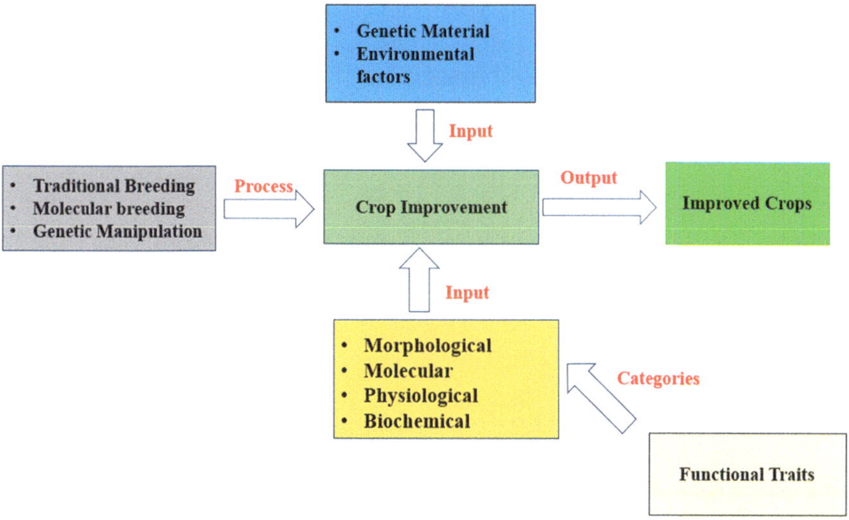

Fig. 2.1 Overview of the role of functional traits in crop improvement

discourse is geared toward offering an understanding of the potential benefits and drawbacks associated with the integration of plant functional traits in the context of crop improvement (Fig. 2.2).

2.2 Role of Quantitative Genetics and Genomics of Plant Functional Traits in Crop Improvement

The quantitative genetics and genomics of plant functional traits play a pivotal role in advancing crop improvement by offering insights into the genetic foundations of intricate traits that contribute significantly to overall plant performance, yield, and adaptability (Singh and Verma 2013a, b). Quantitative genetics aids in the deconstruction of the genetic basis of complex traits through the estimation of the contribution of multiple genes and their interactions. The application of genomic tools, including the identification of quantitative trait loci (QTLs) and marker-assisted selection, empowers breeders to make informed decisions when selecting plants with desirable traits (Tanksley and McCouch 1997). Genomic selection utilizes genomic information to predict the breeding value of individuals, thereby expediting the breeding process (Meuwissen et al. 2001). Functional genomics approaches, such as transcriptomics and metabolomics, contribute to comprehending the molecular mechanisms underlying trait expression (Rensink and Buell 2004). The exploration of the genetics of traits associated with responses to abiotic and biotic stresses is instrumental in developing climate-resilient crop varieties (Varshney et al. 2019). The integration of genomics, transcriptomics, and other omics data provides a holistic understanding of the regulation of functional traits. Additionally, linking genetic

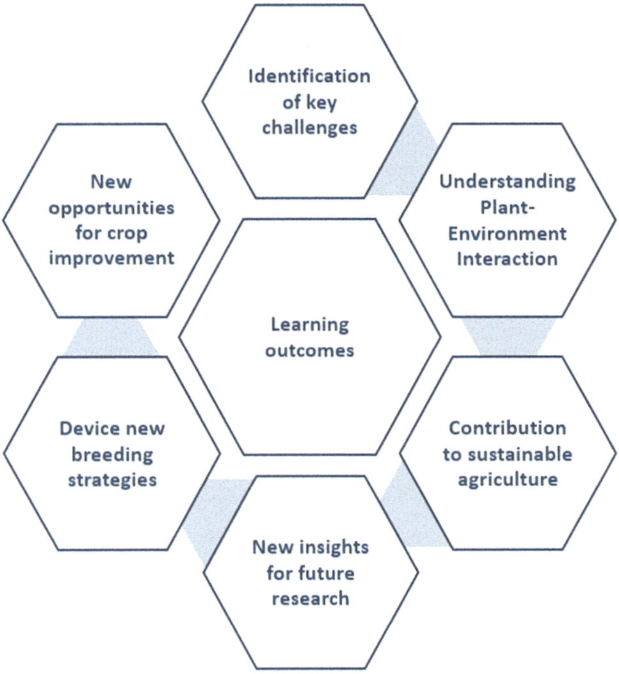

Fig. 2.2 Objectives of the chapter

information with crop models facilitates the prediction of crop performance under diverse environmental conditions.

2.3 Crop Enhancement Through Manipulation of Plant Functional Traits

Modifying plant functional traits to enhance crops involves strategic adjustments to specific characteristics, aiming to improve overall performance, resilience, and yield. These traits, encompassing physiological, morphological, and biochemical attributes, significantly influence plant growth and development and environmental interactions (Singh et al. 2014, 2021, 2024). Targeted manipulation of traits related to water-use efficiency, root architecture, and osmotic regulation shows promise in strengthening a crop's ability to endure water scarcity (Farooq et al. 2009). Similarly, alterations in disease resistance traits, including defense mechanisms and pathogen recognition, offer a means to enhance crop protection (Dangl and Jones 2001).

Interventions in traits associated with nutrient uptake, utilization, and remobilization present a pathway to improve a crop's efficiency in using essential nutrients (Hirel and Lea 2011). Enhancing photosynthesis-related traits, such as chlorophyll content and photosynthetic capacity, holds the potential to elevate overall crop productivity (Long and Ort 2010). Strategically manipulating traits that confer tolerance to diverse abiotic stresses, such as salinity and temperature extremes, is a mechanism to enhance crop resilience (Mittler 2006).

Adjustments in traits governing plant architecture, including branching patterns and height, optimize light interception and resource utilization (Furbank and Tester 2011). Manipulating traits related to seed size, nutritional content, and overall yield significantly contributes to the development of high-quality, high-yielding crops (Sadras and Lawson 2013). Addressing specific climatic conditions through the development of crops with adaptive traits is crucial for aligning agriculture with changing environmental patterns (Lobell and Gourdji 2012).

Utilizing advanced molecular techniques, such as genetic engineering and CRISPR-Cas technology, enables precise manipulation of specific traits, offering a means to achieve targeted crop improvements (Hickey et al. 2019). The strategic manipulation of traits that promote resource-use efficiency and mitigate environmental impact emerges as a critical strategy in advancing sustainable agricultural practices. The exploration of manipulating plant functional traits is a rapidly evolving field, with ongoing progress in biotechnology and genomics offering novel avenues to enhance crop performance and address global food security challenges.

2.4 Challenges in Integrating Plant Functional Trait

Numerous functional traits in plants are governed by a multitude of genes, and the intricacies of their interactions can be challenging to unravel. The interconnection and influence of multiple genes on various plant functional traits add complexity to their understanding and manipulation (Singh et al. 2010a, b). Deciphering the intricate genetic basis of plant functional traits frequently involves navigating complex interactions among multiple genes. Traits such as yield, drought resistance, or nutrient-use efficiency often result from the intricate interplay of numerous genes. The identification of each contributing gene and the comprehension of their collective effects present formidable challenges in the context of trait modification. The genetic landscape becomes more intricate due to complex gene interactions and epistatic effects. In the pursuit of crop improvement through plant functional traits, nonlinear gene interactions may yield unforeseen outcomes. The management and prediction of such gene interactions pose considerable challenges (Mackay 2014). An additional layer of complexity is introduced by phenotypic plasticity, where a single genotype can manifest different phenotypes under varying environmental conditions. The same genetic makeup may lead to different trait expressions depending on the environment. This complicates efforts to consistently identify and modify traits across diverse conditions (Nicotra et al. 2010). QTL mapping is a prevalent approach for identifying genomic regions associated with traits, but it may not fully capture the complexity of the traits to be inherited. While QTL mapping identifies broad genomic regions linked to a considered trait, it may not highlight all contributing genes or their interactions, leading to gaps in understanding the actual trait expression (Mackay et al. 2009). Genetic complexity may vary among populations, making it challenging to generalize trait modifications across different genetic backgrounds. Traits may be influenced by population-specific genetic variations, necessitating specific approaches for each population, further complicating the implementation of crop improvement procedures (Rellstab et al. 2017). Plant functional traits often exhibit phenotypic variability in response to various environmental conditions. The same genotype may express phenotypically distinct traits in varied environments, posing challenges in achieving heritable and universally applicable trait modifications (Violle et al. 2012). The performance of a specific genotype may fluctuate across different environmental conditions, necessitating careful consideration of genotype-environment (GxE) interactions in trait modification strategies. GxE interactions influence trait expression, complicating efforts to identify and modify traits under diverse conditions. For stable gene integration and expression, phenotypic plasticity is required, where traits change in response to variations in environmental conditions. While plasticity enables plants to adapt to changing environments, excessive variability may impede the performance of consistently performing cultivars (Nicotra et al. 2010). Intraspecific variation in response to environmental factors presents challenges in selecting and manipulating consistent trait expressions. Variability within a species may lead to diverse trait responses, complicating efforts to identify and target specific trait modifications (Anderson et al. 2012). Shifts in climate patterns may alter environmental conditions, affecting the performance of functional traits selected under previous climatic conditions. Traits integrated for past climatic conditions may become unsuitable or disadvantageous under changing environmental conditions (Franks and Hoffmann 2012). Unforeseen environmental stress events, such as extreme weather conditions, can disrupt trait expressions and impact crop performance. Various traits may respond differently to sudden and extreme environmental changes, making it more challenging to develop traits resilient to unpredictable stresses (Mittler 2006). Addressing the challenges posed by environmental sensitivity involves robust field trials, the application of agroecological approaches, and the development of strategies that account for GxE interactions. By acknowledging environmental variability, researchers and breeders can enhance the adaptability and stability of plant functional traits in diverse agricultural settings. Numerous functional traits in plants inherently exhibit complexity and diversity, posing challenges in their accurate quantification. Traits like drought tolerance or nutrient-use efficiency involve diverse interacting components, demanding precise and comprehensive measurement techniques to capture their complete intricacy (Perez-Harguindeguy et al. 2013). The spatial and temporal variability of environmental conditions introduces complexity to trait measurement. Traits may exhibit variations across different regions and seasons, necessitating extensive sampling and measurement efforts to acquire representative data. The absence of standardized protocols for trait measurement can result in inconsistencies and hinder cross-study comparisons. Standardization is crucial to ensure that trait measurements are comparable across studies, although achieving consensus on protocols proves challenging (Violle et al. 2007). Certain traits may present challenges in measurement due to technological limitations. Traits involving molecular or physiological processes may require sophisticated technologies, and their measurement can be constrained by the availability and affordability of such tools (Furbank and Tester 2011). Balancing high accuracy and scalability in trait measurement poses a challenge. Some highly accurate measurement techniques may be time-consuming or resource-intensive, limiting their applicability to large-scale field trials. Quantifying belowground traits, such as root architecture, presents additional challenges compared to aboveground traits. Belowground traits, crucial for nutrient and water uptake, often require invasive techniques and specialized equipment (Lynch 2013). Traits often form interconnected networks, and enhancing one trait may unintentionally affect others. Modifying a trait, such as yield, may introduce trade-offs that require careful management (Poorter et al. 2012). Genetic correlations between traits may limit independent improvements. Genes influencing one trait may affect others due to genetic linkage, restricting the selection for specific traits without impacting correlated ones. Limited plant resources may lead to trade-offs when enhancing one trait at the expense of another. Improving traits like fruit size or seed production may divert resources from other crucial functions, resulting in trade-offs in overall plant performance. Traits advantageous in certain environments may be disadvantageous in others, resulting in adaptive trade-offs. Traits evolved for specific ecological niches may not be optimal under different conditions, requiring a balanced approach in trait modifications. Traits may interact differently with the environment, leading to context-dependent trade-offs. A trait beneficial in one environment may become a disadvantage in another, demanding a nuanced understanding of trait-environment interactions (Violle et al. 2012). Evolutionary constraints may hinder the simultaneous enhancement of multiple traits. Evolutionary history may impose limitations that prevent the independent improvement of certain traits without compromising others (Conner and Hartl 2004). Divergent selection pressures on traits may result in conflicting objectives in crop improvement. Traits favored in breeding programs for one goal (e.g., yield) may be disadvantageous for other goals (e.g., stress resistance). Agroecosystems, intricate and dynamic systems, exhibit diverse interactions among crops, organisms, and the environment. Integrating plant functional traits into these systems demands comprehension of these intricate relationships, as modifying traits may yield unforeseen effects on the overall ecosystem (Power 2010). Evaluating the impact of modified traits on biodiversity within agroecosystems requires meticulous consideration. Alterations in plant traits can influence interactions with other species, potentially impacting the diversity of insects, microbes, and other organisms in the agroecosystem. Plant traits wield influence over soil health and microbial communities, affecting nutrient cycling and overall ecosystem functioning. The modification of plant traits may trigger cascading effects on the soil microbiome, with potential repercussions for nutrient availability and plant health (Philippot et al. 2013). Traits optimized for improved performance under specific conditions may lack resilience in the face of changing or extreme environmental stress. Given the variable conditions in agroecosystems, stability-optimized traits are crucial for ensuring long-term productivity (Mittler 2006). Modified traits may impact interactions with pests and pathogens, necessitating careful consideration of plant defense mechanisms. Changes in traits like resistance or tolerance can influence the prevalence and dynamics of pests and diseases within agroecosystems

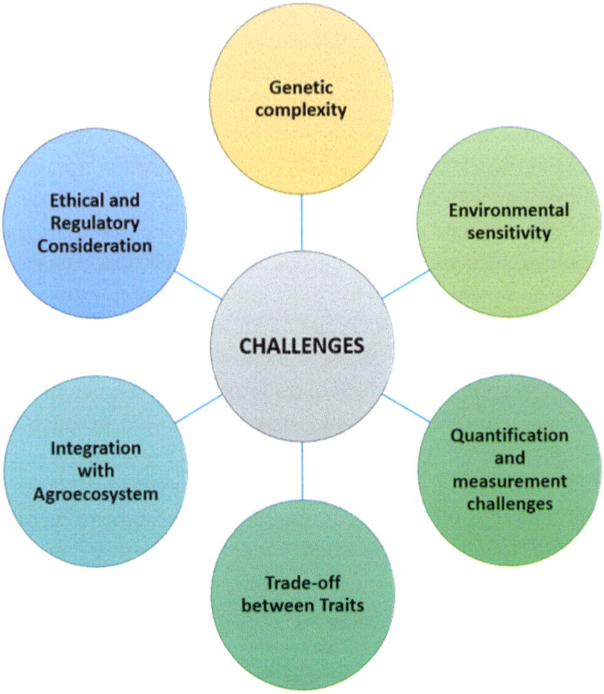

Fig. 2.3 Challenges faced during using plant functional traits in crop

(Gatehouse and Gatehouse 1998). The integration of certain traits may inadvertently increase dependence on agrochemicals. Traits designed for specific pest resistance or nutrient-use efficiency may influence the need for pesticides or fertilizers, thereby influencing the sustainability of agroecosystems (Pretty et al. 2018) (Fig. 2.3).

2.5 Opportunities

The incorporation of plant functional traits into crop improvement programs presents significant opportunities to address key challenges in agriculture. Precision breeding methodologies, including CRISPR-Cas9 and other genome editing tools, afford the precise modification of specific plant functional traits. This capability allows breeders to concentrate on enhancing traits crucial for crop improvement, such as yield, nutrient content, or stress resistance (Jinek et al. 2012). Precision breeding significantly compresses the time needed to develop improved crop varieties. This expedited timeline empowers breeders to respond promptly to emerging challenges, encompassing changing climate conditions or evolving pest and disease threats (Cobb et al. 2013). Precision breeding techniques provide enhanced control over genetic modifications, thereby minimizing off-target effects. This ensures the accurate and predictable integration of introduced traits, addressing concerns associated with unintended consequences (Hsu et al. 2013). Precision breeding facilitates the stacking of multiple desirable traits within a single crop variety. This enables breeders to combine traits such as drought tolerance, disease resistance, and high yield, resulting in crops with enhanced overall performance (Ricroch et al. 2016). Precision breeding can introduce genetic diversity into crops, particularly valuable for overcoming genetic bottlenecks and enhancing adaptability. This makes crops more resilient to environmental changes (Zhang and Voytas 2011). Precision breeding enables the targeted enhancement of nutritional traits, addressing malnutrition and improving the overall nutritional quality of crops. For instance, modifying biosynthesis pathways can elevate the levels of essential nutrients (Beyer et al. 2002). Precision breeding allows for the customization of crops based on local agroecological conditions. Tailoring traits to specific environments enhances the adaptability of crops, improving overall productivity and sustainability (Lin 2011). Precision breeding can contribute to reducing the environmental footprint of agriculture by creating crops with traits that enhance resource-use efficiency. This includes traits related to water use, nutrient uptake, and overall ecosystem impact (Reynolds et al. 2016). Precision breeding technologies offer opportunities to align with evolving regulatory frameworks and address public concerns more effectively. The ability to precisely target and control genetic modifications may enhance public acceptance of trait-modified crops (Lucht 2015). Functional traits associated with enhanced water-use efficiency, deeper root systems, and reduced transpiration rates can contribute to improved drought tolerance in crops. Breeding for these traits helps crops adapt to water scarcity, a crucial factor in the face of changing climate patterns (Bartels and Sunkar 2005). Functional traits related to heat stress, such as heat shock proteins and increased photosynthetic efficiency, can be targeted for crop improvement. Developing heat-resistant crops is essential for mitigating the negative impacts of rising temperatures associated with climate change (Wahid et al. 2007). Modifying flowering time traits can help crops synchronize their reproductive stages with changing climate conditions. This adaptation improves yield stability by avoiding adverse climatic events during critical growth phases. Enhancing traits associated with carbon fixation, such as photosynthetic efficiency, can contribute to increased productivity under elevated carbon dioxide levels. This adaptation helps crops thrive in an environment where atmospheric CO_2 concentrations are rising (Ainsworth and Long 2005). Traits that confer salinity tolerance, such as efficient ion transport and osmoregulation, can be targeted for crop improvement. This adaptation is crucial as changing climate conditions may lead to increased soil salinity in certain regions (Munns 2005). Functional traits associated with resilience to extreme events, such as floods or storms, can be integrated into crop breeding programs. Developing crops with traits that withstand these events contributes to the overall resilience of agriculture (Mittler 2006). Traits associated with resistance to pests and pathogens can be enhanced to mitigate the impact of changing pest dynamics and evolving pathogen strains under climate change. This adaptation reduces reliance on chemical interventions.

Adjusting plant phenology traits, such as flowering and maturation times, allows crops to align their growth cycles with changing climate conditions. This adaptation is crucial for maintaining productivity in the face of shifting seasonal patterns (Parmesan and Yohe 2003). Improving traits related to nitrogen-use efficiency contributes to sustainable agriculture by optimizing nutrient utilization. This adaptation is essential as climate change may affect the availability and dynamics of soil nutrients (Ju et al. 2006). Integrating traits that enhance beneficial biotic interactions, such as symbiotic relationships with microbes or natural predators of pests, can contribute to the overall resilience of crops within the agroecosystem (Tscharntke et al. 2012). Promoting functional trait diversity within agroecosystems enhances their resilience. Utilizing a diverse range of crops with different functional traits improves the stability and adaptability of the entire agricultural system (Lin 2011). Targeting plant functional traits related to water-use efficiency, such as reduced stomatal conductance and improved drought response mechanisms, can lead to crops that thrive with limited water resources. This adaptation is crucial for sustainable agriculture in water-scarce regions (Sinclair and Rufty 2012). Enhancing functional traits associated with nutrient uptake, utilization, and recycling can lead to crops with improved nutrient-use efficiency. This optimization reduces the need for synthetic fertilizers, lowering production costs and minimizing environmental impacts.

Modifying traits related to light interception, such as leaf architecture and chlorophyll content, enhances photosynthetic efficiency. Crops with improved light capture can maximize their energy conversion, leading to increased biomass and yield (Evans 2013). Functional traits influencing carbon-use efficiency, such as efficient allocation of carbon to growth and reduced respiratory losses, can be targeted for crop improvement. This optimization contributes to increased biomass and yield. Modifying traits related to root architecture, such as increased root branching and depth, enhances nutrient exploration and uptake efficiency. This adaptation is critical for crops to access nutrients in diverse soil conditions (Lynch 2013). Traits that confer resilience to abiotic stresses, such as heat or salinity, optimize resource-use efficiency by maintaining crop productivity under challenging conditions. Breeding for these traits ensures stable yields in variable environments (Mittler 2006). Optimizing traits related to efficient nitrogen fixation in leguminous crops enhances their ability to form symbiotic relationships with nitrogen-fixing bacteria. This adaptation reduces the need for external nitrogen inputs, promoting sustainable agriculture. Modifying traits related to xylem efficiency enhances water transport within plants, improving their ability to cope with water scarcity. This adaptation ensures that water is utilized optimally for various physiological processes. Utilizing perennial crops with traits conducive to carbon sequestration in the soil contributes to climate change mitigation. These crops efficiently capture and store carbon, enhancing overall resource-use efficiency (Glover et al. 2010). Developing crops with traits that act as sensors for resource availability, such as nutrient or water sensors, can revolutionize precision agriculture. These traits enable real-time monitoring and adaptive management, optimizing resource-use efficiency (Cobb et al. 2013). Encouraging traits that promote symbiotic interactions with beneficial microbes, such as mycorrhizal fungi, enhance nutrient uptake efficiency. This collaboration optimizes the use of soil nutrients, improving overall resource-use efficiency (van Der Heijden et al. 2015). Disease resistance genes: identifying and enhancing functional traits associated with disease resistance genes can confer broad-spectrum resistance to various pathogens. Breeding for these traits provides durable resistance, reducing the need for chemical pesticides (Dangl and Jones 2001). Modifying traits that result in the production of insect-resistant compounds, such as secondary metabolites or proteins toxic to pests, enhances resistance against insect herbivores. This adaptation reduces crop damage and the reliance on synthetic insecticides (Gatehouse 2002). Developing crops with allelopathic traits, which produce compounds inhibitory to competing plants or pests, can contribute to weed control. These traits enhance the overall resilience of the crop by reducing competition for resources (Inderjit and Duke 2003). Promoting traits associated with induced systemic resistance allows plants to activate their defense mechanisms in response to pathogen attack. Breeding for ISR enhances the overall resilience of crops against a range of pathogens (Pieterse et al. 2014). Antimicrobial peptides: enhancing traits that lead to the production of antimicrobial peptides strengthens the plant's defense against a wide range of pathogens, including bacteria, fungi, and viruses. This adaptation contributes to durable and broad-spectrum resistance. Pathogen recognition proteins: modifying traits associated with pathogen recognition proteins, such as receptor-like kinases, enhances the plant's ability to recognize and respond to specific pathogens. This adaptation facilitates targeted resistance mechanisms (Zipfel 2014). Encouraging traits that facilitate beneficial microbial associations, such as mycorrhizal fungi or rhizobacteria, enhances the plant's ability to resist pathogenic attacks. These symbiotic relationships contribute to a healthier and more resilient crop (Smith and Read 2010). Rapid plant defense responses: targeting traits associated with rapid defense responses, such as the hypersensitive response, enables plants to mount a swift and effective defense against invading pathogens. Breeding for these traits enhances resistance to biotic stress (Coll et al. 2011). Employing systematic gene pyramiding involves combining multiple resistance genes into a single crop variety. This approach enhances resistance durability by countering the ability of pathogens to overcome single-gene resistance (Fukuoka et al. 2014). Inducing traits associated with defense priming allows plants to precondition themselves for a more robust response to future pathogen attacks. Breeding for defense priming enhances the speed and effectiveness of the plant's immune system (Mauch-Mani et al. 2017). Understanding and enhancing host-plant resistance mechanisms, including physical barriers, can contribute to improved resistance against herbivores and pathogens. Breeding for these traits reduces crop vulnerability to biotic stress. Modifying traits related to nutrient uptake efficiency, such as root architecture and nutrient transporters, can improve the plant's ability to absorb essential nutrients from the soil. This adaptation enhances the overall nutritional content of the crop (Lynch 2013). Targeting traits that facilitate the accumulation of essential micronutrients, such as iron, zinc, and vitamin A, enhances the nutritional quality of crops. This biofortification strategy addresses nutrient deficiencies in human diets (Bouis and Saltzman 2017). Increased protein content: Modifying traits associated with protein biosynthesis and accumulation can increase the protein content of crops. This adaptation addresses global protein deficiencies and contributes to improved nutritional outcomes (Shewry and Tatham 2016). Modulating traits related to carbohydrate composition, such as starch and dietary fiber content, influences the nutritional quality and digestibility of crops. This adaptation contributes to addressing issues related to energy availability and dietary fiber intake (Tester and Langridge 2010). Targeting traits that reduce antinutritional compounds, such as phytates and tannins, enhances the bioavailability of essential nutrients in crops. This adaptation improves the overall nutritional value of the harvested produce (Raboy 2009). Modifying traits related to fatty acid composition can result in crops with healthier oil profiles. This adaptation addresses concerns related to diet-related health issues and contributes to the production of healthier cooking oils (Dyer and Mullen 2008). Enhancing traits related to amino acid biosynthesis and accumulation leads to crops with improved amino acid profiles. This adaptation addresses protein quality and contributes to meeting human dietary requirements (Galili and Amir 2013). Modifying traits associated with vitamin biosynthesis and accumulation can lead to crops enriched with essential vitamins. This adaptation addresses micronutrient deficiencies and contributes to improved human health (Paine et al. 2005). Targeting traits related to the production of phytochemicals, such as antioxidants and polyphenols, enhances the health-promoting properties of crops. This adaptation contributes to disease prevention and overall well-being (Pandey et al. 2020). Modifying traits associated with allergenic proteins in crops reduces the risk of allergic reactions in susceptible individuals. This adaptation addresses concerns related to food allergies and expands the inclusivity of certain crops in diets (Koppelman et al. 2005). Modulating traits related to nutrient partitioning within plants ensures that essential nutrients are directed toward edible plant parts. This adaptation maximizes the nutritional content of harvested crops (Fig. 2.4).

2.6 Conclusion

This chapter emphasizes the critical role of plant functional traits in influencing a plant's performance, ecological interactions, and responses to environmental conditions. In the context of crop improvement, these traits, spanning morphological, physiological, biochemical, and molecular attributes, offer substantial potential for enhancing crop yields, quality, and sustainability. However, challenges such as complex trait interactions, limited genetic resources, prolonged breeding cycles, and regulatory constraints impede their integration. These chapter aims to elucidate the significance of plant functional traits in crop enhancement, addressing both opportunities and challenges associated with their integration. It provides insights into the potential benefits and drawbacks of incorporating these traits into crop improvement programs. Classifying traits based on their relevance in crop improvement includes morphological, physiological, biochemical, and molecular traits. Examples range from root architecture and leaf morphology to genetic diversity and gene expression. Manipulating these traits provides opportunities for developing

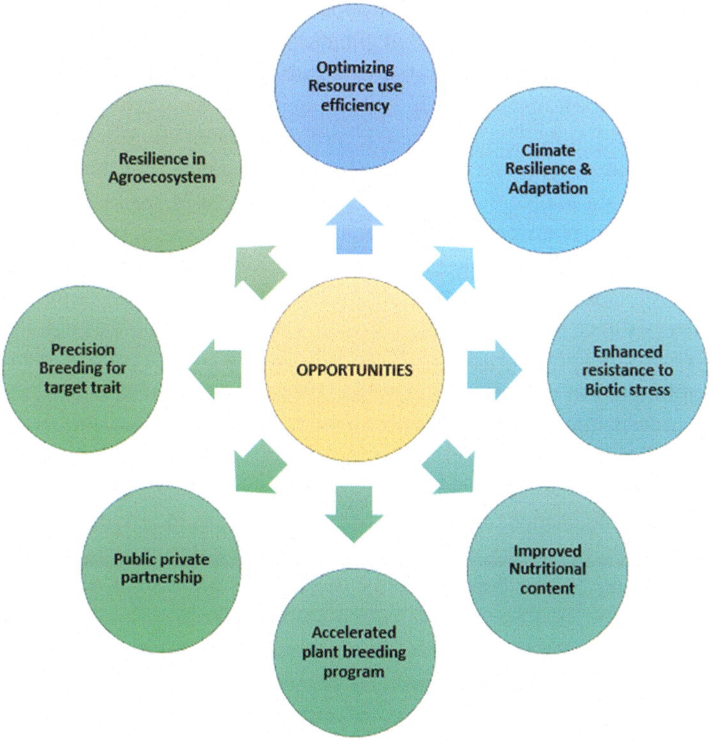

Fig. 2.4 Opportunities of using functional trait in crop improvement

crops with heightened adaptability, resource efficiency, and resilience to environmental challenges.

Quantitative genetics and genomics play a crucial role in understanding the genetic foundations of complex traits. The fusion of these disciplines aids in identifying genes and molecular pathways associated with plant functional traits, facilitating a comprehensive understanding for the selection and breeding of improved crop varieties. Illustrative cases demonstrate successful crop enhancement through the manipulation of plant functional traits, such as optimizing nitrogen-use efficiency in rice and developing drought-tolerant chickpea varieties. These examples highlight tangible benefits and showcase the potential of incorporating plant functional traits into crop improvement strategies. The effective incorporation of plant functional traits faces multifaceted challenges across genetic, environmental, quantification, trait trade-offs, agroecosystem integration, and ethical and regulatory dimensions. Challenges include genetic complexity, environmental responsiveness, quantification and measurement difficulties, trait trade-offs, agroecosystem complexities, and ethical and regulatory considerations. Precision breeding, utilizing CRISPR-Cas9 and genome editing tools, offers transformative opportunities. This approach allows precise modification of essential plant traits, leading to accelerated breeding timelines, mitigation of off-target effects, and stacking of multiple desirable traits. Precision breeding contributes to reduced environmental footprint, aligns with regulatory frameworks, and enhances public acceptance.

Functional traits play a pivotal role in adapting crops to climate change challenges. Breeding for traits associated with drought tolerance, heat stress resistance, optimal flowering time, and improved carbon fixation contributes to climate-resilient agriculture. Targeting plant functional traits related to water- and nutrient-use efficiency, light interception, and resilience to abiotic stress facilitates efficient resource utilization, contributing to sustainable agriculture. Manipulating functional traits associated with disease and insect resistance, allelopathic traits, and beneficial microbial associations offers a promising avenue for sustainable crop improvement, enhancing resistance to biotic stress. Enhancing nutritional traits through the modification of nutrient uptake efficiency, biofortification, increased protein content, and regulation of nutrient partitioning addresses global nutrition challenges, contributing to improved human health. In conclusion, leveraging plant functional traits through precision breeding and targeted modifications presents unprecedented opportunities for sustainable crop improvement, addressing global challenges in agriculture, climate change, resource-use efficiency, resistance to biotic and abiotic stress, and nutritional enhancement. Responsible and ethical application of these technologies can contribute significantly to global food security and agricultural sustainability.

References

Ainsworth EA, Long SP (2005) What have we learned from 15 years of free-air CO2 enrichment (FACE)? A meta-analytic review of the responses of photosynthesis, canopy properties and plant production to rising CO2. New Phytol 165(2):351–372

Anderson JT et al (2012) Evolution and adaptation in the world of plant pathogens: insights from population genomics. Mol Ecol 22(4):1036–1050

Bartels D, Sunkar R (2005) Drought and salt tolerance in plants. Crit Rev Plant Sci 24(1):23–58

Beyer P, Al-Babili S, Ye X, Lucca P, Schaub P, Welsch R, Potrykus I (2002) Golden rice: introducing the β-carotene biosynthesis pathway into rice endosperm by genetic engineering to defeat vitamin A deficiency. J Nutr 132(3):506S–510S

Bouis HE, Saltzman A (2017) Improving nutrition through biofortification: a review of evidence from HarvestPlus, 2003 through 2016. Glob Food Sec 12:49–58

Caligari PD (2001) Plant breeding and crop improvement. In: eLS. Wiley

Cobb JN, DeClerck G, Greenberg A, Clark R, McCouch S (2013) Next-generation phenotyping: requirements and strategies for enhancing our understanding of genotype–phenotype relationships and its relevance to crop improvement. Theor Appl Genet 126:867–887

Coll NS et al (2011) Programs for cell death: the immune defense. Annu Rev Phytopathol 49(1):567–591

Conner JK, Hartl DL (2004) A primer of ecological genetics, vol 425. Sinauer Associates, Sunderland, MA

Dangl JL, Jones JD (2001) Plant pathogens and integrated defence responses to infection. Nature 411(6839):826–833

Dyer JM, Mullen RT (2008) Engineering plant oils as high-value industrial feedstocks for biorefining: the need for underpinning cell biology research. Physiol Plant 132(1):11–22

Evans JR (2013) Improving photosynthesis. Plant Physiol 162(4):1780–1793

Farooq M, Wahid A, Kobayashi NSMA, Fujita DBSMA, Basra SMA (2009) Plant drought stress: effects, mechanisms and management. In: Sustainable agriculture. Springer, pp 153–188
Franks SJ, Hoffmann AA (2012) Genetics of climate change adaptation. Annu Rev Genet 46:185–208
Fukuoka S, Yamamoto SI, Mizobuchi R, Yamanouchi U, Ono K, Kitazawa N et al (2014) Multiple functional polymorphisms in a single disease resistance gene in rice enhance durable resistance to blast. Sci Rep 4(1):4550
Furbank RT, Tester M (2011) Phenomics–technologies to relieve the phenotyping bottleneck. Trends Plant Sci 16(12):635–644
Galili G, Amir R (2013) Fortifying plants with the essential amino acids lysine and methionine to improve nutritional quality. Plant Biotechnol J 11(2):211–222
Gatehouse JA (2002) Plant resistance towards insect herbivores: a dynamic interaction. New Phytol 156(2):145–169
Gatehouse AM, Gatehouse JA (1998) Identifying proteins with insecticidal activity: use of encoding genes to produce insect-resistant transgenic crops. Pestic Sci 52(2):165–175
Glover JD, Reganold JP, Bell LW, Borevitz J, Brummer EC, Buckler ES et al (2010) Increased food and ecosystem security via perennial grains. Science 328(5986):1638–1639
Hickey LT, Hafeez AN, Robinson H, Jackson SA, Leal-Bertioli SC, Tester M et al (2019) Breeding crops to feed 10 billion. Nat Biotechnol 37(7):744–754
Hirel B, Lea PJ (2011) The molecular genetics of nitrogen use efficiency in crops. In: The molecular and physiological basis of nutrient use efficiency in crops. Wiley, pp 139–164
Hsu PD, Scott DA, Weinstein JA, Ran FA, Konermann S, Agarwala V et al (2013) DNA targeting specificity of RNA-guided Cas9 nucleases. Nat Biotechnol 31(9):827–832
Inderjit, Duke SO (2003) Ecophysiological aspects of allelopathy. Planta 217:529–539
Jinek M, Chylinski K, Fonfara I, Hauer M, Doudna JA, Charpentier E (2012) A programmable dual-RNA–guided DNA endonuclease in adaptive bacterial immunity. Science 337(6096):816–821
Ju XT, Kou CL, Zhang FS, Christie P (2006) Nitrogen balance and groundwater nitrate contamination: comparison among three intensive cropping systems on the North China plain. Environ Pollut 143(1):117–125
Koppelman SJ, De Jong GAH, Laaper-Ertmann MD, Peeters KABM, Knulst AC, Hefle SL, Knol EF (2005) Purification and immunoglobulin E-binding properties of peanut allergen Ara h 6: evidence for cross-reactivity with Ara h 2. Clin Exp Allergy 35(4):490–497
Lin BB (2011) Resilience in agriculture through crop diversification: adaptive management for environmental change. Bioscience 61(3):183–193
Lobell DB, Gourdji SM (2012) The influence of climate change on global crop productivity. Plant Physiol 160(4):1686–1697
Long SP, Ort DR (2010) More than taking the heat: crops and global change. Curr Opin Plant Biol 13(3):240–247
Lucht JM (2015) Public acceptance of plant biotechnology and GM crops. Viruses 7(8):4254–4281
Lynch JP (2013) Root phenes for enhanced soil exploration and phosphorus acquisition: tools for future crops. Plant Physiol 156(3):1041–1049
Mackay TF (2014) Epistasis and quantitative traits: using model organisms to study gene– gene interactions. Nat Rev Genet 15(1):22–33
Mackay TF, Stone EA, Ayroles JF (2009) The genetics of quantitative traits: challenges and prospects. Nat Rev Genet 10(8):565–577
Martin AR, Isaac ME (2015) Plant functional traits in agroecosystems: a blueprint for research. J Appl Ecol 52(6):1425–1435
Mauch-Mani B, Baccelli I, Luna E, Flors V (2017) Defense priming: an adaptive part of induced resistance. Annu Rev Plant Biol 68:485–512
Meuwissen TH, Hayes BJ, Goddard M (2001) Prediction of total genetic value using genome-wide dense marker maps. Genetics 157(4):1819–1829
Mittler R (2006) Abiotic stress, the field environment and stress combination. Trends Plant Sci 11(1):15–19
Munns R (2005) Genes and salt tolerance: bringing them together. New Phytol 167(3):645–663

Nicotra AB, Atkin OK, Bonser SP, Davidson AM, Finnegan EJ, Mathesius U et al (2010) Plant phenotypic plasticity in a changing climate. Trends Plant Sci 15(12):684–692
Paine JA, Shipton CA, Chaggar S, Howells RM, Kennedy MJ, Vernon G et al (2005) Improving the nutritional value of Golden Rice through increased pro-vitamin A content. Nat Biotechnol 23(4):482–487
Pandey MK, Pandey AK, Kumar R, Nwosu CV, Guo B, Wright GC et al (2020) Translational genomics for achieving higher genetic gains in groundnut. Theor Appl Genet 133:1679–1702
Parmesan C, Yohe G (2003) A globally coherent fingerprint of climate change impacts across natural systems. Nature 421(6918):37–42
Perez-Harguindeguy N, Diaz S, Garnier E, Lavorel S, Poorter H, Jaureguiberry P et al (2013) New handbook for standardised measurement of plant functional traits worldwide. Aust Bot 61:167–234
Philippot L, Spor A, Hénault C, Bru D, Bizouard F, Jones CM et al (2013) Loss in microbial diversity affects nitrogen cycling in soil. ISME J 7(8):1609–1619
Pieterse CM, Zamioudis C, Berendsen RL, Weller DM, Van Wees SC, Bakker PA (2014) Induced systemic resistance by beneficial microbes. Annu Rev Phytopathol 52:347–375
Poorter H, Niklas KJ, Reich PB, Oleksyn J, Poot P, Mommer L (2012) Biomass allocation to leaves, stems and roots: meta-analyses of interspecific variation and environmental control. New Phytol 193(1):30–50
Power AG (2010) Ecosystem services and agriculture: tradeoffs and synergies. Philos Trans R Soc B Biol Sci 365(1554):2959–2971
Pretty J, Benton TG, Bharucha ZP, Dicks LV, Flora CB, Godfray HCJ et al (2018) Global assessment of agricultural system redesign for sustainable intensification. Nat Sustain 1(8):441–446
Raboy V (2009) Approaches and challenges to engineering seed phytate and total phosphorus. Plant Sci 177(4):281–296
Rellstab C, Fischer MC, Zoller S, Graf R, Tedder A, Shimizu KK, Gugerli F (2017) Local adaptation (mostly) remains local: reassessing environmental associations of climate-related candidate SNPs in Arabidopsis halleri. Heredity 118(2):193–201
Rensink WA, Buell CR (2004) Arabidopsis to rice. Applying knowledge from a weed to enhance our understanding of a crop species. Plant Physiol 135(2):622–629
Reynolds MP, Quilligan E, Aggarwal PK, Bansal KC, Cavalieri AJ, Chapman SC et al (2016) An integrated approach to maintaining cereal productivity under climate change. Glob Food Sec 8:9–18
Ricroch AE, Ammann K, Kuntz M (2016) Editing EU legislation to fit plant genome editing: the use of genome editing technologies in plant breeding requires a novel regulatory approach for new plant varieties that involves farmers. EMBO Rep 17(10):1365–1369
Sadras VO, Lawson C (2013) Nitrogen and water-use efficiency of Australian wheat varieties released between 1958 and 2007. Eur J Agron 46:34–41
Shewry PR, Tatham AS (2016) Improving wheat to remove coeliac epitopes but retain functionality. J Cereal Sci 67:12–21
Sinclair TR, Rufty TW (2012) Nitrogen and water resources commonly limit crop yield increases, not necessarily plant genetics. Glob Food Sec 1(2):94–98
Singh H, Verma A (2013a) Physiological responses of rice cultivars to various nitrogen levels. Int J Agric Environ Biotechnol 6(3):383–388
Singh H, Verma A (2013b) Characterization and screening of high nitrogen efficient rice genotype to curtail environmental pollution. Int J Agric Environ Biotechnol 6(1):23–30
Singh H, Verma A, Shukla A (2010a) Comparative study of yield and yield components of hybrid and inbred genotypes of rice (Oryza Sativa L.). Int J Agric Environ Biotechnol 3:355–360
Singh H, Verma A, Krishnamoorthy M, Shukla A (2010b) Consequence of diverse nitrogen levels on leaf pigments in five rice genotypes under field emergent circumstance. Int J Bioresour Stress Manag 1:189–193
Singh H, Verma A, Rai SK (2013a) Biochemical evaluation of different rice genotypes grown at various nitrogen levels using SDS-PAGE. Curr Adv Agric Sci 5(1):144–146

Singh H, Verma A, Shukla A (2013b) Guttation fluid as a physiological marker for selection of nitrogen efficient rice (Oryza sativa L.) genotypes. Afr J Biotechnol 12(44):6276–6281
Singh H, Verma A, Ansari MW, Shukla A (2014) Physiological response of rice (Oryza sativa L.) genotypes to elevated nitrogen applied under field conditions. Plant Signal Behav 9:e29015. https://doi.org/10.4161/psb.29015
Singh S, Singh H, Sharma SK, Nautiyal R (2021) Seasonal variation in biochemical responses of bamboo clones in the sub-tropical climate of Indian Himalayan foothills. Heliyon 7(4):e06859. https://doi.org/10.1016/j.heliyon.2021.e06859
Singh M, Singh H, Kumar A, Kumar M, Barthwal S, Thakur A (2024) Soil nitrogen availability determines the CO2 fertilization effect on tree species (Neolamarckia cadamba): growth and physiological evidence. Environ Sustain. https://doi.org/10.1007/s42398-023-00300-w
Smith SE, Read DJ (2010) Mycorrhizal symbiosis. Academic Press
Tanksley SD, McCouch SR (1997) Seed banks and molecular maps: unlocking genetic potential from the wild. Science 277(5329):1063–1066
Tester M, Langridge P (2010) Breeding technologies to increase crop production in a changing world. Science 327(5967):818–822
Tscharntke T, Clough Y, Wanger TC, Jackson L, Motzke I, Perfecto I et al (2012) Global food security, biodiversity conservation and the future of agricultural intensification. Biol Conserv 151(1):53–59
van Der Heijden MG, Martin FM, Selosse MA, Sanders IR (2015) Mycorrhizal ecology and evolution: the past, the present, and the future. New Phytol 205(4):1406–1423
Varshney RK, Pandey MK, Bohra A, Singh VK, Thudi M, Saxena RK (2019) Toward the sequence-based breeding in legumes in the post-genome sequencing era. Theor Appl Genet 132(3):797–816
Violle C, Navas ML, Vile D, Kazakou E, Fortunel C, Hummel I, Garnier E (2007) Let the concept of trait be functional! Oikos 116(5):882–892
Violle C, Enquist BJ, McGill BJ, Jiang LIN, Albert CH, Hulshof C et al (2012) The return of the variance: intraspecific variability in community ecology. Trends Ecol Evol 27(4):244–252
Wahid A, Gelani S, Ashraf M, Foolad MR (2007) Heat tolerance in plants: an overview. Environ Exp Bot 61(3):199–223
Zhang F, Voytas DF (2011) Targeted mutagenesis in Arabidopsis using zinc-finger nucleases. In: Plant chromosome engineering: methods and protocols. Humana Press, pp 167–177
Zipfel C (2014) Plant pattern-recognition receptors. Trends Immunol 35(7):345–351

The Interplay Between Plant Functional Traits and Climate Change

3

Chandra Kanta, Abhishek Kumar, Anjali Chauhan, Hukum Singh, and Ishwar Prakash Sharma

Abstract

Global climate change poses extraordinary challenges to ecosystems, profoundly affecting plant species and their functional traits. The chapter explores the dynamic relationships between crucial plant functional traits and various aspects of climate change, including temperature changes, rainfall patterns, and elevated atmospheric CO_2 concentrations. Understanding the complex connection between plant functional traits and climate change becomes paramount for sustainable development as universal heat rises and climatic patterns change. This chapter also highlights the dynamic interactions between plant characteristics and the evolving climate while being true to the SDGs. We investigate how climate change affects the functional traits of plants, such as leaf morphology, photosynthetic efficiency, and reproductive strategies, and how these changes, in turn, affect ecosystems. Our investigation shows that climate-induced modifications in plant functional traits have far-reaching consequences for ecosystem structure and function. Changes in leaf morphology, photosynthesis rates, and reproductive strategies impact individual plant fitness, community structure, and overall ecosystem resilience. It is important to highlight the complex relationships between plant functional traits and climate change to predict the future of ecosystems. To provide a comprehensive understanding of the mechanisms driving plant characteristic responses to changes in climate, further research direc-

C. Kanta (✉) · A. Chauhan
School of Applied and Life Sciences (SALS), Uttaranchal University, Dehradun, Uttarakhand, India

A. Kumar · H. Singh
Forest Research Institute, Dehradun, Uttarakhand, India

I. P. Sharma
Patanjali Herbal Research Division, Patanjali Research Foundation, Haridwar, Uttarakhand, India

N. Kumar, H. Singh (eds.), *Plant Functional Traits for Improving Productivity*,
https://doi.org/10.1007/978-981-97-1510-7_3

tions shall focus on integrating observational and experimental approaches, considering both individual species and community-level dynamics. This synthesis contributes to the broader effort to develop effective strategies to mitigate and adapt to the ecological consequences of ongoing climate change.

Keywords
Climate change · CO_2 · Community · Ecosystem · Plant functional traits · Temperature

3.1 Introduction

Climate change is the greatest environmental issue for the present and upcoming biodiversity (Sala et al. 2000; Bellard et al. 2012; Kumar et al. 2019). The climate system is like a large puzzle with many pieces. It includes the air we breathe in, the land we stand on, the ice, snow, seas, oceans, and all plants and animals. All these parts are interconnected and work together to make our Earth's weather and climate (Hannah and Lovejoy 2003). Climate changes mean a long-lasting and noticeable shift in the average conditions in things which includes temperature, air pressure, rainfall, and many more, and it typically lasts for many years or even longer (Solomon 2007). In the present day, there is no doubt that shifts in the climate are having a strong effect on the existence of humans, and these are typically connected to wider-scale events (Alnsour and Ludwig-Müller 2015; Kumar et al. 2019). Climatic changes are making it harder to ensure there is enough food for everyone in the world, and this is becoming a huge problem for the twenty-first century. As the population keeps growing in a world with changing and challenging situations, it is a strong question of how we can make sure there is plenty of food for everyone (Kang et al. 2009). Appraisement of WHO states that 60% of people worldwide rely on traditional medicines, which are primarily derived from plants. In the contribution toward climatic changes, human interference plays an important role across the planet (Mishra 2016).

Plant functional traits encompass a wide array of morphophysiological and phenological plant features that significantly impact a plant's overall fitness by influencing its ability to grow, reproduce, and survive (Violle et al. 2007; Kumar et al. 2015, 2016; Singh et al. 2020). In the early period of plant functional traits theory, three key traits, i.e., plant height (Ph), leaf surface area (LSA), and seed mass (SM), are identified as the primary factors influencing how plants shape atmospheric functions (Westoby 1998). These features of plants are the product of biological adaptations to the environment, reflecting fundamental trade-offs among particular species, as they dictate how plants acquire, process, and allocate resources, ultimately shaping the plant strategies for survival and reproduction (Chapin et al. 2000; Kumar et al. 2017). The effect of variations in climate on the functional traits of plants is closely connected with several UN sustainable goals such as goal 13 (climate action) and goal 15 (life on land).

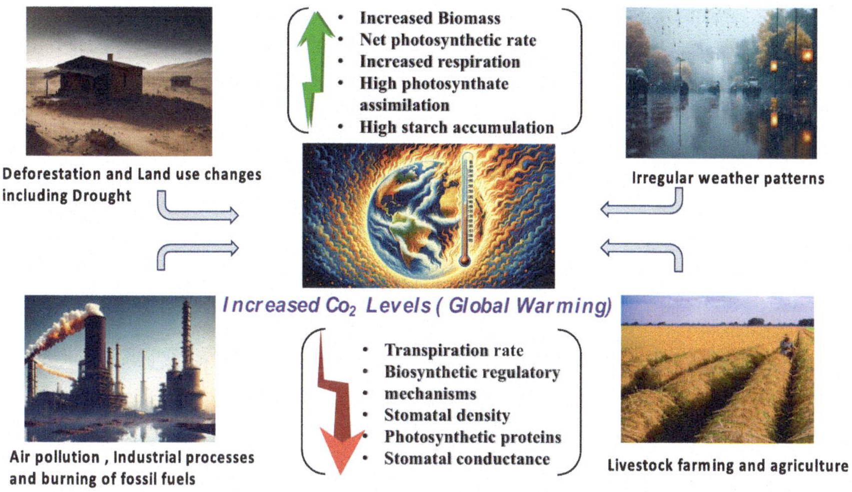

Fig. 3.1 Effects of global climate change drivers resulting in increased CO_2 levels (the green arrow indicates the physiological changes that increase due to the elevated levels and red marks the changes that are downregulated by the increase in the CO_2 level

Understanding the relationship between climate change and plant functional traits and their suggestions for SDGs is essential for designing effective strategies that promote sustainable development while managing the tasks posed by global environmental change. Differences in functional traits have far-reaching effects on ecosystem characteristics such as biodiversity, the cycle of nutrients, and soil properties, as various species adapt to and are shifted by factors like climate, disturbances, and fundamental interactions with other organisms (Lavorel and Garnier 2002). Global climate changes strongly affect land ecosystems, biological science, types of plants, and grazing ecology which are thoroughly connected to climate changes.

This means that florae, which are a significant part of land ecosystems, react to climatic change in various ways. As a result, plants are excellent subjects for studying and pretending the impacts of climatic crisis. This research delves into the interaction between plant useful traits and environment change, seeking to offer a thorough comprehension of how plant features react to changes in the environment and how these reactions impact an ecosystem. Hence, there is an urgent need to investigate the impact of global climatic change on the functional traits of plants to facilitate effective initiatives in this evolving world (Fig. 3.1).

3.2 Climate Influencers

The plant relies on specific factors like sunlight, temperature, precipitation, carbon dioxide, and humidity. These influencers that affect climate are essential not only for plant development but also for the well-being of human nutrition. These factors exhibit variations in their levels across different geographical locations. The growing progress of plants is fully dependent upon temperature, with every species having a distinct optimal temperature range to remain alive and thrive in its specific ecological niche (Hatfield and Prueger 2015). Population is increasing day by day which creates a huge strain on the Earth, emerging as per the key affecting catalysts for climate change. This altering climate exerts a reflective effect on various aspects of the environment including temperature patterns, soil conditions, and rainfall. Some climate influencers and major factors affecting climate are enumerated in Tables 3.1 and 3.2, respectively.

3.3 Acclimatization in Plants: Environmental Responses

A plant experiences several chemical, physiological, and cellular changes when its tolerance and adaptability to heat are surpassed. These changes can have detrimental consequences on growth and development, even resulting in plant mortality. When exposed to high temperature, plant species that are thought to be tolerant of high temperatures can activate their acclimation mechanisms through modifications to their transcriptome, proteome, and metabolome within the plant cell. The composition and integrity of cell membranes alter in response to heat, and this is one of the first acclimatization reactions (Kumar et al. 2020a, b).

Table 3.1 List of some climatic influencers

S. no.	Climate influencers	Impact of climatic changes	References
1.	Drought	Climate change contributes to severe drought, driven by increased emissions of greenhouse gases that increase air temperatures. This increase in temperature, in turn, increases the rate of evaporation, making dry soils less efficient at absorbing moisture from the ground	https://www.climaterealityproject.org/blog/facts-about-climate-change-and-drought
2.	Rainfall	There have been reports of extreme rainfall and snowfall as a result of climatic changes across the globe	Tollefson (2016)
3.	Air pollution	The primary source of atmospheric pollution is CO_2 emissions, but climate change is also influenced by additional air pollutants collectively called as temporary climate-releasing toxins. This group includes methane (CH_4), black carbon, sulfate aerosols, and ground-level ozone emerging as important contributors along with carbon dioxide	https://www.iass-potsdam.de/en/output/dossiers/air-pollution-and-climate-change

Table 3.2 Chief influencers affecting climate

S. no.	Influencers	Description
1.	Prevailing worldwide air pattern	Several air patterns are observed in the northern and southern hemispheres, with airflows linked to seasonal fluctuations
2.	Impacts of geography	The geographical characteristics of a location, including its distance from mountains and extensive bodies of water, play a vital role in determining its climate. The location of an area dictates its predominant wind circulation
3.	Altitude or elevation influences the climate	At higher altitudes, climatic conditions generally become cooler
4.	Shifting in climate conditions across the years	Cold and hot phases are primary fluctuations on the earth. Sometimes, these changes in conditions may be of short duration, but at other times, they may extend over hundreds to thousands of years
5.	Topography	The topographical location of a particular area can have a profound impact on its climate. Mountain ranges are accepted as natural hindrances to the movement of air
6.	Earth's surface	The degree of atmospheric heating depends upon the number of sunrays absorbed or reflected by a surface, with heavily vegetated areas excelling at absorption compared to snowy and icy landscapes, which usually exhibit exceptional reflectivity

Source: https://www.climateandweather.net/global-warming/factors-that-influence-climate/

3.3.1 Stomatal Regulation

As key components of both the hydrological and carbon cycles, stomata control how much carbon dioxide (CO_2) plants take in for photosynthesis as well as how they lose water widely through transpiration to the atmosphere (Berry et al. 2010; Kumar et al. 2021a, b). Plant evolution has relied heavily on the need to balance CO_2 uptake and water loss, a need that will only grow in importance as the globe gets hotter and drier (Swann 2018). A well-acknowledged set of mechanisms by which plants adapt to their surroundings by controlling phenotypic plasticity and life history transitions is represented by epigenetics (Duncan et al. 2014). It is crucial to comprehend how epigenetic mechanisms affect phenotypes and fitness to comprehend how organisms can quickly adapt to changing environmental conditions. Plants adjust photosystem stoichiometry according to the light availability (Schottler and Toth 2014). The photosynthetic yield is impacted by changes in several abiotic variables, including temperature, the availability of nutrients and water, and the intensity and quality of incident light. As a result, regulatory responses have evolved in all photosynthetic species to adapt their photosynthetic capacities to real environmental conditions. Rapid alterations in light quality, such as sun flecks resulting from leaves' motion, are offset by state changeovers wherein the photosystems' antenna structure is reorganized. On the other hand, the majority of dense plant stands usually show persistent differences in light quality. A long-term response (LTR) counteracts these by readjusting the stoichiometry of the photosystem to restore the photosynthetic energy balance. State transitions, which take place on the scale of minutes, are a method via which the relative antenna cross sections of the photosystems vary, redistributing excitation energy among them. A portion of PSII's photosynthesis antenna complex moves laterally to produce the variation. A redox-sensitive kinase is triggered by the reduction of the plastoquinone (PQ) pool, which moves electrons from PSII to the cytochrome b6f (Cyt b6f) complex (Schiphorst et al. 2022). This causes the mobile LHCII to become phosphorylated, detach from PSII, and attach to photosystem I (PSI). The thylakoid membrane's redox signal governs both the short- and long-term acclimation responses, which act at separate periods but have the same functional goal of improving the rate-limiting photosystem's electron transport capacity (Johnson and Wientjes 2020).

3.3.2 Photosynthetic Adjustments (Heat Shock Proteins and Osmoregulation)

In reaction to variations in temperature, carbon dioxide concentration, and light intensity, plants can rapidly adjust their photosynthetic rates (Dusenge et al. 2019). Because of their adaptability, plants can maximize their efficiency in capturing energy and using resources by the local environmental conditions. In addition, plants produce heat shock proteins when exposed abruptly to high temperatures (Timperio et al. 2008). Under stressful circumstances, these proteins defend cellular structures and preserve regular physiological processes. Misfolded proteins build up and attach to heat shock proteins (HSPs) at high temperatures, triggering the release of heat stress transcription factors (HSFs) and the heat strain response (Gomez-Pastor et al. 2018). Furthermore, heat strain triggers MAPK, which controls the production of HSP genes. Heat-induced modifications to membrane fluidity and calcium signaling are linked to MAPK activation and are crucial for the expression of HSP genes and heat resistance (Kumar et al. 2017, 2021).

Plants may experience direct or indirect hyperosmotic pressure as a result of abiotic stressors including scarcity and excessive salinity (Yadav et al. 2020) governed by the mechanism of osmoregulation. Osmoregulation is the process by which plant cells adjust to harsh conditions by increasing the concentration of solutes, mostly through the buildup of different inorganic and organic materials. Osmoregulation-related chemicals can be generically classified into two groups: organic solutes (including proline, betaine, soluble sugars (SSs), and polyols) and inorganic ions (like K^+, Na^+, and Ca^{2+}) (Bhattacharyya 2022). Drought tolerance can be partially achieved by storing proline, SS, and soluble protein (SP) and reducing their osmotic potential. Moreover, autophagy plays a substantial role in plant resistance mechanisms and is widely associated with various aspects of plant growth, development, and responses to stress (Wang et al. 2021). Numerous plants exhibit the autophagy process, which is controlled by the ATG gene family.

3.3.3 Leaf Angles

The pre-Socratic philosophers Anaxagoras and Empedocles were interested in leaf angles (Whippo and Hangarter 2006), but scientists such as Darwin were the first to conduct contemporary scientific research on leaf angles in the nineteenth century. Studying the leaf angle has been important in many interesting ecological studies (Givnish 1987). Foliage approach is essential for several roles of leaf angle, ecology developments, and remote sensing measurements in specific competition which have rarely been explored (Pearcy et al. 2005). With the rapidly evolving technologies, we now understand the importance of the leaf angles' role to some extent (Sack and Scoffoni 2013). Plants undergo two types of environmental adaptation: ecophysiological changes, which occur over shorter periods (hours to months), and genetic modifications, which occur over longer periods (years). This also applies to leaf angle, where variations enhance carbon gain and minimize stresses (Kao and Forseth 1992). MacArthur and Pianka (1966) recently designated eco-evolutionary optimality the evolutionarily-based optimality theory. Harrison et al. (2021) provide a helpful framework to analyze differences in leaf-angle and its interaction with further features. The existing framework, which posits that leaves tend to be more upright in dehydrated and warm settings and more horizontal in moist and light-limited conditions, fails to account for the diverse range of leaf angle variations observed among individuals within the same environment. The initial and crucial task is to systematically document the differences in foliage angle, especially its upright distribution, across a broad spectrum of environments, spanning from humid to boreal environments (Yang et al. 2023).

3.3.4 Root Dynamics

We have come a long way in understanding the components of plants belowground, but we are still in the early stages of investigating the complicated links between root functions and features. The likely belongings of several universal changes in variables on the long-term dynamics of biomass and the combinations of functional traits are not well understood (Wu et al. 2023). In a meta-review analysis, positive plant behaviors to climate and environmental deviations across biomes seem to be best predicted by seven key traits: plant height, lower relative growth rate, greater water-use efficiency (WUE), specific leaf area, larger resprouting ability, higher wood density, and better rooting depth (Kühn et al. 2021). In another study by Freschet et al. (2021), they did a critical assessment of the areas of current strength and knowledge gaps, as well as the identification of research challenges going forward in the arena of root ecology. Importantly, they discovered that the belowground features that are most broadly significant for the functioning of plants and ecosystems are not the ones that are most frequently measured. Understanding the spatial and temporal variations in root traits, along with their associated impacts, is crucial, particularly for making well-informed recommendations regarding when and where measurements should be taken. Utilizing hierarchical spatial sampling and successive election practices would offer valuable insights into the spatial and chronological dynamics of root traits and their influence on system working. De Vries (2021) comprehensively comprehended how interactions between traits and the environment influence natural selection in unfamiliar settings; it is essential to examine specific plants with multifaceted characteristics within dynamic surroundings. This encompasses abiotic gradients, biotic interactions, and their collective impact on crucial plant fitness determinants. Evolutionary functional-structural plant demonstration is specifically considered to simulate character environment connections steering eco-evolutionary dynamics at various scales from individual to community levels. This modeling approach facilitates efficient exploration of extensive, intricate, and dynamic fitness landscapes that arise when in view of plants with multidimensional traits in multidimensional environments.

3.4 Ominous Effect of Plant Functional Traits on Earth's Climatic Shift

Plant functional trait changes can signal shifts in climate patterns. These traits adapt in response to alterations in the environment. For example, when the environmental conditions shift from being humid to experiencing drought, the size of plant leaves becomes smaller, while their nitrogen and phosphorus levels increase. Simultaneously, there is a reduction in the carbon-nitrogen ratio, accompanied by an increase in the rate at which plants assimilate nutrients from their surroundings (Reich et al. 1999). Plants respond to shifts in their surroundings by altering their outward appearance and their inner physiological features while modifying their input in the functions of the ecosystem (Qiuhong et al. 2008). The changes in both the physical and functional traits of plants, encompassing morphological variations, stomatal behavior, modifications in photosynthetic structures and pathways, alterations in plant photosynthesis and respiration rates, and adaptations in water physiology, all represent noticeable indicators of universal changes. The rise in species richness or the expansion of plant communities serves as an indication of global warming. Moreover, the functional characteristics of plants not only mirror specific adaptations but also signify the functional attributes of the ecosystem. Enhancing the methodology of research aimed at elucidating the crucial role of plants in climatic changes can streamline the spatiotemporal dimensions and foster a linkage between global changes and the dynamic shifts occurring at the population and community levels (Zhang and Li 2019).

Plant leaf traits exhibit substantial variation across diverse locales and species worldwide. Moreover, within the same plant species, these qualities display significant variability when exposed to author environments, showing their remarkable plasticity and underscoring the plant's capacity to adapt to and serve as indicators of outer environmental conditions (Guo et al. 2015). The role of plants as indicators of climate change has become increasingly apparent, and ecologists are showing more interest in the functional attributes of plants. However, research on how plants respond to climatic changes remains notably insufficient. Hence, we will gain an understanding of the effect of climatic changes on plant functional traits by examining the interplay between these traits and climate shifts.

3.5 Influence of Different Climatic Features on Plant Growth/Development and Morphology

3.5.1 Impact of Upraised CO_2

The Industrial Revolution initiated a significant change in atmospheric CO_2 level from 280 ppm to more than 400 ppm. The indirect consequences of CO_2 on plants are also important, because of its dual role as a strong greenhouse gas, contributing to global warming and subsequent climate changes (Meehl et al. 2007). Plants experience the direct effects of CO_2 through changes in photosynthetic gas exchange, which subsequently affect various developmental processes. In 12 free air CO_2 enrichment experiments, based on extensive studies involving 40 different species, the stimulation of photosynthetic rate associated with carbon flow by elevated CO_2 reached the mean of 31% (Ainsworth and Long 2005). Ainsworth and Long (2005) reported that exposure to uplifted CO_2 resulted in an average growth of 20% in aboveground biomass among C_3 species. In contrast, Reich et al. (2014) reported that the impact of high CO_2 level on aboveground biomass varied in the field context, reaching up to 33%. However, the extent of this activation depends upon water and nitrogen availability, resulting in a reduced biomass response under dry and low-nutrient conditions (Singh et al. 2013a, b; Reich et al. 2014). It has been also observed that increased CO_2 causes substantial increases in root biomass in several crop species (Madhu and Hatfield 2013; Kumar et al. 2022).

In a lot of species, such as soybean, wheat, rice, peanuts, and beans, increases in shoot biomass are related to significant increases in seed yield (Hatfield et al. 2011; Singh and Verma 2013a, b). Soybean (*Glycine max*) serves as a widely investigated model for understanding how a complete plant's growth and generative result respond to increased CO_2, as shown in Fig. 3.2. In response to increased CO_2, soybean exhibits developmental changes, including larger leaf size and enlarged leaf nodes (Dermody et al. 2006), increased root length, and altered root depth dissemination and nodule pattern (Gray et al. 2012, 2016), with increasing pods number and seed yield (Morgan et al. 2005; Bishop et al. 2015). The increased biomass highlights the importance of CO_2 reactions as a characteristic of agricultural interest, prompting exploration of the complex interactions between the advancement process and carbon dioxide in shaping reproductive output.

3.5.2 Impact of Increased Temperature

The 0.85 °C upsurge in the biosphere's mean surface temperature from 1880 to 2012 can be attributed to the increasing accumulation of carbon dioxide and other greenhouse gases (Hartmann et al. 2013). Global average surface temperatures will

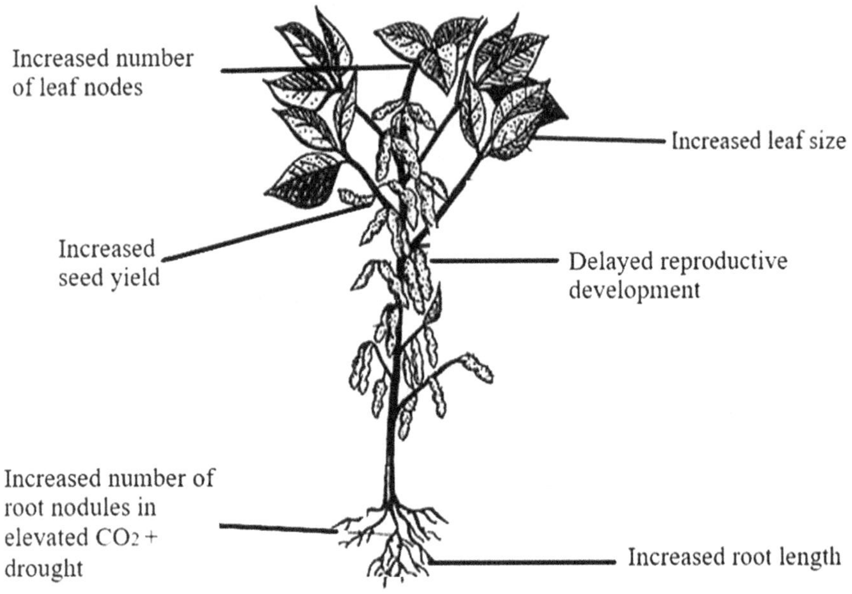

Fig. 3.2 In the diagram, the consequences of increased carbon dioxide on the growth and development of soybean are illustrated (Bishop et al. 2014, 2016; Dermody et al. 2006). (Drawing is adapted from https://www.lhf.org/learning-fields/crops/soybeans/)

probably increase by 1.0–3.7 °C by the termination of the century. In addition to the continued rise in the overall world's temperature, the flora is set to undergo heat stress due to the rising occurrence, strength, and span of heat waves. In contrast to evenly distributed atmospheric carbon dioxide, future projections of global surface temperatures show considerable variation across geographic regions. As a result, plants in different regions will experience these changes differently. For instance, surface temperatures in the Arctic are likely to rise more rapidly than the global mean (Pachauri et al. 2014).

Lobell and Asner (2003) reported that the effect of temperature on corn and soybean yields in the Midwest United States was different from that in the Northern Great Plains. In areas where temperature demonstrated an unfavorable correlation with yield, a 1 °C rise in growing season temperature was connected with an estimated 17% reduction in yield for both corn and soybeans (Lobell and Asner 2003). Despite its thermal stability up to 50 °C, RuBisCo (ribulose-1,5-bisphosphate carboxylase/oxygenase) exhibits variation in its efficiency to catalyze CO_2 fixation in organic compounds under elevated temperatures. This change in plant physiology provides a clear illustration of how increased temperature affects the process of photosynthetic carbon assimilation (Crafts-Brandner and Salvucci 2000). The constraints on photosynthesis under moderately increased temperatures can be attributed to the reduced activity of RuBisCo, an enzyme responsible for eliminating inhibitory molecules from the catalytic site of RuBisCo.

Alternatively, limitations may arise from the reduced revival of ribulose-1,5-bisphosphate (RuBP), the essential 5-carbon sugar that serves as a substrate for the carboxylation reaction of RuBisCo (Sage et al. 2008). The temperature response is also affected by photosynthetic functional type, with C_3 plants showing a lower thermal optimum for photosynthesis than C_4 plants (Sage and Kubien 2007). This functional perspective delivers insight into how plant yield responds to warming, as species-specific temperature plays a key role in determining photosynthetic response to optimal growing temperature, which ultimately controls plant growth and yield.

3.5.3 Impact of Drought Stress

Temperature change will probably increase aquatic stress in plants in various global areas. The incidence of drought has already risen significantly since the mid-twentieth era (Dai 2011), and the frequency of drought is projected to increase by the conclusion of the twenty-first century in areas that are currently arid. Drought appears with differing levels of strength and duration, causing widespread effects on plant development, and is commonly well-defined as a prolonged period of unusually dry weather; drought causes severe hydrological imbalances that can disrupt ecosystems and agricultural systems. For decades, the importance of water availability as an important abiotic factor affecting crop yields has been widely recognized (Boyer 1982). It's important role in plant growth processes underlines its importance. Due to drought stress, there is currently a considerable decrease in plant yield and agricultural produce. The extent of these results varies and depends upon factors such as the developmental stage at which the flora undergoes drought stress (Skirycz et al. 2010; Verelst et al. 2010) and soil-related factors, including soil texture. Furthermore, wide intra- and interspecies differences in the way organisms respond developmentally and physiologically to drought stress result in a diverse sensitivity to climate change-induced drought in different ecosystems covering both natural and agricultural landscapes.

Drought stress triggers specific evolutionary adaptations in various plant parts and tissues. In a general sense, plants respond to drought by reallocating resources toward root tissue, which has deleterious effects on shoot tissue. One way to roughly measure this reallocation is through increased root-to-shoot ratios of biomass (Poorter and Nagel 2000). Additionally, at the level of molecules, changes in resource allotment from shoots-to-roots during scarcity are revealed in the chemical compositions of individual tissue. As Gargallo-Garriga et al. (2014) discussed in their study, the basic content of sugar, amino acid, and nucleoside increases under drought conditions while concurrently decreasing in the young branch tissue of grass species such as *Holcus lanatus* and *Alopecurus pratensis* (Gargallo-Garriga et al. 2014). Plants, according to the theory of functional equilibrium, adjust their allotment of resources between cells to minimize the accession of the most limited resource (Brouwer 1983). During times of water scarcity, directing inputs toward root tissues rather than leaf tissues not only conserves water but also reduces the surface area available for water loss through transpiration. These adjustments can help the plant navigate its developmental stages, thereby helping it achieve generative maturity even in the presence of water shortage challenges.

3.6 Current State of Research

3.6.1 Leaf Functional Trait Induces Different Responses to Changing Climatic Conditions

Several global research studies the leaf traits, revealing relationships between leaf shape and climatic factors. Generally, low temperatures and low annual rainfall are associated with smaller leaf size. Habitat dryness also affects leaf area, with dry conditions causing leaves to become smaller. Moreover, leaf size reduces at elevated altitudes, and plant branches become more vulnerable to embolism in challenging conditions, such as low temperatures. In windy environments, plants may choose to have smaller leaves to avoid wind loss. Conversely, better environmental conditions often result in larger leaves. LMA (leaf mass per area), an important indicator of plant growth, varies with factors such as temperature and rainfall (Wang and Wang 2015). On a global scale, leaf mass per area shows an increase with decreasing annual rainfall, suggesting that drought conditions lead to higher leaf mass. However, there is a research gap concerning the impacts of rising temperatures over time on leaf traits, and the precise time-series response of plant leaf traits remains relatively uncertain.

3.6.2 Leaf Trait Responses to Changing Climatic Conditions

The overall carbon fixation by an individual leaf throughout its life span is intricately linked to both its productivity and the duration it remains actively involved. When studying species characterized by long leaf life and high LMA (leaf mass per area), there is a tendency for reduced leaf productivity and lower leaf mass per unit area. In such situations, the nitrogen concentration per unit of leaf mass typically remains at a relatively low level (Singh et al. 2010a, b). In contrast, species with shorter leaf life and lower LMA display elevated leaf productivity and a greater amount of leaf mass per unit mass. In addition, these species have higher nitrogen content per leaf mass. In summary, interactions between leaf traits, such as life span and LMA, significantly influence the carbon fixation capacity of individual leaves (Poorter and Bongers 2006). Changes in the interactions among leaf traits suggest that the effects of environmental changes vary for different leaf characteristics. Plants exhibit a high degree of adaptation to changing environments, the extent of adjustment being dependent on specific changes in various traits (Singh et al. 2014, 2021, 2024). It describes the direct correlation between leaf traits and environmental constraints, underscoring the intricate and direct relationship between plant traits and their surrounding conditions.

3.6.3 Root Functional Trait Induces Different Responses to Climate Change

The fine roots that serve as the plant's absorptive organs for water and nutrients have properties that encode valuable environmental data, making them reliable indicators for assessing plant growth and distribution (Shi et al. 2011). Fine roots have a significant impact on helping plants grow and adjust to changing climatic conditions, yet current research on the characteristics of plant roots is severely limited. Although the likelihood of global climate change is considered relatively low, ecological impacts appear on a significant and unified scale. As a result, complex ecosystem communities are constantly undergoing restructuring due to the impacts of worldwide climate variations. The effects of short-term disturbances on long-term environmental changes remain unclear, as well as the impact of underlying population growth processes on community viability and sustainability (Post and Forchhammer 2001). Due to the substantial impact of global changes on the functional traits of plants (Meng et al. 2007), presently, research primarily focuses on domains such as plant physiology, phenology, species distribution, and community structure. However, there is a scarcity of long-term studies and large-scale experiments in the field of functional trait research, given that the majority of studies are brief and concentrated on physiological and ecological aspects.

3.7 Conclusion

In conclusion, the interplay between plant functional traits and climate change highlights the complex nature of ecological dynamics in the face of global environmental changes. Current research has mainly delved into areas such as plant physiology, phenology, and community structure, revealing the complex ways in which plants respond to changing climatic conditions. However, the current emphasis on short-period experiments highlights the need for more comprehensive, long-term studies to better understand how plant functional traits evolve in response to climate change. Addressing this research gap is crucial for mitigating the potential impacts of climate change on plant communities and ecosystems in the evolving global climatic conditions. Highlighting the complex relationships between the functional traits of plants and climate change is paramount to predicting future ecosystem trajectories. To gain a deep knowledge of underlying mechanisms driving the specific responses of plants to climatic change, forthcoming research must focus on combining observational and experimental methods. This holistic approach should include both individual species and community-level dynamics. Such a synthesis is essential in the broader effort to design effective approaches or tactics to mitigate and adjust the ecological consequences arising from the continuing challenges of climate change.

References

Ainsworth EA, Long SP (2005) What have we learned from 15 years of free-air CO2 enrichment (FACE)? A meta-analytic review of the responses of photosynthesis, canopy properties and plant production to rising CO_2. New Phytol 165(2):351–372

Alnsour M, Ludwig-Müller J (2015) Potential effects of climate change on plant primary and secondary metabolism and its influence on plant ecological interactions. J Endocytobiosis Cell Res 26:90–99

Bellard C, Bertelsmeier C, Leadley P, Thuiller W, Courchamp F (2012) Impacts of climate change on the future of biodiversity. Ecol Lett 15(4):365–377

Berry JA, Beerling DJ, Franks PJ (2010) Stomata: key players in the earth system, past and present. Curr Opin Plant Biol 13(3):232–239

Bhattacharyya S (2022) Mechanism of temperature stress acclimation and the role of transporters in plants. In: Plant perspectives to global climate changes. Academic Press, pp 413–457

Bishop KA, Betzelberger AM, Long SP, Ainsworth EA (2015) Is there potential to adapt soybean (Glycine max M err.) to future [CO2]? An analysis of the yield response of 18 genotypes in free-air CO2 enrichment. Plant Cell Environ 38(9):1765–1774

Boyer JS (1982) Plant productivity and environment. Science 218(4571):443–448

Brouwer R (1983) Functional equilibrium: sense or nonsense? Netherlands J Agric Sci 31(4):335–348

Chapin FS III, Zavaleta ES, Eviner VT, Naylor RL, Vitousek PM, Reynolds HL, Díaz S (2000) Consequences of changing biodiversity. Nature 405(6783):234–242

Crafts-Brandner SJ, Salvucci ME (2000) Rubisco activase constrains the photosynthetic potential of leaves at high temperature and CO2. Proc Natl Acad Sci 97(24):13430–13435

Dai A (2011) Drought under global warming: a review. Wiley Interdiscip Rev Clim Change 2(1):45–65

Dermody O, Long SP, DeLucia EH (2006) How does elevated CO2 or ozone affect the leaf-area index of soybean when applied independently? New Phytol 169(1):145–155

de Vries J, Evers JB, Kuyper TW, van Ruijven J, Mommer L (2021) Mycorrhizal associations change root functionality: a 3D modelling study on competitive interactions between plants for light and nutrients. New Phytol 231(3):1171–1182

Duncan EJ, Gluckman PD, Dearden PK (2014) Epigenetics, plasticity, and evolution: How do we link epigenetic change to phenotype? J Exp Zool B Mol Dev Evol 322(4):208–220

Dusenge ME, Duarte AG, Way DA (2019) Plant carbon metabolism and climate change: elevated CO2 and temperature impacts on photosynthesis, photorespiration and respiration. New Phytol 221(1):32–49

Freschet GT, Roumet C, Comas LH, Weemstra M, Bengough AG, Rewald B, Stokes A (2021) Root traits as drivers of plant and ecosystem functioning: current understanding, pitfalls and future research needs. New Phytol 232(3):1123–1158

Gargallo-Garriga A, Sardans J, Pérez-Trujillo M, Rivas-Ubach A, Oravec M, Vecerova K, Penuelas J (2014) Opposite metabolic responses of shoots and roots to drought. Sci Rep 4(1):6829

Givnish TJ (1987) Comparative studies of leaf form: assessing the relative roles of selective pressures and phylogenetic constraints. New Phytol 106:131–160

Gomez-Pastor R, Burchfiel ET, Thiele DJ (2018) Regulation of heat shock transcription factors and their roles in physiology and disease. Nat Rev Mol Cell Biol 19(1):4–19

Gray SB, Strellner RS, Puthuval KK, Ng C, Shulman RE, Siebers MH, Leakey AD (2012) Minirhizotron imaging reveals that nodulation of field-grown soybean is enhanced by free-air CO2 enrichment only when combined with drought stress. Funct Plant Biol 40(2):137–147

Gray SB, Dermody O, Klein SP, Locke AM, Mcgrath JM, Paul RE, Leakey AD (2016) Intensifying drought eliminates the expected benefits of elevated carbon dioxide for soybean. Nat Plants 2(9):1–8

Guo R, Wen ZM, Wang HX, Qi DH (2015) Relationships among leaf traits and their expression in different vegetation zones in Yanhe River basin, Northwest China. Ying Yong Sheng Tai Xue Bao (J Appl Ecol) 26(12):3627–3633

Hannah LJ, Lovejoy TE (eds) (2003) Climate change and biodiversity: synergistic impacts (No. 4). Center for Applied Biodiversity Science at Conservation International

Harrison SP, Cramer W, Franklin O, Prentice IC, Wang H, Brannstrom A et al (2021) Eco-evolutionary optimality as a means to improve vegetation and land-surface models. New Phytol 231(6):2125–2141

Hartmann DL, Tank AMK, Rusticucci M, Alexander LV, Brönnimann S, Charabi YAR, Zhai P (2013) Observations: atmosphere and surface. In: Climate change 2013 the physical science basis: working group I contribution to the fifth assessment report of the Intergovernmental Panel on Climate Change. Cambridge University Press, pp 159–254

Hatfield JL, Prueger JH (2015) Temperature extremes: effect on plant growth and development. Weather Clim Extremes 10:4–10

Hatfield JL, Boote KJ, Kimball BA, Ziska LH, Izaurralde RC, Ort D, Wolfe D (2011) Climate impacts on agriculture: implications for crop production. Agron J 103(2):351–370

Johnson MP, Wientjes E (2020) The relevance of dynamic thylakoid organisation to photosynthetic regulation. Biochim Biophys Acta Bioenerg 1861(4):148039

Kang Y, Khan S, Ma X (2009) Climate change impacts on crop yield, crop water productivity and food security—a review. Progress Nat Sci 19(12):1665–1674

Kao W-Y-Y, Forseth IN (1992) Diurnal leaf movement, chlorophyll fluorescence and carbon assimilation in soybean grown under different nitrogen and water availabilities. Plant Cell Environ 15:703–710

Kühn N, Tovar C, Carretero J, Vandvik V, Enquist BJ, Willis KJ (2021) Globally important plant functional traits for coping with climate change. Front Biogeogr 13(4):e53774

Kumar N, Kumar N, Shukla A, Shankhdhar SC, Shankhdhar D (2015) Impact of terminal heat stress on pollen viability and yield attributes of rice (Oryza sativa L.). Cereal Res Commun 43(4):616–626

Kumar N, Shankhdhar SC, Shankhdhar D (2016) Impact of elevated temperature on antioxidant activity and membrane stability in different genotypes of rice (Oryza sativa L.). Indian J Plant Physiol (Now Plant Physiol Rep) 21(1):37–43

Kumar N, Suyal DC, Sharma IP, Verma A, Singh H (2017) Elucidating stress proteins in rice (Oryza sativa L.) genotype under elevated temperature: a proteomic approach to understand heat stress response. 3 Biotech 7:205

Kumar N, Jeena N, Singh H (2019) Elevated temperature modulates rice pollen structure: a study from foothill Himalayan agro-ecosystem in India. 3 Biotech 9:175

Kumar A, Dwivedi GK, Tewari S, Paul J, Anand R, Kumar N, Kumar P, Singh H, Kaushal R (2020a) Carbon mineralization and inorganic nitrogen pools under Terminalia chebula Retz.-based agroforestry system in Himalayan Foothills, India. For Sci 66(5):634–643

Kumar A, Kumar P, Singh H, Kumar N (2020b) Adaptation and mitigation potential of roadside trees with bio-extraction of heavy metals under vehicular emissions and their impact on physiological traits during seasonal regimes. Urban For Urban Green 58:126900

Kumar N, Jeena N, Kumar A, Khairakpam R, Singh H (2021) Comparative response of rice cultivars to elevated air temperature in Bhabar region of Indian Himalaya: status on yield attributes. Heliyon 7:e07474

Kumar A, Kumar P, Singh H, Bisht S, Kumar N (2021a) Relationship of physiological plant functional traits with soil carbon stock in temperate forest of Garhwal Himalaya. Curr Sci 120(8):1368–1373

Kumar A, Kumar P, Singh H, Kumar N (2021b) Modulation of plant functional traits under essential plant nutrients during seasonal regime in natural forests of Garhwal Himalayas. Plant Soil 465:197–212

Kumar A, Singh H, Kumari G, Bisht S, Malik A, Kumar N, Singh M, Raturi A, Barthwal S, Thakur A, Kaushal R (2022) Adaptive resilience of roadside trees to vehicular emissions via leaf enzymatic, physiological, and anatomical trait modulations. Environ Pollut 313:120191

Lavorel S, Garnier E (2002) Predicting changes in community composition and ecosystem functioning from plant traits: revisiting the Holy Grail. Funct Ecol 16(5):545–556
Lobell DB, Asner GP (2003) Climate and management contributions to recent trends in US agricultural yields. Science 299(5609):1032–1032
MacArthur RH, Pianka ER (1966) On optimal use of a patchy environment. Am Nat 100(916):603–609
Madhu M, Hatfield JL (2013) Dynamics of plant root growth under increased atmospheric carbon dioxide. Agron J 105(3):657–669
Meehl GA, Stocker TF, Collins WD, Friedlingstein P, Gaye AT, Gregory JM, Raper SCB, Watterson IG, Weaver AJ, Zhao ZC (2007) Global climate projections. In: Solomon S, Qin D, Manning M (eds) Climate change 2007: the physical science basis. Contribution of working group I to the fourth assessment report of the Intergovernmental Panel on Climate Change. Cambridge University Press, pp 247–845
Meng TT, Ni J, Guohong W (2007) Plant functional traits and environmental and ecosystem functions. J Plant Ecol 1:150–165
Mishra T (2016) Climate change and production of secondary metabolites in medicinal plants: a review. Int J Herb Med 4(4):27–30
Morgan PB, Bollero GA, Nelson RL, Dohleman FG, Long SP (2005) Smaller than predicted increase in aboveground net primary production and yield of field-grown soybean under fully open-air [CO2] elevation. Glob Chang Biol 11(10):1856–1865
Pachauri RK, Team CW, Meyer LA (2014) IPCC, 2014: climate change 2014: synthesis report. Contribution of working groups I, II and III to the fifth assessment report of the Intergovernmental Panel on Climate Change. IPCC, Geneva, p 151
Pearcy RW, Muraoka H, Valladares F (2005) Crown architecture in sun and shade environments: assessing function and trade-offs with a three-dimensional simulation model. New Phytol 166(3):791–800
Poorter L, Bongers F (2006) Leaf traits are good predictors of plant performance across 53 rain forest species. Ecology 87(7):1733–1743
Poorter H, Nagel O (2000) The role of biomass allocation in the growth response of plants to different levels of light, CO2, nutrients and water: a quantitative review. Funct Plant Biol 27(12):1191–1191
Post E, Forchhammer MC (2001) Pervasive influence of large-scale climate in the dynamics of a terrestrial vertebrate community. BMC Ecol 1(1):1–7
Qiuhong F, Zuomin S, Lili D (2008) Response and application of plant functional traits to environment. For Sci 4:125–131
Reich PB, Ellsworth DS, Walters MB, Vose JM, Gresham C, Volin JC, Bowman WD (1999) Generality of leaf trait relationships: a test across six biomes. Ecology 80(6):1955–1969
Reich PB, Hobbie SE, Lee TD (2014) Plant growth enhancement by elevated CO2 eliminated by joint water and nitrogen limitation. Nat Geosci 7(12):920–924
Sack L, Scoffoni C (2013) Leaf venation: structure, function, development, evolution, ecology and applications in the past, present and future. New Phytol 198(4):983–1000
Sage RF, Kubien DS (2007) The temperature response of C3 and C4 photosynthesis. Plant Cell Environ 30(9):1086–1106
Sage RF, Way DA, Kubien DS (2008) Rubisco, Rubisco activase, and global climate change. J Exp Bot 59(7):1581–1595
Sala OE, Stuart Chapin FIII, Armesto JJ, Berlow E, Bloomfield J, Dirzo R, Wall DH (2000) Global biodiversity scenarios for the year 2100. Science 287(5459):1770–1774
Schiphorst C, Achterberg L, Gomez R, Koehorst R, Bassi R, van Amerongen H, Wientjes E (2022) The role of light-harvesting complex I in excitation energy transfer from LHCII to photosystem I in Arabidopsis. Plant Physiol 188(4):2241–2252
Schottler MA, Toth SZ (2014) Photosynthetic complex stoichiometry dynamics in higher plants: environmental acclimation and photosynthetic flux control. Front Plant Sci 5:188

Shi Y, Wen Z, Gong S (2011) Comparisons of relationships between leaf and fine root traits in hilly area of the Loess Plateau, Yanhe River basin, Shaanxi Province, China. Shengtai Xuebao (Acta Ecol Sin) 31(22):6805–6814
Singh H, Verma A (2013a) Physiological responses of rice cultivars to various nitrogen levels. Int J Agric Environ Biotechnol 6(3):383–388
Singh H, Verma A (2013b) Characterization and screening of high nitrogen efficient rice genotype to curtail environmental pollution. Int J Agric Environ Biotechnol 6(1):23–30
Singh H, Verma A, Shukla A (2010a) Comparative study of yield and yield components of hybrid and inbred genotypes of rice (*Oryza Sativa L.*). Int J Agric Environ Biotechnol 3:355–360
Singh H, Verma A, Krishnamoorthy M, Shukla A (2010b) Consequence of diverse nitrogen levels on leaf pigments in five rice genotypes under field emergent circumstance. Int J Bioresour Stress Manag 1:189–193
Singh H, Verma A, Rai SK (2013a) Biochemical evaluation of different rice genotypes grown at various nitrogen levels using SDS-PAGE. Curr Adv Agric Sci 5(1):144–146
Singh H, Verma A, Shukla A (2013b) Guttation fluid as a physiological marker for selection of nitrogen efficient rice (*Oryza sativa* L.) genotypes. Afr J Biotechnol 12(44):6276–6281
Singh H, Verma A, Ansari MW, Shukla A (2014) Physiological response of rice (*Oryza sativa* L.) genotypes to elevated nitrogen applied under field conditions. Plant Signal Behav 9:e29015. https://doi.org/10.4161/psb.29015
Singh H, Yadav M, Kumar N, Kumar A, Kumar M (2020) Assessing adaptation and mitigation potential of roadside trees under the influence of vehicular emissions: a case study of Grevillea robusta and Mangifera indica planted in an urban city of India. PLoS One 15(1):e0227380
Singh S, Singh H, Sharma SK, Nautiyal R (2021) Seasonal variation in biochemical responses of bamboo clones in the sub-tropical climate of Indian Himalayan foothills. Heliyon 7(4):e06859. https://doi.org/10.1016/j.heliyon.2021.e06859
Singh M, Singh H, Kumar A, Kumar M, Barthwal S, Thakur A (2024) Soil nitrogen availability determines the CO_2 fertilization effect on tree species (*Neolamarckia cadamba*): growth and physiological evidence. Environ Sustain. https://doi.org/10.1007/s42398-023-00300-w
Skirycz A, De Bodt S, Obata T, De Clercq I, Claeys H, De Rycke R, Inzeo D (2010) Developmental stage specificity and the role of mitochondrial metabolism in the response of Arabidopsis leaves to prolonged mild osmotic stress. Plant Physiol 152(1):226–244
Solomon S (ed) (2007) Climate change 2007-the physical science basis: working group I contribution to the fourth assessment report of the IPCC, vol 4. Cambridge University Press
Swann AL (2018) Plants and drought in a changing climate. Curr Clim Change Rep 4:192–201
Timperio AM, Egidi MG, Zolla L (2008) Proteomics applied on plant abiotic stresses: role of heat shock proteins (HSP). J Proteomics 71(4):391–411
Tollefson J (2016) Global warming already driving increases in rainfall extremes. Nature:443–444. https://doi.org/10.1038/nature.2016.19508
Verelst W, Skirycz A, Inze D (2010) Abscisic acid, ethylene and gibberellic acid act at different developmental stages to instruct the adaptation of young leaves to stress. Plant Signal Behav 5(4):473–475
Violle C, Navas ML, Vile D, Kazakou E, Fortunel C, Hummel I, Garnier E (2007) Let the concept of trait be functional! Oikos 116(5):882–892
Wang CS, Wang SP (2015) A review of research on responses of leaf traits to climate change. Chin J Plant Ecol 39(2):206–216
Wang P, Wang T, Han J, Li M, Zhao Y, Su T, Ma C (2021) Plant autophagy: an intricate process controlled by various signalling pathways. Front Plant Sci 12:754982
Westoby M (1998) A leaf-height-seed (LHS) plant ecology strategy scheme. Plant Soil 199:213–227
Whippo CW, Hangarter RP (2006) Phototropism: bending towards enlightenment. Plant Cell 18(5):1110–1119
Wu A, Xiong X, González-M R, Li R, Li A, Liu J, Zhang Q (2023) Climate change reshapes plant trait spectrum to explain biomass dynamics in an old-growth subtropical forest. Front Plant Sci 14:1260707

Yadav S, Modi P, Dave A, Vijapura A, Patel D, Patel M (2020) Effect of abiotic stress on crops. In: Sustainable crop production. IntechOpen, p 3
Yang X, Li R, Jablonski A, Stovall A, Kim J, Yi K, Lerdau M (2023) Leaf angle as a leaf and canopy trait: rejuvenating its role in ecology with new technology. Ecol Lett 26:1005
Zhang J, Li P (2019) Response of plant functional traits to climate change. In: IOP conference series: earth and environmental science, Jul 2019, vol 300(3). IOP Publishing, p 032078

Scientific Advancement in Trait Measurement and Quantification

4

Jharna Maiti and Amit Joshi

Abstract

The field of trait measurement and quantification has witnessed significant scientific advancements in recent years, driven by the growing demand for precise and comprehensive assessments of plants' characteristics and attributes. However, contemporary research has embraced innovative approaches that incorporate cutting-edge technologies and interdisciplinary methodologies to enhance the accuracy and reliability of trait assessment. One noteworthy advancement involves the integration of advanced statistical techniques, such as item response theory and latent variable modeling, to refine the measurement precision of psychological traits. These techniques allow researchers to extract nuanced information from data, improving the granularity of trait quantification. Furthermore, the advent of physiological monitoring has opened new avenues for trait measurement. Likewise, wearable devices and biosensors have facilitated continuous and ecologically valid trait monitoring, providing real-time data that enhances trait quantification. Additionally, the incorporation of big data and machine learning techniques has revolutionized trait measurement. Large-scale datasets and sophisticated algorithms enable the identification of subtle patterns and relationships among traits, paving the way for more accurate predictive models and personalized trait assessments. Moreover, cross-disciplinary collaborations have emerged as a hallmark of progress in trait measurement. Physiologist, geneticists, and computer scientists are joining forces to develop holistic models that encompass both genetic and environmental influences on traits, resulting in a more comprehensive understanding of plants nature. Scientific advancements in trait measurement and quantification have ushered in a new era of precision and objectivity. Innovations in statistical methods, wearable technology, machine

J. Maiti · A. Joshi (✉)
Department of Biochemistry, Kalinga University, Naya-Raipur, Chhattisgarh, India
e-mail: amit.joshi@kalingauniversity.ac.in

N. Kumar, H. Singh (eds.), *Plant Functional Traits for Improving Productivity*,
https://doi.org/10.1007/978-981-97-1510-7_4

learning, and interdisciplinary cooperation have collectively contributed to the refinement and expansion of our capacity to assess plants traits. These developments hold promise not only for academic research but also for applications in psychology, education, and beyond.

Keywords
Trait measurement · Scientific advancements · Biosensors · Real-time data · Physiological traits

4.1 Introduction

The comprehension and enhancement of plant growth, productivity, and resilience have been made possible by scientific breakthroughs in the measurement and quantification of plant characteristics. The lengthy history of botanical research, which dates back to the groundbreaking work of scientists like Gregor Mendel, Charles Darwin, and Nikolai Vavilov, is the source of this advancement (Le et al. 2023). Addressing urgent global issues including food security, climate change, and sustainable agriculture is why this field is important (Kumar et al. 2015, 2016; Wei et al. 2023). Through precise measurement of plant characteristics such as growth rate, photosynthetic efficiency, nutrient uptake, and stress tolerance, agronomists and researchers may create crops that have lower environmental impact and higher yields. This might then help to lessen the effects of climate change and guarantee a sufficient supply of food for the world's expanding population (Devi et al. 2023). Modern methods of measuring plant traits make use of cutting-edge technology such as molecular biology, high-throughput phenotyping, and remote sensing. These methods provide high-precision, noninvasive evaluations of plant traits, encouraging data-driven decisions in ecosystem management and agriculture (Kumar et al. 2017; Zhang et al. 2023).

Our understanding of plant evolution, ecology, and adaptation is enriched by the continuous improvements in trait quantification, which broadens our perspective of the natural world. Scientific advancement in this area is essential for promoting sustainable solutions and guaranteeing the welfare of ecosystems and people at large, particularly in light of the tremendous problems facing our planet (Katamadze et al. 2023).

4.1.1 Scope and Objectives

The scope and goals of scientific progress in plant characteristic assessment and quantification are diverse and cover a wide variety of objectives meant to increase our knowledge of plant biology and tackle pressing global issues (Brusset et al. 2023).

Enhanced crop productivity: Enhancing agricultural yields and quality through the identification and selection of plant features that lead to improved performance is one of the main goals. This can include characteristics linked to nutrient intake, disease resistance, drought resistance, and general plant health (Kumar et al. 2019; Goswami and Pariyar 2023).

Sustainable agriculture: Enhancing agricultural yields and quality through the identification and selection of plant features that lead to improved performance is one of the main goals (Kumar et al. 2021; Woo et al. 2023). This can include characteristics linked to nutrient intake, disease resistance, drought resistance, and general plant health.

Climate resilience: Another important goal is to comprehend the characteristics of plants that allow for adaptation to changing climatic conditions. Crop varieties that are climate resilient are developed by identifying features associated with heat tolerance, CO_2 assimilation, and resilience to harsh weather events (Singh et al. 2020; Kumar et al. 2021a, b; Elnagar et al. 2023).

Biodiversity conservation: Scientific study in this area contributes to the preservation of natural ecosystems by assessing plant traits that influence the fitness and competitiveness of species (Lenzi et al. 2023).

Ecosystem management: Selecting native plant species that can improve biodiversity and ecosystem functioning is made easier with the help of plant trait studies, which are crucial for managing and restoring ecosystems (Kumar et al. 2020a, b; Paul et al. 2023).

The necessity for sustainable agriculture, climate adaptability, biodiversity protection, and better ecosystem management has shaped the scope and goals of scientific advances in plant characteristic measurement and quantification (Viñals et al. 2023). These developments have the potential to improve our understanding of plant biology and ecological systems while tackling urgent global issues.

4.2 Historical Perspective

Early plant trait quantification techniques were straight forward and mostly dependent on basic measurements and eye observation. This led to the scientific advancement of plant trait quantification (Ramzan and Li 2023). The more sophisticated and exact procedures that are currently employed in plant research were made possible by these techniques. In the past, basic morphological factors like plant height, leaf size, and the quantity of branches or flowers were frequently used to measure plant attributes (Karantonis et al. 2023). Basic measurement instruments and manual observation were used to evaluate these observable characteristics. Physiological evaluations like water retention and gas exchange measures were also included in the early plant trait measurement process. Scientists frequently employed techniques such as gas exchange chambers to measure photosynthetic rates or weighed plant samples both before and after drying to ascertain the water content (Lukaski and Talluri 2023). Understanding plant attributes required an understanding of plant germination and growth rate studies. Scientists would sow seeds and record the length of time it took for the seeds to sprout, the rate at which the seedlings grew, and the plant's overall progress (Wang et al. 2023). Systems of visual grading were commonly used for characteristics that were difficult to measure. Based on visual observations, traits including disease resistance, leaf color, and flowering period were frequently given numerical rankings (Dobrolinska et al. 2023). Simple instruments like scales, rulers, and calipers were also used to measure certain fundamental biometric parameters including plant weight, leaf area, and root length (Avisdris et al. 2023).

Even though these early techniques were somewhat basic, they helped us comprehend plant biology and offered crucial insights into plant features. Their accuracy, reproducibility, and scalability were constrained, nevertheless. Because of this, contemporary plant trait assessment now makes use of cutting-edge instruments and technology including remote sensing, high-throughput phenotyping, molecular markers, and complex imaging methods (Shen et al. 2023). Researchers are able to take on challenging problems in agriculture, ecology, and environmental science because of the increased precision, automation, and capacity to examine a wider variety of features provided by these more advanced techniques (Lamichhane et al. 2023). However, the essential ideas behind these early approaches to trait measurement continue to be important for understanding plant features. Significant advancements in our knowledge of plant biology have been made possible by turning plant trait quantification into scientific milestones. This has made ecological management, environmental conservation, and agricultural efficiency all more possible (Raymond-Léonard et al. 2023). The trajectory of plant trait assessment has been significantly shaped by a number of significant turning points:

An important turning point was reached in the early twentieth century with the rediscovery of Gregor Mendel's work. This established the framework for comprehending heredity, which paved the way for genetic mapping and the identification of particular genes in charge of certain plant characteristics (Pollen et al. 2023). The plant breeding programs were established in the twentieth century with the goal of creating crop varieties with desired features. High-yield crop types were developed as a result of the Green Revolution, which was sparked by Norman Borlaug's work and revolutionized agriculture worldwide (Healy and Dawson 2023). Quantitative trait loci (QTL) analysis emerged as a potent method for determining the genetic areas linked to plant attributes in the late twentieth century (Aguet et al. 2023). With the help of this method, it is possible to precisely map the genes in charge of quantitative traits, facilitating focused breeding to produce desired features. Plant trait quantification has quickened with the introduction of high-throughput phenotyping technology including automated imaging systems and sensor-based measurements (Wen et al. 2023). These technologies make it possible to quickly evaluate multiple traits at once, which improves the effectiveness of trait selection in breeding programs. Large-scale analyses of plant characteristics in both natural and agricultural environments have been made possible by remote sensing technologies, such as satellite and drone-based imagery (Di and Yu 2023). This has been useful in tracking the health of the plants, calculating crop yields, and evaluating the effects of environmental changes. The genome-wide association studies (GWAS) and next-generation sequencing are two examples of sophisticated genomic methods that have made it easier to identify genetic markers linked to particular plant features (Waszczuk et al. 2023). As a result, breeding efforts have become more focused and accurate. A comprehensive understanding of plant traits has been made possible by the integration of data from genomes, transcriptomics, proteomics, and metabolomics. This has helped to clarify the molecular mechanisms underpinning trait expression and response to environmental stimuli (Kumar et al. 2017, 2022; Sathyanarayanan et al. 2023; Taylor et al. 2023).

These landmarks in the quantification of features have greatly improved our capacity to quantify and comprehend plant characteristics, ranging from stress resistance and agricultural output to ecosystem dynamics. In a period of constantly changing environmental conditions, they continue to shape the sector by providing promising possibilities for biodiversity conservation, sustainable agriculture, and environmental management.

4.3 Advances in Biological Trait Measurement

4.3.1 Biomarkers in Trait Assessment

The potential of biomarkers to provide light on a variety of characteristics pertaining to plant health, stress response, and adaptation is being investigated in the field of plant science.

For example, the identification of particular chemical substances or metabolites in plant tissues can function as biomarkers for illnesses, exposure to environmental stressors, or deficits in certain nutrients. These biomarkers can direct treatments to maximize plant productivity and health by offering early and accurate signals of plant conditions (Table 4.1).

Additionally, a plant's growth, resilience, or environmental adaptability may be indicated by genetic indicators, such as certain gene expression patterns or epigenetic alterations (Menéndez-Miguélez et al. 2023). These genetic biomarkers can be found by scientists using molecular methods such as DNA methylation analysis and RNA sequencing, which gives them a better grasp of plant characteristics (Terracciano et al. 2023).

In addition to improving our ability to track and maximize plant performance, the use of biomarkers in plant trait evaluation helps to promote more environmentally friendly farming practices and ecosystem management. It makes it possible to precisely select traits, identify problems early, and create more resilient and productive plant varieties, all of which are essential for tackling global issues like climate change and food security. As this field of study develops, biomarkers are turning into useful instruments for the quest of a more sustainable and environmentally friendly future (Huang and Marefati 2023).

Table 4.1 Biomarkers in trait assessment

Type	Biomarker	Trait assessed	Application
Plants	Gene expression	Drought tolerance	Identifying plants resilient to water scarcity
	Chlorophyll content	Photosynthetic efficiency	Monitoring plant health and response to environmental changes
	Metabolite profiles	Nutrient uptake efficiency	Enhancing crop yield and quality
Animals	DNA sequence	Disease susceptibility	Predicting predisposition to certain illnesses
	Hormone levels	Reproductive fitness	Assessing fertility and breeding success
	Protein markers	Muscle development	Improving livestock quality
Bacteria	Antibiotic resistance	Drug susceptibility	Determining effective treatment options
	Quorum sensing molecules	Communication	Understanding bacterial behavior and interactions
	Metabolic pathways	Environmental adaptation	Studying bacterial response to varying conditions

4.3.2 SNP Chip Genotyping

Single nucleotide polymorphism (SNP) chip genotyping has transformed the way scientists examine and improve a variety of plant properties, and it has become an essential tool in plant genetics. With the use of this state-of-the-art technology, plant genomes may be thoroughly analyzed to provide insights into the genetic differences underlying characteristics like yield, disease resistance, and environmental adaptability (Estopa et al. 2023). By locating and monitoring particular genetic differences linked to desired traits, SNP chip genotyping enables precision breeding. With a more focused strategy, new plant varieties with enhanced traits—like higher yield, more nutritional value, and greater resistance to pests and diseases—are developed more quickly. SNP genotyping offers important insights into genetic markers linked to stress tolerance and adaptability, which are important as plants grapple with the effects of climate change. This information is essential for creating crops that can survive in a variety of changing environmental settings (Viana and Souza 2023).

Plant genotyping using SNP chips is based on similar concepts to human genomics. Plant samples are used to extract genomic DNA, which is then broken up and tagged with fluorescent dyes. After that, the tagged DNA is hybridized onto an SNP chip that has probes unique to plant SNPs. Fluorescence at each probe site is indicative of the existence of certain SNP variations linked to different characteristics. Because SNP genotyping makes it possible for breeders to efficiently choose plants with desired features, crop development initiatives can move more quickly. Increasing crop yield, quality, and total agricultural productivity are directly impacted by this. Genome-wide association studies (GWAS) make significant use of SNP chips to find genetic variations linked to a variety of disorders (Yang et al. 2023).

4.3.3 Genome-Wide Association Studies (GWAS)-Related Data Analysis

Plant trait assessment and quantification has advanced significantly because of the use of tools like PLINK and R-Studio for data processing related to genome-wide association studies (GWAS). These approaches have been crucial in helping to understand the intricate genetic bases of many plant features, which has advanced crop breeding, ecological research, and conservation initiatives. PLINK is a flexible software program that is frequently used in GWAS for genetic association studies, data administration, and quality assurance. It is an important method for finding genetic markers linked to particular plant attributes since it enables researchers to process large-scale genetic information quickly. On the other hand, R-Studio is a potent platform for statistical analysis and data visualization that makes it possible to integrate genetic data with other pertinent information and explore GWAS results in-depth (Ashwath et al. 2023).

By leveraging GWAS data analysis, plant scientists can uncover the genetic variations that contribute to traits like yield, disease resistance, and stress tolerance. This knowledge facilitates precision breeding efforts, enabling the development of crop varieties with enhanced traits for improved agricultural productivity and sustainability. Furthermore, GWAS-related data analysis aids in understanding plant adaptation to different environments and ecological interactions. This information is essential for ecosystem management, biodiversity conservation, and the development of strategies to mitigate the impact of climate change on plant populations.

Utilizing programs like PLINK and R-Studio for GWAS data analysis has been essential in developing the science of plant trait assessment and quantification (see Fig. 4.1). These methods address global issues like food security and climate change while advancing agriculture, ecology, and environmental research by enabling a greater understanding of the genetic basis of phenotypes (Johnson et al. 2023).

4.3.4 Network Biology Approach for Phenotype Analysis

Plant trait quantification and measurement have advanced scientifically, thanks to the use of a network biology approach for phenotype analysis. This method provides a comprehensive understanding of plant features by utilizing the concepts of systems biology to comprehend intricate relationships between genes, proteins, and metabolic pathways. In plant research, network biology entails building many kinds of biological networks, such as metabolic, protein-protein interaction, and gene regulatory networks. These networks provide a framework for understanding how proteins and genes interact together to produce particular features. The ability of network biology to decipher the complex relationships underlying phenotypic features makes it important for trait analysis in plants. Researchers can find important genes, proteins, and pathways that affect characteristics like growth, stress response, and disease resistance by charting these connections. This discovery offers a thorough comprehension of trait mechanisms, surpassing the level of individual gene correlations (Forrest et al. 2023).

The network biology method also makes it easier to find "hub" genes or proteins, which are essential for the expression of certain traits. By focusing on these centers, breeding and trait manipulation techniques may become more successful, thereby increasing crop resilience and yield. Moreover, a thorough examination of trait-related processes is made possible by network biology's integration of multi-omics data, such as those from genomes, transcriptomics, proteomics, and metabolomics. This multifaceted method improves the precision and depth of trait measurement, allowing researchers to more thoroughly evaluate qualities. The methodology of network biology represents a revolutionary development in the measurement and examination of plant characteristics. It provides new insights into plant biology by clarifying the molecular mechanisms underlying these features. Network biology addresses the urgent global concerns of food security and environmental sustainability by offering a systems-level viewpoint that aids in the production of more robust, productive, and sustainable crop varieties.

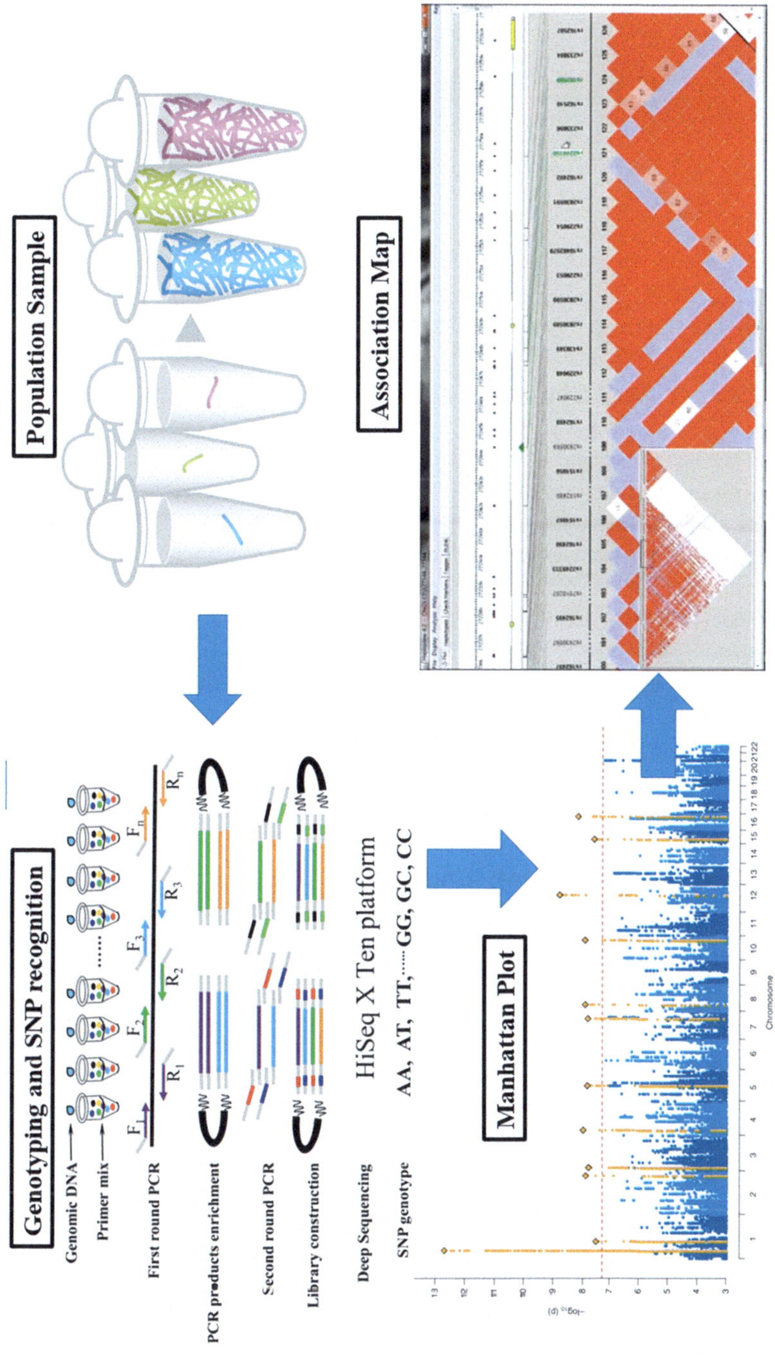

Fig. 4.1 Genome-wide association studies (GWAS) for the population SNP data

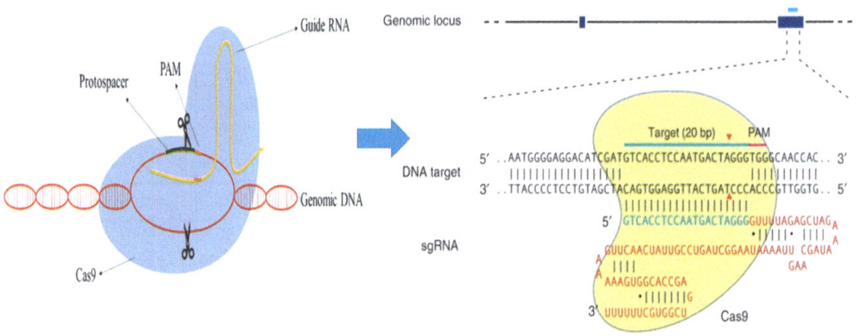

Fig. 4.2 Fundamental functionality of CRISPR-Cas9 for genomic modification

4.3.5 Crisper-Cas9 Genomic Editing

Plant trait measurement and quantification have seen a significant scientific breakthrough with the introduction of CRISPR-Cas9 genome editing (Fig. 4.2). With the use of this groundbreaking technology, it is now possible to precisely and specifically alter a plant's genetic code, producing uniquely accurate customized features (Grinin et al. 2023). Researchers can directly edit or insert genes linked to particular qualities using CRISPR-Cas9, such as enhanced yield and nutritional content, disease resistance, and stress tolerance. The time and resources needed to grow crops with desired features have decreased, thanks to the transformation of traditional breeding methods brought about by the capacity to edit plant genomes with such precision.

The ability of CRISPR-Cas9 to address global concerns in agriculture and food security is one of its most important advantages. It offers the possibility of improved crop yield, decreased pesticide use, and the creation of hardier plant types that can withstand shifting environmental conditions by fine-tuning plant features. Moreover, functional genomics research can make use of CRISPR-Cas9 technology to help researchers better understand the role of particular genes and pathways in trait expression. Our understanding of plant biology is strengthened by this information, which makes trait measurement and quantification more accurate. In plant science, CRISPR-Cas9 genome editing is a revolutionary technology. It provides the accuracy and speed required to propel advances in trait assessment and quantification, addressing urgent global issues linked to environmental sustainability and food security while promoting more productive and sustainable agriculture (Zhao et al. 2023).

4.3.6 RNA-Seq Analysis for Trait Measurement

As a key scientific development in the measurement of plant attributes, RNA-Seq analysis has shown promise in providing previously unattainable insights into the genetic and molecular underpinnings of a wide range of features (see Table 4.2). The extensive measurement of gene expression levels is made possible by this state-of-the-art technology to fully comprehend how genes influence various properties, including growth, disease resistance, and stress responses in plants.

Table 4.2 RNA-Seq tools for trait measurement

Trait assessed	RNA-Seq application	Description/use cases	Latest software tools
Gene expression profiles	Differential expression analysis	Comparing gene expression between conditions and phenotypes	DESeq2, edgeR, limma, sleuth
	Co-expression network analysis	Identifying genes acting together in biological pathways	WGCNA, Cytoscape
	Alternative splicing events	Studying different mRNA isoforms and their functions	SUPPA, rMATS, MAJIQ
	Transcriptome assembly and annotation	Creating reference transcriptomes for non-model species	Trinity, StringTie, HISAT2
Trait-specific biomarkers	Biomarker discovery	Identifying RNA biomarkers associated with specific traits	FeatureCounts, HTSeq, Ballgown
	Signature analysis	Characterizing unique RNA signatures for certain traits	LIONESS, MetaSignature, GSVA
	Marker-assisted selection	Selecting organisms based on RNA markers for desired traits	SelectionTools, MaAsLin2
	Pathway enrichment analysis	Understanding functional implications of gene expression	GOSeq, KOBAS, Reactome
Environmental responses	Stress response profiling	Analyzing gene expression changes under different stressors	heatmaps, GSEA, clusterProfiler
	Adaptation and acclimation studies	Understanding how organisms respond and adapt to environments	CladeLMM, PCAdapt, ecoevolity
	Environmental impact assessment	Assessing how environmental changes affect gene expression	ecolSeq, ecoPCR, ecostructure
Disease diagnosis	Biomarker identification for diseases	Discovering RNA markers indicative of specific illnesses	DEsingle, sSeq, multiClust
	Monitoring disease progression	Tracking changes in RNA expression related to disease	SeqGSEA, GSCA, RNApeg
	Stratifying diseases based on molecular profiles	Classifying diseases into subtypes using RNA-Seq data	MethylMix, SNF, CancerSubtypes

Because it can capture the dynamic and subtle characteristics of gene activity, RNA-Seq is important for measuring plant traits. It measures the quantity of mRNA molecules as well as noncoding RNAs, alternative splicing, and posttranscriptional modifications, all of which are essential for the expression of certain traits. Identification of important genes, regulatory pathways, and biological processes linked to particular traits is made easier with the help of RNA-Seq analysis data. Researchers can use this knowledge to pinpoint the genes that produce desired features in crops, leading to more accurate and productive breeding programs (Wu et al. 2023).

Furthermore, RNA-Seq has several uses in stress biology, allowing for the study of how plants react to temperature fluctuations, pathogen attacks, and drought. Through the exploration of the molecular pathways behind stress tolerance, scientists hope to create crop types that are more tolerant of the effects of climate change. A revolutionary technique in the measurement of plant traits is RNA-Seq analysis. It provides a systems-level perspective on gene expression, which advances our knowledge of plant biology and opens up new avenues for productive and sustainable agriculture that tackles global issues of environmental sustainability and food security.

4.3.7 Machine Learning and ANN in Trait Analysis

Plant trait analysis has undergone a revolution, thanks to machine learning and artificial neural networks (ANNs), which have made considerable scientific progress in trait assessment and quantification possible. Large datasets may contain intricate patterns and linkages that these computational methods can uncover, improving our knowledge of plant characteristics and speeding up attempts at breeding and conservation.

Genomic, transcriptomic, proteomic, and phenotypic data are just a few of the many biological data sets from which useful insights are extracted using machine learning techniques. The identification of genetic markers, patterns of gene expression, and metabolic pathways linked to particular traits is made possible by this data-driven methodology (Jamali et al. 2023).

A type of machine learning called artificial neural networks (ANNs) has shown remarkable potential in modeling and trait prediction. These artificial neural networks can recognize complex correlations between input data and desired characteristic outcomes. It has been used to forecast plant production, growth, and responses to stress using field observations, climate variables, and multi-omics datasets. Quantifying complex, diverse features that arise from the interaction of multiple genetic and environmental factors is a special application of machine learning and artificial neural networks (ANNs). They provide a predictive, data-driven method for analyzing features that makes it possible to identify plant varieties with the best qualities for particular environmental circumstances and to design more exact breeding plans.

The field of measuring and quantifying plant traits has undergone a paradigm shift, thanks to machine learning and ANN. They give us the tools to quickly and accurately examine enormous datasets, which ultimately advances our understanding of plant biology and improves our capacity to create resilient, high-yielding crops and maintain natural ecosystems in a world that is changing quickly. The worldwide issues of food security, climate change, and biodiversity conservation are greatly aided by these developments (Attri et al. 2023).

4.4 Future Trends and Challenges

With the incorporation of emerging technology, the measurement and quantification of plant traits is undergoing a wave of innovation and scientific improvement. With the use of these technologies, plant scientists should be able to analyze traits with more accuracy, efficiency, and depth, which will lead to improvements in ecology, agriculture, and environmental management. High-Throughput Phenotyping (HTP): Plant trait assessments may be completed quickly, nondestructively, and on a wide scale, thanks to HTP platforms, which include drones, advanced imaging systems, and remote sensing technologies. These technologies offer useful data for better managing natural resources, tracking ecosystem dynamics, and comprehending agricultural performance (Bacha and Iqbal 2023).

Next-Generation Sequencing (NGS): In plant science, NGS has completely changed the fields of functional genomics and genomics research. It makes it possible for scientists to investigate the complete transcriptome, epigenome, and genome of plants, which makes it easier to find genetic markers linked to a variety of characteristics, such as stress tolerance and disease resistance. Metabolomics and Proteomics: Plant metabolic and proteomic profiles may now be measured with greater accuracy, thanks to sophisticated analytical instruments. By providing light on the molecular underpinnings of characteristics, these technologies aid in the disentanglement of intricate relationships that affect development, adaptation, and reactions to external cues (Bennett et al. 2023).

Gene-Editing Technologies: Trait modification in plants has been revolutionized by CRISPR-Cas9 and other gene-editing approaches. They provide exact techniques for introducing desirable features, including enhanced yield, disease resistance, or nutrient content, into the plant DNA (Jeong et al. 2023). The environment for measuring and quantifying plant traits is changing as a result of these new technologies. They offer improved ecosystem management, a greater comprehension of plant biology, and more productive, sustainable agriculture. Plant science is well-positioned to more precisely and effectively address pressing global issues like food security, climate change, and biodiversity conservation as these technologies advance.

4.5 Conclusion

Important advances have been made in the scientific study of plant trait measurement and quantification. Predictive modeling, genetic insights, and data collecting have all been transformed by high-throughput phenotyping, next-generation sequencing, and machine learning. More accurate trait analysis is made possible by these advancements, which is advantageous for ecological study, environmental management, and agriculture. New technologies are opening up new avenues for agricultural development and sustainable practices, which makes them crucial. Multidisciplinary cooperation has promoted comprehensive methods of assessing traits. To guarantee that these developments are applied responsibly and ethically, ethical and societal problems, such as data privacy and equitable resource allocation, must be carefully addressed. The development of plant trait measuring and quantification science has significant effects on both practice and research.

References

Aguet F, Alasoo K, Li YI, Battle A, Im HK, Montgomery SB, Lappalainen T (2023) Molecular quantitative trait loci. Nat Rev Methods Primers 3(1):4

Ashwath MN, Lavale SA, Santhoshkumar AV, Mohapatra SR, Bhardwaj A, Dash U et al (2023) Genome-wide association studies: an intuitive solution for SNP identification and gene mapping in trees. Funct Integr Genomics 23(4):297

Attri I, Awasthi LK, Sharma TP, Rathee P (2023) A review of deep learning techniques used in agriculture. Eco Inform 77:102217

Avisdris N, Link Sourani D, Ben-Sira L, Joskowicz L, Malinger G, Yagel S et al (2023) Improved differentiation between hypo/hypertelorism and normal fetuses based on MRI using automatic ocular biometric measurements, ocular ratios, and machine learning multi-parametric classification. Eur Radiol 33(1):54–63

Bacha SAS, Iqbal B (2023) Advancing agro-ecological sustainability through emerging genetic approaches in crop improvement for plants. Funct Integr Genomics 23(2):145

Bennett HM, Stephenson W, Rose CM, Darmanis S (2023) Single-cell proteomics enabled by next-generation sequencing or mass spectrometry. Nat Methods 20(3):363–374

Brusset X, Ivanov D, Jebali A, La Torre D, Repetto M (2023) A dynamic approach to supply chain reconfiguration and ripple effect analysis in an epidemic. Int J Prod Econ 263:108935

Devi S, Kumar N, Kumar A, Dhansu P, Pazhany AS, Mann A et al (2023) Potential use of nanofertilizers in alleviating stresses in plants. In: Salinity and drought tolerance in plants: physiological perspectives. Springer Nature, Singapore, pp 521–535

Di L, Yu E (2023) Remote sensing. In: Remote sensing big data. Springer International Publishing, Cham, pp 17–43

Dobrolinska MM, Lazarenko SV, van der Zant FM, Does L, van der Werf N, Prakken NH et al (2023) Performance of visual, manual, and automatic coronary calcium scoring of cardiac 13N-ammonia PET/low dose CT. J Nucl Cardiol 30(1):239–250

Elnagar E, Zeoli A, Rahif R, Attia S, Lemort V (2023) A qualitative assessment of integrated active cooling systems: a review with a focus on system flexibility and climate resilience. Renew Sust Energ Rev 175:113179

Estopa RA, Paludeto JGZ, Müller BSF, de Oliveira RA, Azevedo CF, de Resende MDV et al (2023) Genomic prediction of growth and wood quality traits in Eucalyptus benthamii using different genomic models and variable SNP genotyping density. New For 54(2):343–362

Forrest MP, Dos Santos M, Piguel NH, Wang YZ, Hawkins NA, Bagchi VA et al (2023) Rescue of neuropsychiatric phenotypes in a mouse model of 16p11.2 duplication syndrome by genetic correction of an epilepsy network hub. Nat Commun 14(1):825

Goswami B, Pariyar B (2023) Nutrient analysis of soil and manures to enhance crop productivity in organic farming—an insight. Int J Res Biosci Agric Technol 1:55–60

Grinin L, Grinin A, Malkov S (2023) Sociopolitical transformations: a difficult path to cybernetic society. In: Reconsidering the limits to growth: a report to the Russian Association of the Club of Rome. Springer International Publishing, Cham, pp 169–189

Healy GK, Dawson JC (2023) Participatory plant breeding and social change in the Midwestern United States: perspectives from the seed to kitchen collaborative. In: Rethinking food system transformation. Springer Nature, Cham, pp 61–71

Huang W, Marefati M (2023) Development, exergoeconomic assessment and optimization of a novel municipal solid waste-incineration and solar thermal energy based integrated power plant: an effort to improve the performance of the power plant. Process Saf Environ Prot 172:562–578

Jamali M, Soufizadeh S, Yeganeh B, Emam Y (2023) Wheat leaf traits monitoring based on machine learning algorithms and high-resolution satellite imagery. Eco Inform 74:101967

Jeong BR, Jang J, Jin E (2023) Genome engineering via gene editing technologies in microalgae. Bioresour Technol 373:128701

Johnson KH, Krishna TA, Dash M, Thiyageshwari S, Ceasar SA, Selvi D (2023) Food and nutritional security: innovative approaches for improving micronutrient use efficiency in soybean (Glycine max (L.) Merrill) under hostile soils. J Soil Sci Plant Nutr 23(1):56–70

Karantonis JA, Carruthers SP, Burdick KE, Pantelis C, Green M, Rossell SL et al (2023) Brain morphological characteristics of cognitive subgroups of schizophrenia-spectrum disorders and bipolar disorder: a systematic review with narrative synthesis. Neuropsychol Rev 33(1):192–220

Katamadze A, Vergara-Díaz O, Uberegui E, Yoldi-Achalandabaso A, Araus JL, Vicente R (2023) Evolution of wheat architecture, physiology, and metabolism during domestication and further cultivation: lessons for crop improvement. Crop J 11(4):1080–1096

Kumar N, Kumar N, Shukla A, Shankhdhar SC, Shankhdhar D (2015) Impact of terminal heat stress on pollen viability and yield attributes of rice (Oryza sativa L.). Cereal Res Commun 43(4):616–626

Kumar N, Shankhdhar SC, Shankhdhar D (2016) Impact of elevated temperature on antioxidant activity and membrane stability in different genotypes of rice (Oryza sativa L.). Indian J Plant Physiol (Now Plant Physiol Rep) 21(1):37–43

Kumar N, Suyal DC, Sharma IP, Verma A, Singh H (2017) Elucidating stress proteins in rice (Oryza sativa L.) genotype under elevated temperature: a proteomic approach to understand heat stress response. 3 Biotech 7:205

Kumar N, Jeena N, Singh H (2019) Elevated temperature modulates rice pollen structure: a study from foothill Himalayan agro-ecosystem in India. 3 Biotech 9:175

Kumar A, Dwivedi GK, Tewari S, Paul J, Anand R, Kumar N, Kumar P, Singh H, Kaushal R (2020a) Carbon mineralization and inorganic nitrogen pools under Terminalia chebula Retz.-based agroforestry system in Himalayan foothills, India. For Sci 66(5):634–643

Kumar A, Kumar P, Singh H, Kumar N (2020b) Adaptation and mitigation potential of roadside trees with bio-extraction of heavy metals under vehicular emissions and their impact on physiological traits during seasonal regimes. Urban For Urban Green 58:126900

Kumar N, Jeena N, Kumar A, Khairakpam R, Singh H (2021) Comparative response of rice cultivars to elevated air temperature in Bhabar region of Indian Himalaya: status on yield attributes. Heliyon 7:e07474

Kumar A, Kumar P, Singh H, Bisht S, Kumar N (2021a) Relationship of physiological plant functional traits with soil carbon stock in temperate forest of Garhwal Himalaya. Curr Sci 120(8):1368–1373

Kumar A, Kumar P, Singh H, Kumar N (2021b) Modulation of plant functional traits under essential plant nutrients during seasonal regime in natural forests of Garhwal Himalayas. Plant Soil 465:197–212

Kumar A, Singh H, Kumari G, Bisht S, Malik A, Kumar N, Singh M, Raturi A, Barthwal S, Thakur A, Kaushal R (2022) Adaptive resilience of roadside trees to vehicular emissions via leaf enzymatic, physiological, and anatomical trait modulations. Environ Pollut 313:120191

Lamichhane JR, Alletto L, Cong WF, Dayoub E, Maury P, Plaza-Bonilla D et al (2023) Relay cropping for sustainable intensification of agriculture across temperate regions: crop management challenges and future research priorities. Field Crop Res 291:108795

Le QT, Truong HA, Nguyen DT, Yang S, Xiong L, Lee H (2023) Enhanced growth performance of abi5 plants under high salt and nitrate is associated with reduced nitric oxide levels. J Plant Physiol 286:154000

Lenzi D, Balvanera P, Arias-Arévalo P, Eser U, Guibrunet L, Martin A et al (2023) Justice, sustainability, and the diverse values of nature: why they matter for biodiversity conservation. Curr Opin Environ Sustain 64:101353

Lukaski HC, Talluri A (2023) Phase angle as an index of physiological status: validating bioelectrical assessments of hydration and cell mass in health and disease. Rev Endocr Metab Disord 24(3):371–379

Menéndez-Miguélez M, Álvarez-Álvarez P, Pardos M, Madrigal G, Ruiz-Peinado R, López-Senespleda E et al (2023) Development of tools to estimate the contribution of young sweet chestnut plantations to climate-change mitigation. For Ecol Manag 530:120761

Paul SS, Hasselquist EM, Jarefjäll A, Ågren AM (2023) Virtual landscape-scale restoration of altered channels helps us understand the extent of impacts to guide future ecosystem management. Ambio 52(1):182–194

Pollen AA, Kilik U, Lowe CB, Camp JG (2023) Human-specific genetics: new tools to explore the molecular and cellular basis of human evolution. Nat Rev Genet 24:1–25

Ramzan M, Li H (2023) An analytical framework to link factors affecting agricultural trade intensity in the world: pathways to sustainable agricultural development 2030 agenda. Environ Dev Sustain:1–50. https://doi.org/10.1007/s10668-023-03908-4

Raymond-Léonard LJ, Cortet J, Handa IT (2023) Towards a standardization of new functional trait measurements by assessing intraspecific variation: the case of springtail mandibles. Eur J Soil Biol 117:103495

Sathyanarayanan A, Mueller TT, Moni MA, Schueler K, Baune BT, Lio P et al (2023) Multi-omics data integration methods and their applications in psychiatric disorders. Eur Neuropsychopharmacol 69:26–46

Shen S, Zhan C, Yang C, Fernie AR, Luo J (2023) Metabolomics-centered mining of plant metabolic diversity and function: past decade and future perspectives. Mol Plant 16(1):43–63

Singh H, Yadav M, Kumar N, Kumar A, Kumar M (2020) Assessing adaptation and mitigation potential of roadside trees under the influence of vehicular emissions: a case study of Grevillea robusta and Mangifera indica planted in an urban city of India. PLoS One 15(1):e0227380

Taylor G, Garassino F, Aarts MG, Harbinson J (2023) Improving C3 photosynthesis by exploiting natural genetic variation: Hirschfeldia incana as a model species. Food Energy Secur 12(1):e420

Terracciano A, Walker K, An Y, Luchetti M, Stephan Y, Moghekar AR et al (2023) The association between personality and plasma biomarkers of astrogliosis and neuronal injury. Neurobiol Aging 128:65–73

Viana JMS, Souza CAS (2023) Efficiency of mapping epistatic quantitative trait loci. Heredity 131:1–8

Viñals E, Maneja R, Rufí-Salís M, Martí M, Puy N (2023) Reviewing social-ecological resilience for agroforestry systems under climate change conditions. Sci Total Environ 869:161763

Wang J, Cheng JH, Sun DW (2023) Enhancement of wheat seed germination, seedling growth and nutritional properties of wheat plantlet juice by plasma activated water. J Plant Growth Regul 42(3):2006–2022

Waszczuk MA, Jonas KG, Bornovalova M, Breen G, Bulik CM, Docherty AR et al (2023) Dimensional and transdiagnostic phenotypes in psychiatric genome-wide association studies. Mol Psychiatry:1–11. https://doi.org/10.1038/s41380-023-02142-8

Wei D, Wang X, Luo N, Zhu Y, Wang P, Meng Q (2023) Alleviating groundwater depletion while realizing food security for sustainable development. J Clean Prod 393:136351

Wen T, Li JH, Wang Q, Gao YY, Hao GF, Song BA (2023) Thermal imaging: the digital eye facilitates high-throughput phenotyping traits of plant growth and stress responses. Sci Total Environ 899:165626

Woo SL, Hermosa R, Lorito M, Monte E (2023) Trichoderma: a multipurpose, plant-beneficial microorganism for eco-sustainable agriculture. Nat Rev Microbiol 21(5):312–326

Wu L, Yang Y, Wang X, Weng Z, Hua S, Li D et al (2023) Genome-wide QTL mapping and RNA-seq reveal the genetic variation influencing growth traits in giant grouper (Epinephelus lanceolatus). Aquaculture 563:738944

Yang Z, Wang S, Wei L, Huang Y, Liu D, Jia Y et al (2023) BnIR: a multi-omics database with various tools for Brassica napus research and breeding. Mol Plant 16(4):775–789

Zhang H, Wang L, Jin X, Bian L, Ge Y (2023) High-throughput phenotyping of plant leaf morphological, physiological, and biochemical traits on multiple scales using optical sensing. Crop J 11:1303

Zhao F, Ding X, Liu Z, Yan X, Chen Y, Jiang Y et al (2023) Application of CRISPR/Cas9-based genome editing in ecotoxicology. Environ Pollut 336:122458

Application of Functional Traits in Modelling Productivity and Resilience Under Climate Change

5

Apurva Malik, Kishan Kumar, Garima Kumari, Narendra Kumar, and Hukum Singh

Abstract

The changing climatic trends have increased global temperatures and drought risk, thus accelerating catastrophes that threaten food, nutrition, and environmental security. Functional traits, inherent characteristics of organisms influencing their interactions with the environment, are indispensable tools in understanding and predicting ecosystem responses to climate fluctuations. The chapter explores integration of functional traits into various modelling approaches, including species distribution models (SDMs) and trait-based community assembly models, showcasing their power to anticipate shifts in species distributions and community compositions under climate change. It also highlights how functional traits underpin essential ecosystem processes and resilience assessments, emphasizing their role in sustaining productivity and adaptability. Practical applications in agriculture, forestry, aquatic systems, invasive species management, and restoration ecology underscore real-world relevance. The chapter acknowledges challenges in data collection and charts future research directions, ultimately emphasizing the critical importance of functional traits in navigating the complex challenges posed by climate change in ecological systems.

A. Malik (✉) · G. Kumari
Forest Ecology and Climate Change Division, Forest Research Institute, Dehradun, India

K. Kumar · H. Singh
Plant Physiology Discipline, Forest Genetics and Tree Improvement Division, Forest Research Institute, Dehradun, India

N. Kumar
College of Agriculture, Central Agriculture University (I), Krydemkulai, Meghalaya, India

N. Kumar, H. Singh (eds.), *Plant Functional Traits for Improving Productivity*,
https://doi.org/10.1007/978-981-97-1510-7_5

Keywords

Climate change · Ecological models · Functional traits · Resilience assessment · Functional diversity

5.1 Introduction

Climate change, driven primarily by human activities such as the burning of fossil fuels, deforestation, and industrial processes, results in the release of greenhouse gases like carbon dioxide (CO_2), methane (CH_4), and nitrous oxide (N_2O) into the atmosphere. These gases trap heat, leading to rising global temperatures and a suite of environmental impacts. Ecosystems are significantly affected, with direct consequences on plant and animal species. Rising temperatures influence species distribution and behavior, posing challenges for those adapted to specific temperature ranges. The melting of glaciers and polar ice caps contributes to sea level rise, endangering coastal habitats and species dependent on them for breeding and feeding. Altered precipitation patterns, including more frequent and intense droughts, floods, and storms, disrupt ecosystems by affecting water availability, soil moisture, and overall balance. Ocean acidification, resulting from the absorption of excess CO_2 by oceans, harms marine life with calcium carbonate structures, such as corals and shellfish. Shifting species distribution occurs as ecosystems change, with some species migrating to higher latitudes or elevations, impacting local biodiversity (Chakraborty et al. 2014; Singh et al. 2020). Climate change poses a significant threat to biodiversity overall, as some species struggle to adapt or migrate quickly enough, leading to potential disruptions and cascading effects throughout food webs. The increased frequency and intensity of extreme weather events, including hurricanes, wildfires, and heat waves, have immediate and severe impacts on ecosystems, causing habitat destruction, loss of life, and disruptions to natural balances. Addressing these challenges requires global efforts to reduce greenhouse gas emissions, adapt to changing conditions, and protect vulnerable species and habitats (Akpodiogaga-a and Odjugo 2010; Kumar et al. 2019).

Understanding and predicting ecosystem responses to climate change is vital for various reasons. Ecosystems, housing diverse plant and animal species, necessitate proactive conservation efforts based on insights into their reactions to environmental shifts. This knowledge aids in identifying vulnerable species and habitats, safeguarding biodiversity, and preventing critical species loss. Additionally, ecosystems deliver essential services like clean water, crop pollination, climate regulation, and fisheries support. Predicting their responses to climate change is crucial for evaluating potential disruptions to these services, guiding strategies for sustainable management, and ensuring the well-being of human societies. Ecosystems directly contribute to human well-being by providing food, medicine, and resources, and understanding their response to climate change helps anticipate disruptions, informing adaptive measures in agriculture, water resource management, and community impact mitigation (Marshall et al. 2008; Kumar et al. 2016). The information is fundamental for developing effective adaptation and mitigation strategies, whether protecting coastal communities from sea-level rise, implementing sustainable agriculture practices, or restoring degraded ecosystems. Ecosystem responses are integral to global climate models, enhancing their accuracy in forecasting climate change impacts and aiding policymakers in informed decision-making on mitigation and adaptation. Furthermore, studying ecosystem responses contributes to a broader understanding of ecological processes, advancing scientific models and theories, and enabling innovative solutions for mitigating climate change impacts. Lastly, understanding and predicting ecosystem responses are essential for developing early warning systems, particularly for extreme weather events, facilitating timely community preparedness and response to disasters such as floods, wildfires, or hurricanes (Cleverly et al. 2019).

Functional traits are fundamental in ecological research, offering crucial insights into the strategies and adaptations of organisms within ecosystems. These specific characteristics influence an organism's performance, behavior, and ecological interactions. For instance, in species ecology and niche differentiation, functional traits like leaf morphology or root architecture contribute to understanding how species interact, coexist, and compete. They also play a key role in comprehending ecosystem functioning, with traits related to nutrient cycling, carbon sequestration, and water use efficiency providing insights into the roles different species play in vital processes (Kumar et al. 2020a). As sensitive indicators of environmental changes, functional traits help predict shifts in species composition due to factors like climate change. They also contribute to explaining community assembly and dynamics, shedding light on mechanisms such as environmental filtering, competition, and facilitation (Carlucci et al. 2020). In the realm of biodiversity-ecosystem functioning relationships, functional trait analyses help identify key traits driving ecosystem processes, emphasizing the importance of diverse functional traits for overall ecosystem health. Functional traits are valuable tools in conservation and restoration ecology, aiding in assessing ecosystem resilience, identifying key species for restoration, and predicting the success of introduced or invasive species. In the context of global change ecology, functional traits provide a framework for predicting species responses to environmental changes and assessing ecosystem vulnerability. Moreover, the integration of functional traits into ecological models enhances our understanding of ecosystem dynamics, fostering the development of more accurate and predictive models for ecological processes (McGill et al. 2006; Kumar et al. 2020b).

5.2 Functional Traits: Key Drivers of Ecosystem Processes and Functioning

5.2.1 Definition and Significance of Functional Traits

Functional traits, encompassing morpho-physio-phenological characteristics, indirectly influence organism fitness by affecting growth, reproduction, and survival (Violle et al. 2007). Categorized as response and effect traits, they serve as indicators of adaptive strategies and ecosystem impact. Quantifiable and diverse, these traits provide insights into ecological strategies, adaptations, and community roles (Violle et al. 2007; Kumar et al. 2021b, c). Evaluations using functional traits offer a means to understand fluctuations in community structure and ecosystem operation.

5.2.2 Establishing Links Between Traits and Ecosystem Functioning

Functional traits, such as leaf area, wood density, and water-related characteristics, play crucial roles in ecosystem functioning. These traits are pivotal in processes like carbon sequestration, water use efficiency, and reproductive dynamics. Indicative of ecosystem resistance, resilience, and stability, they provide valuable insights into the intricate interplay of invasive species' traits with native species, contributing to a deeper understanding of biodiversity. Additionally, these traits form a comprehensive framework for unravelling ecological strategies and predicting the impacts of organisms on ecosystem structure, function, and resilience. Moreover, the essential characteristics of organisms, categorized as functional traits, significantly influence key ecosystem processes. These traits have a far-reaching impact, dictating nutrient cycling efficiency, primary productivity, and water cycling. They regulate various aspects of ecosystem dynamics, including pollination, drought resistance, and defense, which, in turn, affect reproductive success and ecosystem recovery post-disturbance. Traits such as wood density and growth form are intricately linked to carbon sequestration, shaping the contributions of ecosystems to the global carbon cycle (Kumar et al. 2015; Chen et al. 2018; Lone et al. 2019; Wang et al. 2020). Collectively, understanding the role of functional traits provides a holistic perspective on how organisms influence and shape essential processes within ecosystems (Table 5.1).

5.2.3 Importance of Trait-Based Ecology in Understanding Ecosystem Dynamics

The context-dependent nature of trait expression introduces variations in broadly defined patterns, a concept illustrated by Kraft et al. (2015) in their study on plant functional traits. This context dependency allows for the elucidation of differences in fitness and ecological niches among species. Combinations of multiple traits, as highlighted by, effectively capture niche differences, underscoring the potential of functional traits in clarifying species interactions, coexistence, and overall ecosystem function. This contextual understanding of functional traits establishes a seamless link to the importance of trait-based ecology in comprehending how these characteristics drive ecological processes. Trait-based ecology, transcending traditional taxonomy, serves as a vital framework for predicting biodiversity-function relationships, community assembly, and species interactions. As emphasized by de Bello et al. (2010), these approaches offer valuable insights into community dynamics, particularly under the influence of environmental changes, thereby supporting

Table 5.1 Linking functional traits to specific ecosystem processes

Functional trait	Linked ecosystem process	Explanation
Larger leaf area	Photosynthesis	Larger leaf area enhances the surface for capturing sunlight, contributing to higher photosynthetic rates
High specific leaf area (SLA)	Nutrient cycling	High SLA influences leaf decomposition rates, affecting nutrient release and contributing to cycling
Fine roots and mycorrhizae	Nutrient uptake	Fine roots and mycorrhizal associations enhance nutrient uptake efficiency, contributing to nutrient cycling
Stomatal conductance	Water cycling	Stomatal conductance influences water vapor exchange, impacting water cycling within ecosystems
Flowering phenology	Pollination	Flowering phenology influences the timing of reproductive events, affecting interactions with pollinators
Leaf chemistry	Decomposition	Leaf chemistry influences litter decomposition rates, impacting nutrient release and organic matter cycling
High wood density	Carbon sequestration	High wood density is associated with slower decomposition, contributing to long-term carbon sequestration
Drought resistance traits	Resilience to drought	Succulence and deep roots enhance resistance and resilience to drought, contributing to ecosystem stability
Chemical defense traits	Regulation of herbivory	Chemical defense traits deter herbivores, influencing herbivore population dynamics and regulating herbivory
Fire adaptation traits	Response to disturbance (fire)	Fire-adapted traits, like fire-resistant bark, influence ecosystem response to fire disturbances

ecosystem resilience. This comprehensive framework is indispensable for addressing challenges in global change ecology, managing the impacts of invasive species, and guiding effective conservation and restoration efforts. The seamless integration of insights derived from the context-dependent nature of trait expression with trait-based ecology enhances our understanding of species interactions and ecological dynamics.

5.2.4 Implications for Ecosystem Productivity and Resilience

Understanding the implications of functional traits for ecosystem productivity and resilience is pivotal. Enhanced ecosystem productivity results from species optimizing resource utilization, and functional diversity correlates with increased stability. Functional traits contribute to species adaptation, ensuring ecosystem persistence amid challenges like climate change. Resilience to disturbances is bolstered by the diversity of functional strategies, impacting ecosystem recovery (Mori et al. 2013).

Diverse and resilient functional traits contribute to human well-being by supporting essential ecosystem services (Kumar et al. 2017; Barnett et al. 2019; Hevia et al. 2017; Lundberg and Moberg 2003; Kumar et al. 2019, 2021a, b, c).

5.3 Modelling Approaches

Trait-based modelling in ecology is a method that focuses on understanding and predicting ecological patterns and processes by examining an organism's functional features. These features encompass traits such as size, morphology, physiology, behavior, and life history, collectively influencing an organism's ability to function and interact within its environment. The approach, as outlined by Violle et al. (2007), seeks to provide a deeper mechanistic understanding of ecosystem dynamics by integrating diverse functional features within and across species. Functional traits play a crucial role in an organism's adaptation to its surroundings, resource competition, and overall contribution to ecosystem health. McGill et al. (2006) emphasize that trait-based modelling elucidates the relationships between characteristics, ecological strategies, and community dynamics, contributing to the broader field of trait-based community ecology. This subdiscipline aims to use an organism's functional traits to explain species coexistence, community structure, and ecosystem functioning, enabling models to predict community responses to environmental changes (Lavorel and Garnier 2002). Trait-based models also explore how functional features impact ecosystem-level activities like primary productivity and nutrient cycling, enhancing predictions of ecosystem responses to environmental shifts (Mouillot et al. 2013). Despite its potential, challenges faced by trait-based modelling include data availability, integration with other modelling techniques, and understanding the evolutionary aspects of traits (Díaz et al. 2016).

5.3.1 Overview of Modelling Techniques in Ecology

Ecological modelling, a versatile approach combining computer tools, systems analysis, and mathematical techniques, plays a pivotal role in unravelling intricate ecological dynamics. Striking the right balance between environmental intricacies, core issues, and data constraints is imperative for effective modelling (Jorgensen 2003; Jorgensen and Bendoricchio 2001). Various modelling techniques employed in ecology are included in Table 5.2.

5.3.2 Incorporation of Functional Traits into Ecological Models

Functional traits serve as key components across various ecological modelling frameworks, enhancing our understanding of complex dynamics. In trait-based community ecology models, functional features like specific leaf area, nitrogen content, and seed mass are integrated to elucidate species coexistence, biodiversity

Table 5.2 Modelling techniques used in ecology

Modelling technique	Description	Examples/references
Population models	Focus on population dynamics, including size, age distribution, and interactions	Environmental protection, agricultural yields, disease spread
Community models	Investigate interspecies relationships such as mutualism, competition, and predation	Relationships within ecological communities (Tilman 1982)
Ecosystem models	Abstract, mathematical depictions of ecological systems to enhance understanding	Representation of entire biomes or ecological communities (Hall and Day 1977)
Individual-based models (IBMs)	Examine how interactions between individuals and their environment contribute to population and ecosystem ecology	Study of population and ecosystem dynamics (Grimm and Railsback 2005)
Agent-based models (ABMs)	Similar to IBMs but with an emphasis on modelling interactions among autonomous agents in a spatially explicit context	Modelling interactions with autonomous agents (Railsback and Grimm 2019)
Spatial models	Consider the spatial distribution of ecological entities in modelling	Spatial aspects in ecological systems
Stochastic models	Introduce randomness to simulate variability in ecological systems	Simulating variability in ecological processes
Process-based models	Mathematical descriptions of one or more processes that describe the operation of well-defined biological systems	Modelling well-defined biological processes
Bayesian models	Use Bayesian statistical techniques to draw conclusions about ecological phenomena that are probabilistic	Probabilistic conclusions about ecological phenomena (Hobbs and Hooten 2015)
Meta-analysis models	Combine information from multiple studies to make generalized conclusions about ecological trends and patterns	Deriving generalized insights from multiple studies (Gurevitch and Hedges 1999)

patterns, and community dynamics (Kraft et al. 2008; Singh et al. 2010a, b, 2013a, b). Ecosystem models incorporate traits, such as root depth and nutrient uptake efficiency, to explore how different species contribute to processes like nitrogen cycling and overall ecosystem production (Singh and Verma 2013a, b; Singh et al. 2024; Suding et al. 2015). Individual-based models (IBMs) take a more granular approach, including functional traits at the level of individual organisms to simulate realistic population dynamics and community interactions, providing insights into how trait variation impacts ecological processes (DeAngelis and Mooij 2005). In the context of plant competition models, traits linked to resource acquisition, growth rates, and stress tolerance are frequently considered, contributing to predictions of community composition and species dominance (Adler et al. 2013). This holistic integration of functional traits across these modelling frameworks enriches our understanding of ecological phenomena, spanning from intricate community dynamics to broader ecosystem-level processes.

5.4 Species Distribution Models (SDMs) and Functional Traits

Functional traits play a crucial role in species distribution models (SDMs) across various contexts:

Trait-based climate change impact model: Functional trait-based SDMs are valuable tools for assessing the impact of climate change on different species. These models facilitate the prediction of traits that confer resilience or vulnerability in the face of changing environmental conditions, offering insights into species' responses to climate change (Engler et al. 2011).

Phenotypic plasticity and SDMs: SDMs incorporating functional features associated with phenotypic plasticity account for a species' capacity to modify traits in response to changing environmental circumstances. This inclusion enhances the models' ability to capture the adaptive capacity of species in dynamic environments (Valladares et al. 2014).

Trait-based approaches to invasion ecology: Functional trait-based SDMs are instrumental in monitoring and anticipating the spread of invasive species. Invasion ecology models leverage characteristics related to dispersal, reproductive tactics, and ecological preferences to better understand and manage invasive species dynamics (Gallien et al. 2010).

Functional diversity and SDMs: SDMs can predict patterns of biodiversity and community composition by incorporating functional diversity, a measure of the variety of functional traits within a community. This approach provides a more comprehensive understanding of the ecological dynamics shaping species distributions (Mouillot et al. 2013).

Trait-based niche modelling: SDMs utilize functional traits alongside environmental variables to determine a species' niche, offering a more nuanced understanding of how different species respond to their surroundings. This trait-based niche modelling enhances the complexity of species-environment relationships in SDMs (Fitzpatrick and Hargrove 2009).

Trait-environment relationships in SDMs: SDMs establish correlations between species and environmental variables by applying functional traits. For example, plant characteristics like temperature preference or drought tolerance are integral in predicting species distribution, highlighting the importance of trait-environment relationships in these models (Thuiller et al. 2011). The incorporation of functional traits in SDMs thus enables a more comprehensive and nuanced exploration of species distribution dynamics across various ecological scenarios.

5.5 Trait-Based Community Assembly Models

Trait-based community assembly models offer a paradigm shift from traditional taxonomic approaches by focusing on functional traits to understand and predict patterns of species coexistence and community composition. Departing from treating species as interchangeable entities, these models delve into the characteristics—physiological, morphological, reproductive, and life history—that influence how species interact with their environment and each other. Central to trait-based community assembly is the concept of trade-offs, where species with similar trait values may compete strongly for resources, while those with dissimilar traits can coexist by utilizing different niches, thereby influencing community structure (Larson and Funk 2016). These models distinguish between environmental filtering and biotic interactions as mechanisms driving community assembly, with environmental filtering selecting for traits suited to prevailing conditions and biotic interactions involving species sorting based on interactions. Trait-based models fall into niche-based and neutral categories, emphasizing environmental factors and stochastic processes, respectively. Functional diversity metrics quantify the range of functional traits within a community, with higher diversity indicating enhanced resilience to environmental changes. These models also find application in studying successional dynamics, predicting community responses to disturbances, and are empirically validated using real-world data to assess their accuracy in representing community dynamics (Jung et al. 2010).

5.5.1 Concept of Community Assembly

Community assembly encompasses the intricate processes governing the composition and structure of ecological communities over time, considering historical and contemporary factors (Kraft et al. 2015; Mittelbach and Schemske 2015). Historical events, such as geological changes and climate fluctuations, along with evolutionary processes, leave a lasting impact on species composition, contributing to the pool available for colonization and establishment. Abiotic factors like climate and soil conditions, coupled with biotic factors such as competition, predation, and mutualism, interact to determine species persistence (Weiher et al. 2011). The traits of individual species, influencing competition and coexistence dynamics, play a central role in community assembly (Weiher et al. 2011). Functional diversity, reflecting a variety of traits within a community, contributes significantly to stability and resilience. Community assembly theories fall into neutral and niche-based categories, emphasizing stochastic processes or the role of ecological niches and interactions (Schlägel et al. 2020). Climate change exerts profound impacts on ecological communities, causing shifts in composition and functioning (Van Vuuren et al. 2012). Rising temperatures induce species migration, disrupting established distribution patterns and leading to upslope movements in mountainous regions and poleward shifts at higher latitudes. Climate-driven alterations influence the timing of crucial life events, creating mismatches that cascade through ecosystems, affecting interactions between species. Climate-induced stressors, such as prolonged drought or extreme weather events, reshape community structure by favoring certain species and impacting mutualistic relationships. These disruptions extend to polar ecosystems, vulnerable to climate change, experiencing shifts in marine and terrestrial communities due to warming temperatures and ice melt (Moline et al. 2008; Ledger et al. 2012). The transformations underscore the urgent need for comprehensive strategies to mitigate the effects of climate change on ecological communities.

5.5.2 Functional Traits Play a Crucial Role in Guiding Community Assembly Models

Functional traits serve as pivotal elements in informing community assembly models, offering a mechanistic lens into the intricate dynamics of species interactions and their responses to the environment. These specific organism characteristics unravel the ecological strategies that underpin species coexistence and competition within communities. In the realm of community assembly models, functional traits are instrumental in simulating trait-based filtering, where species with distinct traits are more likely to persist under specific environmental conditions (van der Plas et al. 2015). For instance, in environments prone to drought, species endowed with water use efficiency traits take precedence, shaping a community composition filtered by functional traits. The concept of trade-offs, inherent to these models, elucidates how species, limited by their inability to excel in all traits simultaneously, coexist by exploiting diverse ecological niches. A trade-off, such as between competitive ability and stress tolerance, facilitates the harmonious existence of multiple species within a community. Beyond this, functional traits contribute to calculating community functional diversity, signaling the spectrum of traits present and fortifying the community's adaptability to environmental changes (Martin 2014). As models integrate these diversity metrics, they become valuable tools for assessing the stability and resilience of ecosystems. Functional traits are integral to modelling shifts in community composition along environmental gradients, predicting trait distributions across diverse conditions. They provide critical insights into community responses to disturbances, with species resilient to fire or drought dominating in susceptible ecosystems (Dubuis et al. 2013). Niche-based community assembly models leverage functional traits to simulate the intricate determination of species' ecological niches, allowing for a nuanced representation of how environmental conditions interact with traits to shape community structure. Moreover, functional traits prove indispensable in forecasting community responses to changing environmental conditions linked to climate change, enabling models to simulate potential shifts in composition based on altered temperatures, precipitation patterns, or nutrient availability. Empirical validation of functional traits within these models, by comparing predictions with observed community patterns, enhances the robustness and reliability of trait-based assumptions (Herrera 2016).

5.6 Resilience Assessment and Functional Diversity

Resilience assessment in ecosystems is intricately tied to the concept of functional diversity, representing the array of functional traits present within a community. This diversity serves as a key determinant of ecosystems' capacity to endure disturbances, adapt to evolving conditions, and recover from perturbations. Assessing resilience involves considering the diversity of response traits among different species, as unique functional characteristics influence their reactions to disturbances (Chillo et al. 2011). Functional redundancy, where multiple species share similar traits, enhances resilience by enabling compensation for negative impacts on specific species, thereby maintaining crucial ecosystem processes. The combination of high functional redundancy and diverse traits contributes to stability in the face of disturbances, highlighting the system's ability to uphold essential functions despite external pressures. Specific functional traits act as indicators of resilience, with attributes linked to stress tolerance, recovery rates, and adaptive capacity providing insights into ecosystems' potential for recovery. Functional traits also play a crucial role in evaluating the efficiency of ecosystem services, influencing their reliability and stability over time (Espelta et al. 2020). Response diversity, encompassing a variety of response strategies within a community, contributes to adaptive capacity, a vital component of resilience assessment. Temporal stability in ecosystem functions is linked to functional diversity changes over time, with consistent diversity indicative of greater stability (Parsons et al. 2016). Invasive species pose a challenge, altering functional diversity and impacting resilience, prompting the assessment of ecosystems' ability to resist and recover from such intrusions. Functional diversity acts as an early warning system for changes in ecosystem health, offering insights into potential stressors and enabling timely intervention. Understanding the nexus between resilience and functional diversity informs conservation and management strategies, emphasizing the need to protect and restore functional diversity to ensure ecosystems can recover from disturbances and continue providing essential services (Renault et al. 2022).

5.6.1 Role of Functional Diversity and Redundancy in Enhancing Resilience

Functional diversity and redundancy are integral components of biodiversity that significantly contribute to enhancing ecosystem resilience. Functional diversity, representing the variety of functional traits and strategies across species, plays a crucial role in adaptive capacity by offering diverse responses to environmental changes. It promotes complementarity, occupying various ecological niches to reduce competition and enhance resource use efficiency, fostering stability and robustness in ecosystems (Biggs et al. 2020). On the other hand, functional redundancy, denoting similarity in functional roles among different species, acts as insurance against losses by allowing functionally similar species to compensate for disturbances, maintaining essential ecosystem processes. This redundancy contributes to flexibility and robustness, ensuring that ecosystems can adapt to changing conditions. Together, the synergy of functional diversity and redundancy creates a powerful combination, with diverse traits providing adaptability and redundancy ensuring the persistence of essential functions even in the face of species loss. This dual mechanism enhances resilience to novel or unpredictable conditions, allowing

Table 5.3 Case studies illustrating resilience assessment using trait-based approaches

Case study	Objective	Trait-based approach	Findings	Reference
Coral reefs in the Caribbean	Assess resilience to climate change, specifically coral bleaching	Examined traits related to thermal tolerance, growth, and reproduction	Species with traits conferring thermal tolerance identified as potentially more resilient	Baker et al. (2008)
Grassland resilience to drought	Assess resilience to prolonged drought conditions	Focused on water use efficiency, root depth, and seed traits	Species with deep roots and high water use efficiency associated with greater resilience	Mehran et al. (2015)
Mangrove forests and storm surges	Evaluate resilience to storm surges and sea-level rise	Investigated traits related to seed dispersal, root structure, and salt tolerance	Species with buoyant seeds and salt tolerance identified as contributors to resilience	Rezaie et al. (2020)
Alpine plants and warming temperatures	Assess resilience to climate warming in alpine plant communities	Examined traits related to temperature preferences, growth forms, and phenology	Species with traits allowing for a phenological shift were seen as potentially more resilient	Winkler et al. (2019)
Temperate forests and fire disturbances	Evaluate resilience to recurrent fire disturbances	Studied traits related to flammability, bark thickness, and resprouting ability	Species with thick bark and resprouting ability associated with greater postfire resilience	Bryant et al. (2019)

ecosystems to better cope with unforeseen challenges. Understanding and conserving both functional diversity and redundancy are crucial for effective conservation strategies, and in restoration efforts, balancing these aspects is key to promoting ecosystem resilience (Montoya et al. 2012) (Table 5.3).

5.7 Applied Applications

5.7.1 Crop Modelling and Breeding Strategies

1. Crop growth models simulate various physiological processes of nutrient uptake and crop growth. They aid in estimating yield under various circumstances and predict stages of crop development.
2. The dynamics of disease and pest populations in agricultural ecosystems can be better understood with the use of ecological models. The development of efficient management techniques, such as planning the application of pest control treatments and the choice of resistant crop varieties.

3. It is possible to simulate how crop development and yield may be affected by climate change using ecological models. This facilitates the interpretation of the potential effects of changing climate on crop production.
4. By simulating water flow in agricultural landscapes, ecological models can help optimize irrigation techniques for sustainability and efficient use of water.

5.7.2 Forestry and Sustainable Forest Management

1. Modelling aids in the estimation of rates of carbon sequestration, evaluation of the effects of forest management techniques on carbon stocks, and comprehension of the function of forests in reducing global warming.
2. Models such as FARSITE and BehavePlus help anticipate fire spread, severity, and potential effects on forest ecosystems by simulating the behavior of fire while taking weather, topography, and vegetation into account.
3. They assist in determining important habitats, assessing how forest management techniques affect biodiversity, and guiding conservation plans.
4. Understanding water flow, sediment transport, and nutrient cycling in forested watersheds is aided by modelling the hydrological cycle.

5.7.3 Aquatic Ecosystems and Fisheries Management

1. The models aid in the estimation of fish population abundance, growth rates, and death rates. Also, the best harvest levels and maximum sustainable production for fisheries are estimated with the use of models.
2. Various models aid in the identification of diseases that spread throughout fish populations. They support the development of disease management plans for aquaculture and wild fisheries.
3. To evaluate the economic viability, models can be incorporated into ecological and economic considerations by optimizing fishing operations.
4. It is possible to predict changes in fish habitats by modelling the effects of climate change on ocean temperature, currents, and the distribution of marine species.

5.7.4 Invasive Species Prediction and Management

1. Based on environmental factors, SDMs aid in the prediction of invasive species' prospective dispersal. These models help prioritize management efforts and identify regions that are vulnerable to invasion.
2. To effectively manage invasive species, models aid in the timing and combination of techniques including chemical, biological, and cultural practices.

3. Models help in the creation of early warning systems to detect possible invasive species before they have a chance to establish themselves. Rapid reaction and containment efforts are made possible by early discovery.
4. Models evaluate how invasive species affect the economy by considering things like yield losses, environmental degradation, and control expenses.

5.7.5 Restoration Ecology and Ecosystem Restoration

1. Models enable the identification and prioritization of restoration sites according to variables including ecosystem services, biodiversity, and habitat quality.
2. The course of vegetation and ecosystem development throughout time is simulated by models. Comprehending successional dynamics assists planners in projecting post-restoration ecosystem changes and making appropriate planning decisions.
3. The models help choose appropriate species and restoration tactics that are resilient to shifting environmental circumstances.
4. Models can evaluate the cost-effectiveness of various restoration projects by considering both ecological and economic aspects.

5.8 Challenges and Future Directions

5.8.1 Challenges in Collecting and Applying Functional Trait Data

Functional trait data collection often faces challenges due to the lack of standardized protocols. Inconsistencies in measurement methods, units, and reporting can hinder the comparability of data across studies and limit the robustness of meta-analyses. Traits can exhibit scale-dependent variability, posing challenges in choosing the appropriate spatial and temporal scales for data collection. The mismatch between the scale of trait measurements and the scale of ecological processes can affect the accuracy of predictions and applications. Also, there is considerable intraspecific variability in functional traits, and capturing this variation within and among populations is challenging. Overlooking intraspecific variability can lead to inaccurate assessments of community dynamics and ecosystem functioning. Understanding trade-offs and redundancies among functional traits is complex. Traits may be correlated, and identifying the specific traits that drive ecological processes can be challenging. Functional trait data are often biased toward common species, and there is limited information available for rare or understudied species. This bias can affect the comprehensiveness of trait databases and hinder a holistic understanding of ecosystem dynamics.

5.8.2 Emerging Trends in Trait-Based Research

Integrating molecular data with traditional functional traits provides a more comprehensive understanding of species interactions and responses to environmental changes. Genomic approaches can reveal underlying genetic mechanisms influencing traits. Advancement in remote sensing and high-throughput technologies enable the collection of trait data at larger spatial and temporal scales. It facilitates the measurement of traits such as leaf area index, while high-throughput technologies streamline the process of trait data collection. Trait-based research is expanding to microbial ecology, acknowledging the role of microbial functional traits in ecosystem processes. These researches are also increasingly being applied to aquatic ecosystems. Identifying functional traits of aquatic organisms helps in understanding the ecological processes in freshwater and marine environments.

5.8.3 Future Directions and the Role of Functional Traits in Addressing Climate Change Challenges

Functional traits can aid in identifying species and ecosystems that are more resilient or adaptive to climate change. This information is valuable for developing conservation and management strategies to mitigate the impacts of climate change on biodiversity. Integrating functional traits into climate models can enhance predictions of species distribution shifts and ecosystem responses to climate change. This trait-based modelling approach provides a more mechanistic understanding of how species and ecosystems will respond to changing environmental conditions. Functional traits can guide ecosystem restoration efforts by helping select species with traits that enhance ecosystem resilience. This is crucial for restoring ecosystems that have been degraded due to climate change-induced events such as wildfires, droughts, or floods.

Establishing global-scale trait monitoring networks can contribute to a better understanding of how functional traits vary across different regions and ecosystems. This information is essential for implementing effective global conservation and climate change mitigation strategies.

5.9 Conclusion

Understanding and predicting ecosystem responses to climate change is imperative for various reasons, and functional traits play a pivotal role in advancing ecological research. The exploration of functional traits provides vital insights into the strategies and adaptations of organisms within ecosystems, influencing their performance, behavior, and ecological interactions. Trait-based modelling in ecology emerges as a promising and insightful approach, offering a mechanistic understanding of how species interact with their environment (Singh et al. 2014, 2021). Notably, trait-based models contribute to community ecology by explaining species coexistence and community structure while providing valuable predictions regarding ecosystem responses to environmental changes.

The integration of functional trait analysis with species distribution modelling enhances our perspective on the variables influencing species presence and abundance across different settings. In a rapidly changing world, this integrated approach has the potential to strengthen biodiversity management, interpretation, and prediction. It offers a comprehensive framework for navigating the complexities of ecosystems in the era of climate change.

Furthermore, incorporating functional traits into ecological research and resilience assessments establishes a robust foundation for addressing environmental shifts. This understanding not only aids in predicting and mitigating the impacts of climate change but also guides conservation efforts. It ensures the persistence of essential ecosystem functions, promoting the well-being of both ecosystems and the societies they support. The continued exploration of functional traits and resilience will be instrumental in developing adaptive strategies. The ongoing researches are crucial for fostering a sustainable coexistence between human activities and the intricate web of life on Earth. By incorporating functional traits into our ecological understanding, a harmonious balance that supports the resilience of ecosystems in the face of ongoing environmental challenges can be developed.

References

Adler PB, Fajardo A, Kleinhesselink AR, Kraft NJ (2013) Trait-based tests of coexistence mechanisms. Ecol Lett 16(10):1294–1306

Akpodiogaga-a P, Odjugo O (2010) General overview of climate change impacts in Nigeria. J Hum Ecol 29(1):47–55

Baker, A. C., Glynn, P. W., & Riegl, B. (2008). Climate change and coral reef bleaching: An ecological assessment of long-term impacts, recovery trends and future outlook. Estuarine, coastal and shelf science, 80(4), 435-471.

Barnett LA, Jacobsen NS, Thorson JT, Cope JM (2019) Realizing the potential of trait-based approaches to advance fisheries science. Fish Fish 20(5):1034–1050

Biggs CR, Yeager LA, Bolser DG, Bonsell C, Dichiera AM, Hou Z et al (2020) Does functional redundancy affect ecological stability and resilience? A review and meta-analysis. Ecosphere 11(7):e03184

Bryant T, Waring K, Sánchez Meador A, Bradford JB (2019) A framework for quantifying resilience to forest disturbance. Front For Glob Chang 2:56

Carlucci MB, Brancalion PH, Rodrigues RR, Loyola R, Cianciaruso MV (2020) Functional traits and ecosystem services in ecological restoration. Restor Ecol 28(6):1372–1383

Chakraborty S, Pattanayak A, Mandal S, Das M, Roychowdhury R (2014) An overview of climate change: causes, trends and implications. In: Crop improvement in the era of climate change. IK International Publishing House, New Delhi, pp 1–29

Chen W, Koide RT, Eissenstat DM (2018) Nutrient foraging by mycorrhizas: from species functional traits to ecosystem processes. Funct Ecol 32(4):858–869

Chillo V, Anand M, Ojeda RA (2011) Assessing the use of functional diversity as a measure of ecological resilience in arid rangelands. Ecosystems 14:1168–1177

Cleverly J, Eamus D, Edwards W, Grant M, Grundy MJ, Held A et al (2019) TERN, Australia's land observatory: addressing the global challenge of forecasting ecosystem responses to climate variability and change. Environ Res Lett 14(9):095004 de Bello F, Lavorel S, Díaz S, Harrington R, Cornelissen JH, Bardgett RD et al (2010) Towards an assessment of multiple ecosystem processes and services via functional traits. Biodivers Conserv 19:2873–2893

DeAngelis DL, Mooij WM (2005) Individual-based modeling of ecological and evolutionary processes. Annu Rev Ecol Evol Syst 36:147–168

Díaz S, Kattge J, Cornelissen JH, Wright IJ, Lavorel S, Dray S et al (2016) The global spectrum of plant form and function. Nature 529(7585):167–171

Dubuis A, Rossier L, Pottier J, Pellissier L, Vittoz P, Guisan A (2013) Predicting current and future spatial community patterns of plant functional traits. Ecography 36(11):1158–1168

Engler R, Randin CF, Thuiller W, Dullinger S, Zimmermann NE, Araujo MB et al (2011) 21st century climate change threatens mountain flora unequally across Europe. Glob Chang Biol 17(7):2330–2341

Espelta JM, Cruz-Alonso V, Alfaro-Sánchez R, Hampe A, Messier C, Pino J (2020) Functional diversity enhances tree growth and reduces herbivory damage in secondary broadleaf forests, but does not influence resilience to drought. J Appl Ecol 57(12):2362–2372

Fitzpatrick MC, Hargrove WW (2009) The projection of species distribution models and the problem of non-analog climate. Biodivers Conserv 18(8):2255–2261

Gallien L, Münkemüller T, Albert CH, Boulangeat I, Thuiller W (2010) Predicting potential distributions of invasive species: where to go from here? Divers Distrib 16(3):331–342

Grimm V, Railsback SF (2005) Individual-based modeling and ecology. Princeton University Press

Gurevitch J, Hedges LV (1999) Statistical issues in ecological meta-analyses. Ecology 80(4):1142–1149

Hall CA, Day JW Jr (1977) Ecosystem modeling in theory and practice: an introduction with case histories. Wiley

Herrera JP (2016) Interactions between plants and primates shape community diversity in a rainforest in Madagascar. J Anim Ecol 85(4):982–993

Hevia V, Martín-López B, Palomo S, García-Llorente M, de Bello F, González JA (2017) Trait-based approaches to analyze links between the drivers of change and ecosystem services: synthesizing existing evidence and future challenges. Ecol Evol 7(3):831–844

Hobbs N, Hooten M (2015) Bayesian models: a statistical primer for ecologists. Princeton University Press, Princeton

Jwgensen, S. E. (2003). The recent development toward a consistent ecosystem theory. WIT Transactions on Ecology and the Environment, 63.

Jorgensen SE, Bendoricchio G (eds) (2001) Fundamentals of ecological modelling, 3rd edn. Amsterdam, Elsevier, 530 pp

Jung V, Violle C, Mondy C, Hoffmann L, Muller S (2010) Intraspecific variability and trait-based community assembly. J Ecol 98(5):1134–1140

Kraft NJ, Valencia R, Ackerly DD (2008) Functional traits and niche-based tree community assembly in an Amazonian forest. Science 322(5901):580–582

Kraft NJ, Adler PB, Godoy O, James EC, Fuller S, Levine JM (2015) Community assembly, coexistence and the environmental filtering metaphor. Funct Ecol 29(5):592–599

Kumar N, Kumar N, Shukla A, Shankhdhar SC, Shankhdhar D (2015) Impact of terminal heat stress on pollen viability and yield attributes of rice (*Oryza sativa* L.). Cereal Res Commun 43(4):616–626

Kumar N, Shankhdhar SC, Shankhdhar D (2016) Impact of elevated temperature on antioxidant activity and membrane stability in different genotypes of rice (*Oryza sativa* L.). Indian J Plant Physiol (Now Plant Physiol Rep) 21(1):37–43

Kumar N, Suyal DC, Sharma IP, Verma A, Singh H (2017) Elucidating stress proteins in rice (Oryza sativa L.) genotype under elevated temperature: a proteomic approach to understand heat stress response. 3 Biotech 7:205

Kumar N, Jeena N, Singh H (2019) Elevated temperature modulates rice pollen structure: a study from foothill Himalayan agro-ecosystem in India. 3 Biotech 9:175

Kumar A, Dwivedi GK, Tewari S, Paul J, Anand R, Kumar N, Kumar P, Singh H, Kaushal R (2020a) Carbon mineralization and inorganic nitrogen pools under *Terminalia chebula* Retz.-based agroforestry system in Himalayan foothills, India. For Sci 66(5):634–643

Kumar A, Kumar P, Singh H, Kumar N (2020b) Adaptation and mitigation potential of roadside trees with bio-extraction of heavy metals under vehicular emissions and their impact on physiological traits during seasonal regimes. Urban For Urban Green 58:126900

Kumar N, Jeena N, Kumar A, Khairakpam R, Singh H (2021a) Comparative response of rice cultivars to elevated air temperature in Bhabar region of Indian Himalaya: status on yield attributes. Heliyon 7:e07474

Kumar A, Kumar P, Singh H, Bisht S, Kumar N (2021b) Relationship of physiological plant functional traits with soil carbon stock in temperate forest of Garhwal Himalaya. Curr Sci 120(8):1368–1373

Kumar A, Kumar P, Singh H, Kumar N (2021c) Modulation of plant functional traits under essential plant nutrients during seasonal regime in natural forests of Garhwal Himalayas. Plant Soil 465:197–212

Larson JE, Funk JL (2016) Regeneration: an overlooked aspect of trait-based plant community assembly models. J Ecol 104(5):1284–1298

Lavorel S, Garnier E (2002) Predicting changes in community composition and ecosystem functioning from plant traits: revisiting the Holy Grail. Funct Ecol 16(5):545–556

Ledger ME, Harris RM, Armitage PD, Milner AM (2012) Climate change impacts on community resilience: evidence from a drought disturbance experiment. In: Advances in ecological research, vol 46. Academic Press, pp 211–258

Lone PA, Dar JA, Subashree K, Raha D, Pandey PK, Ray T et al (2019) Impact of plant invasion on physical, chemical and biological aspects of ecosystems: a review. Trop Plant Res 6(3):528–544

Lundberg J, Moberg F (2003) Mobile link organisms and ecosystem functioning: implications for ecosystem resilience and management. Ecosystems 6:87–98

Marshall JD, Blair JM, Peters DP, Okin G, Rango A, Williams M (2008) Predicting and understanding ecosystem responses to climate change at continental scales. Front Ecol Environ 6(5):273–280

Martin PR (2014) Trade-offs and biological diversity: integrative answers to ecological questions. In: Integrative organismal biology. Wiley, pp 291–308

McGill BJ, Enquist BJ, Weiher E, Westoby M (2006) Rebuilding community ecology from functional traits. Trends Ecol Evol 21(4):178–185

Mehran A, Mazdiyasni O, AghaKouchak A (2015) A hybrid framework for assessing socioeconomic drought: linking climate variability, local resilience, and demand. J Geophys Res Atmos 120(15):7520–7533

Mittelbach GG, Schemske DW (2015) Ecological and evolutionary perspectives on community assembly. Trends Ecol Evol 30(5):241–247

Moline MA, Karnovsky NJ, Brown Z, Divoky GJ, Frazer TK, Jacoby CA et al (2008) High latitude changes in ice dynamics and their impact on polar marine ecosystems. Ann N Y Acad Sci 1134(1):267–319

Montoya D, Rogers L, Memmott J (2012) Emerging perspectives in the restoration of biodiversity-based ecosystem services. Trends Ecol Evol 27(12):666–672

Mori AS, Furukawa T, Sasaki T (2013) Response diversity determines the resilience of ecosystems to environmental change. Biol Rev 88(2):349–364

Mouillot D, Graham NA, Villéger S, Mason NW, Bellwood DR (2013) A functional approach reveals community responses to disturbances. Trends Ecol Evol 28(3):167–177

Parsons M, Glavac S, Hastings P, Marshall G, McGregor J, McNeill J et al (2016) Top-down assessment of disaster resilience: a conceptual framework using coping and adaptive capacities. Int J Disaster Risk Reduct 19:1–11

Railsback SF, Grimm V (2019) Agent-based and individual-based modeling: a practical introduction. Princeton University Press

Renault D, Hess MC, Braschi J, Cuthbert RN, Sperandii MG, Bazzichetto M et al (2022) Advancing biological invasion hypothesis testing using functional diversity indices. Sci Total Environ 834:155102

Rezaie AM, Loerzel J, Ferreira CM (2020) Valuing natural habitats for enhancing coastal resilience: wetlands reduce property damage from storm surge and sea level rise. PLoS One 15(1):e0226275

Schlägel UE, Grimm V, Blaum N, Colangeli P, Dammhahn M, Eccard JA et al (2020) Movement-mediated community assembly and coexistence. Biol Rev 95(4):1073–1096

Singh H, Verma A (2013a) Physiological responses of rice cultivars to various nitrogen levels. Int J Agric Environ Biotechnol 6(3):383–388

Singh H, Verma A (2013b) Characterization and screening of high nitrogen efficient rice genotype to curtail environmental pollution. Int J Agric Environ Biotechnol 6(1):23–30

Singh H, Verma A, Shukla A (2010a) Comparative study of yield and yield components of hybrid and inbred genotypes of rice (Oryza Sativa L.). Int J Agric Environ Biotechnol 3:355–360

Singh H, Verma A, Krishnamoorthy M, Shukla A (2010b) Consequence of diverse nitrogen levels on leaf pigments in five rice genotypes under field emergent circumstance. Int J Bioresour Stress Manag 1:189–193

Singh H, Verma A, Rai SK (2013a) Biochemical evaluation of different rice genotypes grown at various nitrogen levels using SDS-PAGE. Curr Adv Agric Sci 5(1):144–146

Singh H, Verma A, Shukla A (2013b) Guttation fluid as a physiological marker for selection of nitrogen efficient rice (Oryza sativa L.) genotypes. Afr J Biotechnol 12(44):6276–6281

Singh H, Verma A, Ansari MW, Shukla A (2014) Physiological response of rice (Oryza sativa L.) genotypes to elevated nitrogen applied under field conditions. Plant Signal Behav 9:e29015. https://doi.org/10.4161/psb.29015

Singh H, Yadav M, Kumar N, Kumar A, Kumar M (2020) Assessing adaptation and mitigation potential of roadside trees under the influence of vehicular emissions: a case study of Grevillea robusta and Mangifera indica planted in an urban city of India. PLoS One 15(1):e0227380

Singh S, Singh H, Sharma SK, Nautiyal R (2021) Seasonal variation in biochemical responses of bamboo clones in the sub-tropical climate of Indian Himalayan foothills. Heliyon 7(4):e06859. https://doi.org/10.1016/j.heliyon.2021.e06859

Singh M, Singh H, Kumar A, Kumar M, Barthwal S, Thakur A (2024) Soil nitrogen availability determines the CO2 fertilization effect on tree species (Neolamarckia cadamba): growth and physiological evidence. Environ Sustain. https://doi.org/10.1007/s42398-023-00300-w

Suding K, Higgs E, Palmer M, Callicott JB, Anderson CB, Baker M et al (2015) Committing to ecological restoration. Science 348(6235):638–640

Thuiller W, Lavergne S, Roquet C, Boulangeat I, Lafourcade B, Araujo MB (2011) Consequences of climate change on the tree of life in Europe. Nature 470(7335):531–534

Tilman D (1982) Resource competition and community structure. Princeton University Press

Valladares F, Matesanz S, Guilhaumon F, Araújo MB, Balaguer L, Benito-Garzón M et al (2014) The effects of phenotypic plasticity and local adaptation on forecasts of species range shifts under climate change. Ecol Lett 17(11):1351–1364

van der Plas F, Janzen T, Ordonez A, Fokkema W, Reinders J, Etienne RS, Olff H (2015) A new modeling approach estimates the relative importance of different community assembly processes. Ecology 96(6):1502–1515

Van Vuuren DP, Riahi K, Moss R, Edmonds J, Thomson A, Nakicenovic N et al (2012) A proposal for a new scenario framework to support research and assessment in different climate research communities. Glob Environ Chang 22(1):21–35

Violle C, Navas ML, Vile D, Kazakou E, Fortunel C, Hummel I, Garnier E (2007) Let the concept of trait be functional! Oikos 116(5):882–892

Wang Y, Yang XD, Ali A, Lv GH, Long YX, Wang YY et al (2020) Flowering phenology shifts in response to functional traits, growth form, and phylogeny of woody species in a desert area. Front Plant Sci 11:536

Weiher E, Freund D, Bunton T, Stefanski A, Lee T, Bentivenga S (2011) Advances, challenges and a developing synthesis of ecological community assembly theory. Philos Trans R Soc B Biol Sci 366(1576):2403–2413

Winkler DE, Lubetkin KC, Carrell AA, Jabis MD, Yang Y, Kueppers LM (2019) Responses of alpine plant communities to climate warming. In: Ecosystem consequences of soil warming. Academic Press, pp 297–346

6 Enhancing Resource Use Efficiency in Crops Through Plant Functional Traits

Hukum Singh, Narendra Kumar, and Amit Kumar

Abstract

In the scenarios of global challenges such as climate change, burgeoning populations, limited agriculture resources, food security, etc., enhancing the resource use efficiency in crops becomes a crucial need for sustainable agriculture. This chapter explores the potential role of plant functional traits (PFTs) in augmenting resource use efficiency and facilitating crop adaptation amid the aforementioned global challenges. We emphasize the complex interactions between plant functional traits and their influence on crops' resource use efficiency. This chapter explores the potential role of various functional traits, such as physiological, morphological, root architecture, biochemical, and molecular, which govern the plant's ability to harness soil (water, nutrients) and atmospheric (sunlight) resources. Further, this chapter paves the direction to develop climate-resilient crop varieties capable of maintaining productivity under changing environmental conditions by exploring the potential role of PFTs in stress tolerance and climate adaptability. This chapter not only highlights the intrinsic potential of plant functional traits in resource use efficiency but also emphasizes their capacity to transform agricultural practices in the face of global challenges, fostering sustainable and resilient agriculture in the future. This insight may benefit academicians and researchers in developing efficient crop varieties in resource acquisition and utilization by exploring the potential of plant functional traits to ensure food security and achieve sustainable development goals.

H. Singh
Plant Physiology Discipline, Forest Research Institute, Dehradun, Uttarakhand, India

N. Kumar
College of Agriculture, Central Agriculture University (I), Kyrdemkulai, Meghalaya, India

A. Kumar
Department of Forestry, North Eastern Hill University, Tura Campus, Tura, Meghalaya, India

N. Kumar, H. Singh (eds.), *Plant Functional Traits for Improving Productivity*,
https://doi.org/10.1007/978-981-97-1510-7_6

6.1 Introduction

Crops, including agroforestry, face numerous challenges such as climate change and resource availability (nutrients, water, lands, etc.) globally demanding innovative solutions to enhance productivity (Singh et al. 2024). Further, ever-increasing pollution intensifies the demand for food production, thereby impacting food security globally (Singh 2021).

Global climate change affects growing seasons and significantly impacts crop growth, development, and productivity. Climate change and precipitation patterns with increased pests and disease incidence threaten global food security via altering agrarian yields. Concurrently, the depletion of crucial resources, particularly water, land, and soil nutrients, hinders agricultural productivity (Sharma and Singh 2021). Soil health degradation due to soil erosion, deforestation, declined soil fertility, and environmental stressors are other pressures on agricultural productivity (Kumar et al. 2015; Joshi et al. 2020; Joshi and Singh 2020: Joshi et al. 2021).

Such scenarios require urgent scientific interventions and sustainable solutions that enhance resource use efficiency in cropping systems, resulting in better productivity to maintain global food security and sustainability (Singh et al. 2010a, b; Kumar et al. 2016). Various practices are being adopted globally to boost resource use efficiency and productivity of cropping systems with due consideration of the long-term impact of climate change.

Furthermore, exploring the role of plant functional traits (PFTs) is a better approach to enhancing resource use efficiency and productivity of cropping systems. The focus on PFTs is integral to sustainable agriculture production. Breeding crops with functional traits that enhance adaptability and resilience to climate change and associated abiotic and biotic stress, including resisting pests and diseases and thriving in diverse environmental conditions, contribute to long-term sustainability (Kumar et al. 2019a, 2021a). Such approaches boost the adaptability of crops and lessen the dependency on external resource inputs, fostering a more robust and self-sustaining agricultural ecosystem, ensuring food security, and mitigating climate change consequences.

6.2 Significance of Plant Functional Traits

Plant functional traits (PFTs) are crucial in determining and shaping resource extraction, uptake, and utilization by the cropping system. The central role of PFTs in utilizing resources lies in their ability to influence how cropping systems interact with surrounding environments and climate and the availability of resources. These PFTs cover a wide range of plant attributes, particularly root and canopy architecture, leaf morphology and anatomy, and physiological processes, which together determine the crop's response to environmental and climatic circumstances (Yadav et al. 2019a, b). Researchers, including breeders, agronomists, physiologists, molecular biologists, and other plant scientists, may screen and develop crop varieties with enhanced ability to efficiently use resources like water and nutrients by understanding and manipulating these traits.

Crops with deep root systems may be able to access water in deeper soil layers, contributing to drought resistance. Further, crops with a higher leaf area index and specific leaf area capture more solar radiation, resulting in higher photosynthesis and nutrient absorption rates, ultimately improving biological and economic yield (Gupta et al. 2018a, b, 2019a, b). Understanding the potential of plant functional traits assists researchers in developing crop varieties tailored to specific agroclimatic conditions, eventually contributing to sustainable agriculture by minimizing resource inputs, enhancing resilience, and fostering more efficient use of limited resources.

Manipulating PFTs is a novel approach to advancing sustainable and efficient agricultural ecosystem management to enhance productivity. This approach is a better way to optimize resource use, improve crop resilience to future changing climatic variability, and mitigate their impact on the agricultural system. Researchers have screened crop varieties with improved water and nitrogen use efficiency, contributing to water conservation and minimizing dependency on chemical fertilizers (Singh et al. 2014). Furthermore, harnessing the potential of plant functional traits facilitates the development of crop varieties with increased resistance to pests and diseases, reducing reliance on chemical interventions. The paradigm shift in crop improvement programs may boost crop productivity and promote the long-term sustainability of food production systems, aligning with the imperative to balance human needs with global environmental and climate change challenges.

6.3 Fundamentals of Plant Functional Traits

6.3.1 Physiological Traits

Physiological functional traits play a pivotal role in determining growth and development, directly influencing productivity. Physiological traits are integral components of complex mechanisms of adaptation that collectively impact the uptake, utilization, and allocation of resources in plant systems under changing environmental conditions. Therefore, optimizing these traits is vital for shaping growth and development and significantly impacts overall productivity. Understanding the physiological traits that govern plant resource dynamics is essential to explore their role in adapting to climate change and environmental stresses (Singh 2024).

The plant physiological traits that govern their resource dynamics, with root architecture play a pivotal role in resource uptake. The length, density, and morphology of a plant's root system determine its ability to access water and soil nutrients (Kumar et al. 2020a, b). The adaptive nature of root architecture is crucial to plants facing diverse environmental conditions such as water, nutrients, heat, cold, etc. Moreover, water use efficiency (WUE) is a critical physiological trait influencing how plants optimize water resources. Stomatal regulation and transpiration rate are integral components of WUE, enabling plants to balance water conservation and effective photosynthesis (Kumar et al. 2021d, e, f). Hence, understanding these physiological mechanisms sheds light on a plant's ability to navigate water scarcity and provides valuable insights for agricultural practices aimed at enhancing crop resilience in resource-limited environments, thereby improving productivity (Kumar et al. 2021g, h).

The allocation and utilization of resources are finely tuned physiological traits that dictate a plant's growth and survival. Carbon allocation, a dynamic process that determines the distribution of carbon resources to different plant parts such as root, leaf, stem, and branch, plays a crucial role in shaping a plant's adaptive strategy (Kumar et al. 2021b). Efficient allocation patterns, influenced by environmental stress, impact the growth of leaves, stems, and roots. Moreover, nutrient acquisition and utilization represent another set of physiological traits critical for plant health. Efficient nutrient uptake, transport, and assimilation are essential for thriving in nutrient-limited soils. Exploring the role of the physiological traits not only assists in unraveling the adaptive strategies of plants in diverse ecosystems but also guides sustainable agricultural practices by optimizing resource utilization and minimizing environmental impact.

6.3.2 Morphological Traits

The impact of morphological traits on plant resource capture and utilization is profound, influencing their ability to thrive in diverse environmental conditions. Morphological traits, encompassing the physical structure and form of plants, play a crucial role in determining how efficiently plants acquire and utilize essential resources such as water, nutrients, and sunlight (Sharma et al. 2018).

Leaf morphology is a critical factor impacting resource utilization and capture of sunlight for photosynthesis. The size, shape, and arrangement of leaves affect the surface area exposed to sunlight. Plants with optimized leaf area index maximize light absorption, enhancing their photosynthetic efficiency. This influences the plant's capacity to convert solar energy into chemical energy, which is vital for growth and development.

The stem morphological traits contribute to resource allocation and structural support. Stems serve as conduits for nutrient transport between roots and leaves, and their architecture influences the distribution of resources throughout the plant (Sharma et al. 2017). Additionally, the structural integrity of stems plays a role in supporting leaves and flowers, optimizing their exposure to sunlight, and facilitating efficient resource utilization.

Morphological adaptations in leaves, stems, roots, etc. occur in response to abiotic stresses such as water scarcity or high salinity. Morphological traits, like reduced leaf size, waxy coatings, or modified stomatal structures, help plants conserve water and mitigate the impact of stressful conditions. These adaptations reflect the plant's ability to balance resource capture with conservation strategies.

6.3.3 Phenological Traits

Plant phenological traits significantly impact resource use efficiency and are crucial for adaptation to climatic variability. Exploring these vital traits is central to enhancing resource efficiency and facilitating plant adaptation to changing environmental conditions. The timing of various phenophases, such as flowering and fruiting, intricately influences how plants utilize resources like water and nutrients, profoundly affecting resource use efficiency and adaptation in multiple ecosystems including agroecosystem.

The synchronization of phenophases, especially flowering and fruiting, significantly influences a plant's ability to align with environmental conditions, impacting nutrient and water uptake. Efficient resource use is critical for plant survival and growth, profoundly influencing ecosystem dynamics.

Plant species endowed with well-adapted phenological traits demonstrate enhanced resilience to environmental changes, including shifts in temperature, precipitation patterns, and soil moisture. This synchronization allows for effective energy allocation and reproductive success. Conversely, mismatches between phenological events and climatic conditions lead to declines in resource use efficiency and reduced adaptative response.

Scientific understanding of the influences of phenological traits is crucial for predicting how plant communities will respond to climate change (Kumar et al. 2018, 2019b; Rawat et al. 2020). Altered phenological patterns may potentially disrupt ecological relationships, affecting pollinators, herbivores, and other species dependent on specific timing cues, thereby creating cascading effects on ecosystems.

Emphasizing the study of phenological traits is essential for developing strategies to mitigate the impacts of climate change on plant communities. Breeding and conservation efforts should prioritize screening and cultivating plant varieties with adaptive phenological traits to enhance overall resource use efficiency under changing climatic conditions (Singh et al. 2018). Ultimately, comprehending the impacts of phenological traits on resource use efficiency and adaptation is fundamental for conserving biodiversity and maintaining ecosystem stability in the face of climate change.

6.3.4 Biochemical Traits

Biochemical traits constitute a significant group of functional traits in plants that play a fundamental role in resource use efficiency, influencing how plants acquire, allocate, and utilize resources for growth and development. These traits encompass a range of plant chemical processes, including nutrient uptake, production of secondary metabolites, etc. (Kumar et al. 2016, 2017; Singh et al. 2020, 2021a). Hence, understanding the interplay of these biochemical traits is essential for optimizing resource use efficiency in diverse ecological contexts.

One of the critical biochemical traits influencing resource use efficiency is photosynthesis. Plants' efficiency in converting sunlight into chemical energy dictates their ability to assimilate carbon from the atmosphere. Variations in photosynthetic rates among plant species are linked to differences in biochemical pathways, enzyme activity, and chloroplast structure. Improving photosynthetic efficiency enhances a plant's capacity to convert available resources into biomass, ultimately contributing to greater resource use efficiency (Flexas et al. 2016).

Further, biochemical traits are pivotal in nutrient acquisition, with plants deploying specific strategies to optimize nutrient use efficiency. Root exudates, for example, play a role in nutrient mobilization and uptake by influencing soil microbial communities. Additionally, activating specific metabolic pathways, such as mycorrhizal associations and nitrogen-fixing nodules, enhances the efficiency with which plants acquire essential nutrients from the soil. Understanding and manipulating these biochemical processes are key aspects of sustainable agriculture and ecosystem management (Marschner et al. 2011).

Beyond primary metabolism, plant biochemical traits include the production of secondary metabolites, which serve various ecological functions. Some secondary metabolites act as defense compounds, protecting plants against herbivores and pathogens. Allocating resources to synthesizing these compounds represents a strategic step in resource use efficiency, as it enhances a plant's ability to withstand biotic stresses and allocates resources more efficiently toward growth and reproduction (Wink 1999). Understanding the regulation of these biochemical pathways provides insights into how plants balance resource allocation between growth and defense.

The role of plant biochemical traits in resource use efficiency is multifaceted, encompassing processes from photosynthesis to nutrient uptake and defense mechanisms. Efforts to enhance resource use efficiency in plants must consider the intricate biochemical networks governing these traits. Insights into these processes' molecular and metabolic basis pave the way for developing strategies to optimize resource utilization in agriculture, conservation, and sustainable agroecosystem management.

6.4 Impact of PFT on Resource Acquisition

6.4.1 Impact on Soil Nutrient Acquisition

Plant functional traits are crucial in shaping nutrient acquisition strategies, influencing a plant's survival ability in various environments. Root traits, mainly root morphology, architecture, density, and mycorrhizal associations, significantly impact nutrient uptake efficiency from the soil system (Bardgett et al. 2014; Prakash et al. 2022). For example, deep-rooted plants are often better equipped to access nutrients from deeper soil layers. At the same time, mycorrhizal associations enhance nutrient absorption by facilitating nutrient transfer between the soil and plant roots (van der Heijden et al. 2015). These functional traits contribute to the plant's adaptive capacity to varying soil conditions and nutrient availabilities.

Further, leaf functional traits are pivotal in nutrient acquisition, affecting the plant's photosynthetic capacity and nutrient use efficiency. Leaf functional traits, particularly leaf nitrogen content (LNC), specific leaf area (SLA), and leaf longevity, are key functional traits that influence a plant's ability to acquire and utilize nutrients (Reich et al. 1997). Higher leaf nitrogen content is often associated with increased photosynthetic rates and enhanced nutrient uptake, contributing to enhanced resource use efficiency (Aerts and Chapin 2000). Henceforth, understanding the role of leaf functional traits is essential for predicting plant responses to nutrient availability and environmental changes.

The relationship between plant functional traits extends beyond nutrient acquisition to encompass water use efficiency and carbon allocation. For instance, root traits, including root-to-shoot ratio and fine root density, influence water uptake efficiency and plant water use strategies (Comas et al. 2013a). Moreover, carbon allocation to different plant parts, influenced by functional traits such as leaf area (LA) and specific root length (SRL), has implications for nutrient acquisition and overall plant performance and productivity (Mokany et al. 2006). These interconnected functional traits collectively shape a plant's resource acquisition strategies and responses to environmental conditions.

Therefore, it is evident that long-term research on plant functional traits and their impact on nutrient acquisition have practical applications in agriculture, ecology, and conservation. Furthermore, understanding the influence of plant functional traits on nutrient uptake and acquisition may help breeders in screening and characterizing crop varieties, including agriculture and agroforestry, with enhanced nutrient use efficiency and improved resilience to changing climatic and environmental circumstances (Freschet et al. 2015; Kumar et al. 2022). Moreover, considering the role of plant functional traits in crop growth models allows for more accurate predictions of nutrient cycling and ecosystem dynamics, aiding in developing sustainable agriculture management practices (Díaz et al. 2016). This interdisciplinary approach emphasizes the consideration of plant functional traits in the context of nutrient and resource acquisition for addressing global challenges related to food security and ecosystem sustainability (Singh et al. 2022).

6.4.2 Impact on Atmospheric Resource Acquisition

It is well established that plant functional traits determine how plants interact with the surrounding environmental conditions. The traits influence resource acquisition mechanisms and the uptake and absorption of essential elements from the atmosphere. Key functional traits like leaf morphology, stomatal conductance, and biochemical pathways contribute to a plant's ability to efficiently acquire resources from the surrounding environments, influencing the overall growth and development of the plant system.

Stomatal conductance is a critical functional trait that regulates the exchange of gases between plants and the atmosphere. It directly influences water uptake from the soil and atmosphere and releases water during transpiration. Plants with optimized stomatal conductance exhibit enhanced water use efficiency, allowing for better adaptation to varying atmospheric conditions like temperature, CO_2, precipitation, etc. (Raturi et al. 2022; Singh 2024).Stomatal conductance is pivotal for plants in arid environments and water-limited conditions.

Leaf morphological traits, i.e., size, shape, thickness, and surface characteristics, significantly affect a plant's ability to assimilate carbon dioxide from the atmosphere through photosynthesis. It is reported that variations in leaf traits impact the carbon assimilation rate, ultimately influencing plant productivity (Singh 2024). Further, they stressed to recognize the importance of leaf functional traits in understanding a plant's carbon acquisition strategy under climate change and environmental stresses.

Plant biochemical pathways contribute to nutrient acquisition from the atmosphere. For example, nitrogen-fixing plants utilize nodules containing nitrogen-fixing bacteria to convert atmospheric nitrogen into forms accessible to the plants (Singh 2024). Understanding these biochemical traits is essential for improving crop nutrient acquisition strategies and optimizing agricultural productivity.

Roots primarily interact with the soil. Certain root functional traits impact the plant's ability to acquire resources indirectly from the atmosphere. Root exudates influence soil microbial communities, creating a dynamic interface between the soil and the atmosphere (Bardgett and van der Putten 2014). This interaction displays the importance of plant functional traits in resource acquisition from the soil and atmosphere systems.

The impact of plant functional traits on resource acquisition has broader implications for agroecosystem functioning and responses to global climate change. It is therefore recognized that exploring the role of plant functional traits in resource acquisition is crucial for predicting how crop species at individual and ecosystem levels will respond to changing climatic and environmental conditions such as rising atmospheric CO_2 concentrations, temperature, and water- and nutrient-limited conditions (Díaz et al. 2016). Understanding the linkage between plant functional traits and resource acquisition would be an effective instrument to develop strategies for sustainable agriculture and agroecosystem management practices.

6.5 Environmental Adaptation and Climate Resilience

6.5.1 Climate and Plant Functional Traits

Exploring plants' responses to diverse climatic conditions is crucial for predicting the impact of climate change on ecosystems. Plant functional traits, encompassing leaf morphology, physiology, and reproductive strategies, are dynamic and exhibit distinct responses to temperature, precipitation, and other climatic factors. Collectively, these traits determine plant adaption to changing climatic conditions (Singh et al. 2017a).

Leaf morphological traits, such as leaf size, shape, and structure, are highly responsive to climatic variations (Dhyani et al. 2021). For instance, plants in arid environments often exhibit adaptations like reduced leaf size and increased leaf thickness to minimize water loss through transpiration (Reich et al. 1999).

Physiological functional traits, like stomatal conductance and photosynthetic rates, are sensitive indicators of climatic conditions (Nirmal et al. 2021). Climatic conditions such as elevated temperatures and increased CO_2 levels impact stomatal behavior, altering gas exchange, resource uptake, and efficiency, ultimately affecting productivity (Ainsworth and Rogers 2007; Apurva et al. 2017; Kumar et al. 2017; Dhyani et al. 2021).

Phenological traits, such as flowering and fruiting times, are highly responsive to climate variations. Warming temperatures have been linked to shifts in phenological events, with earlier flowering observed in many plant species (Parmesan and Yohe 2003; Gupta and Singh 2017). Further, the change in phenological behavior may have cascading effects on ecosystem dynamics and interactions with pollinators and other species.

Biochemical traits are critical and vital in a plant's response to changing climatic conditions (Singh and Verma 2013a, b; Singh et al. 2013a). Plants exhibit alterations in their secondary metabolite production in response to increased UV radiation or temperature changes, influencing their resistance to herbivores or pathogens (Wink 1999).

Besides, root traits are crucial for nutrient and water acquisition. Root traits respond to variations in soil moisture and temperature. In water-limited environments, plants have been reported to develop deeper and more extensive root systems to access water stored at greater depths (Comas et al. 2013b).

Drought is a significant climatic factor. The plants have been found to evolve various traits to cope with water scarcity and associated stresses (Kumari and Singh 2018). Plant traits such as succulence, deep root systems, and the ability to close stomata under water stress contribute to a plant's drought tolerance (Chaves et al. 2003).

Increasing atmospheric CO_2 levels are a crucial factor responsible for climate change. Plants respond to elevated CO_2 through changes in photosynthesis, leaf nitrogen content, and water use efficiency (Ainsworth and Long 2005; Singh et al. 2024). Understanding the response of plant functional traits is essential for predicting ecosystem dynamics and carbon sequestration.

Plants undergo genetic and evolutionary changes in response to long-term climate change. The long-term climate change conditions may favor plants with traits conferring better fitness under altered conditions, contributing to adaptive evolution (Franks et al. 2014).

Hence, understanding plants' responses to various climatic conditions is invaluable to formulating conservation and management practices. Conservation efforts should consider preserving genetic diversity to allow for adaptive responses, while land management strategies must account for changing plant traits in the wake of climate change.

6.5.2 Soil Interactions with Plant Functional Traits

The relationship between soil properties and plant functional traits is essential in agroecology. Soil properties, including texture, nutrient content, and pH, profoundly influence plant growth, development, and productivity. Researchers worldwide have reported complex associations between soil characteristics and plant functional traits.

Soil texture, defined by sand, silt, and clay proportions, significantly impacts root architecture and development. Comas et al. (2013c) reported that plants adapt their root functional traits, such as root length and branching, in response to soil texture to optimize nutrient and water acquisition.

Further, soil nutrient availability is critical in shaping leaf functional traits. Plants adjust their leaf morphology, such as size and thickness, based on nutrient levels in the soil. Poorter et al. (2012) demonstrated how nutrient-rich soils contribute to developing leaves with higher photosynthetic capacity.

Soil pH is an important soil property that influences microbial communities, which, in turn, impact plant functional traits. Acidic or alkaline soils alter nutrient cycling and availability, influencing traits related to nutrient uptake and utilization by plants (Fierer and Jackson 2006).

The soil's water-holding capacity influences plant traits, including drought tolerance. Plants with low water-holding capacity develop adaptive traits, such as deep root systems and reduced leaf area, to conserve water during periods of drought (Bartlett et al. 2014).

Soil compaction affects root functional traits by limiting root penetration and exploration of the soil profile. Compacted soils often lead to altered root morphology, reduced nutrient absorption, and compromised plant performance (Bengough et al. 2011).

Heavy metal contamination in the soil is a significant problem worldwide. Soils contaminated with heavy metals prompt adaptive responses in plant functional traits. Ali et al. (2013) suggested that plants develop defense mechanisms, altering traits, such as root exudation and metal uptake, to cope with heavy metal stress (Meena et al. 2011; Ali et al. 2013; Singh et al. 2017b).

The diversity and activity of soil microbial communities influence plant functional traits through complex interactions. Soil microbes contribute to nutrient cycling, affecting plant traits related to nutrient acquisition and overall growth (van der Heijden et al. 2008).

Hence, it is evident that the influence of soil characteristics on plant functional traits has broader implications for ecosystem functioning. Therefore, it is essential to understand these complex relationships to predict how ecosystems respond to environmental changes, including land-use alterations and climate change (De Deyn et al. 2008).

Examining soil characteristics in relation to plant functional traits provides valuable insights into the mechanisms governing plant-soil interactions (Kumar et al. 2018a). Long-term and systematic studies on the complex relationship between plant functional traits and soil properties are essential to advancing scientific understanding of agroecosystem dynamics and developing strategies for sustainable management practices.

6.6 Advancements in Genetic Manipulation for Enhanced Traits

6.6.1 Genetic Engineering for Trait Enhancement

Genetic engineering plays a pivotal role in improving plant resource use efficiency through trait enhancement. It allows researchers to manipulate plants' genetic composition precisely, introducing specific traits contributing to better resource use efficiency. This includes improving traits related to nutrient uptake, water utilization, and resistance to environmental stressors. One key aspect is the development of crops with enhanced tolerance to abiotic stressors such as drought, salinity, and nutrient deficiencies. Genes responsible for stress resistance, enabling plants to thrive in challenging environmental conditions, are being introduced in plants. This has significant implications for sustainable agriculture, as it helps maximize crop yield in regions prone to such stressors.

Additionally, genetic engineering facilitates the enhancement of photosynthetic efficiency in plants. There is a need to introduce genes responsible for enhancing photosynthesis efficiency, which will help improve the plant's ability to convert sunlight into energy more effectively. These genes may boost overall plant productivity and contribute to resource efficiency by optimizing the utilization of sunlight, a crucial factor in plant growth. Advances in understanding the molecular mechanisms of photosynthesis have enabled targeted genetic modifications for this purpose, promising more resilient and resource-efficient crops. Furthermore, improving nutrient use efficiency is a critical goal in genetic engineering for sustainable agriculture.

Efforts are in progress globally to develop crops with enhanced nutrient uptake and utilization capabilities. This involves introducing genes that improve the efficiency of nutrient absorption from the soil, reducing the need for excessive fertilization. Genetically modified crops contribute to more sustainable and environmentally friendly agricultural practices by minimizing nutrient loss and maximizing the utilization efficiency of the crops. The introduced genes responsible for increased nutrient uptake or improved nutrient utilization in crops like maize and rice have shown promising results (Lopes and Araus 2019).

Drought is a significant challenge in agriculture, and genetic engineering offers solutions to enhance plant water use efficiency. Introducing genes for improved water retention or reduced crop transpiration may induce adaptation to water-limited conditions (Tardieu and Tuberosa 2010).

Optimizing photosynthesis is crucial for increasing resource use efficiency. This approach allows researchers to introduce genes associated with enhancing the efficiency of photosynthetic processes in plants, leading to improved biomass production. Parry et al. (2009) modified the key enzymes involved in photosynthesis to maximize resource use efficiency and productivity.

Besides, the structure of plant roots significantly influences their ability to absorb water and nutrients, thereby enhancing resource extraction, uptake, and use efficiency. Genetic engineering helps researchers manipulate functional traits associated with root architecture, enhancing the plant's capacity to explore a larger soil volume for resources. This approach improves resource uptake in various crops (York 2019).

Further, recent advancements in molecular biology have provided sophisticated tools for trait enhancement in plants. CRISPR-Cas9, for instance, allows precise and targeted modifications in the plant genome, facilitating the development of crops with specific traits related to resource use efficiency (Hsu et al. 2014).

Genetic engineering holds immense promise for improving resource use efficiency in plants. Ethical considerations and regulatory frameworks are paramount in genetic engineering. Hence, striking a balance between innovations and ensuring environmental and human safety is crucial for applying genetic engineering in agriculture (Kuiper et al. 2001).

6.6.2 Breeding Strategies for Trait Selection

Crop improvement and breeding strategies are pivotal in enhancing desirable traits in plants, contributing to improved resource use, crop performance, and agricultural sustainability. Various approaches have been employed to select and propagate specific traits, catering to the evolving needs of modern agriculture.

Conventional breeding methods have been used for controlled cross-pollination to achieve desired traits for enhancing resource utilization efficiency and productivity. This approach includes the selection of parent plants with desirable characteristics, leading to the development of improved crop varieties (Acquaah 2012).

The marker-assisted selection (MAS) approach is being used, which involves using molecular markers linked to target traits to enable more precise trait selection. Collard and Mackill (2008) used this approach in breeding crops with suitable traits to enhance crop productivity.

Genomic selection (GS) is another strategy that leverages genomic information to predict the breeding value of plants without direct phenotypic assessment. Jannink et al. (2010) and Heffner et al. (2009) have contributed significantly to the understanding and application of genomic selection-specific traits in plant breeding for enhanced productivity.

Gene-editing techniques (GET) have emerged as an advanced technology, like CRISPR-Cas9, which has revolutionized trait selection by allowing precise modifications to the plant genome. Puchta (2017) and Hille and Charpentier (2016) explored the potential use of gene editing in plant breeding for specific traits that are reasonable for the adaptive capacity of crops toward abiotic stresses.

Quantitative trait loci (QTL) mapping is widely used to identify regions of the genome associated with specific traits, aiding in selecting plants with favorable genetic combinations. Phenomics and high-throughput screening advancements in phenotyping technologies enable the rapid and accurate assessment of plant traits (Verma et al. 2020, 2021). Tardieu et al. (2017) have explored the application of high-throughput phenotyping in plant breeding for specific traits to enhance the performance and productivity of crop species.

Integrating breeding strategies for sustainable agriculture has been considered an efficient approach in breeding and crop improvement research for desired traits. This holistic approach combines various breeding strategies to ensure the development of resilient and high-yielding crop varieties. Cobb et al. (2019) and Collard et al. (2005) explore this integrative approach to address the challenges of global food security and environmental sustainability.

6.7 Technological Innovations in Monitoring and Optimization PFTs

Advanced technological innovations significantly advance the monitoring and optimization of plant functional traits, allowing researchers to gain deeper insights into crop performance and make informed decisions for crop management. These innovations encompass a range of technologies, including remote sensing, high-throughput phenotyping, sensor networks, precision agriculture techniques, etc.

High-throughput phenotyping (HTP) is an advanced imaging technology that includes drones, satellites, and ground-based sensors. These technologies allow for rapid and nondestructive monitoring of plant traits on a landscape level. These technologies enable researchers to collect data on plant growth, morphology, and physiological traits efficiently, which helps monitor crop performance (Singh et al. 2013b, 2014).

Remote sensing technology uses satellite and aerial imagery and unmanned aerial vehicles (UAVs or drones) to collect valuable information for monitoring plant traits over large areas. These technologies assess various characteristics of vegetation such plant functional types, plant health, nutrient status, and stress levels, aiding in optimizing cultivation and management practices (Singh et al. 2021b; Kumar et al. 2021c).

Ground-based sensor technology, including proximal and remote sensors, measures various plant traits such as chlorophyll content, canopy temperature, and nutrient levels. These sensors provide real-time data, allowing for quick adjustments in management practices.

Advancements in genomic tools, including next-generation sequencing and genome-wide association studies (GWAS), contribute to understanding the genetic basis of plant functional traits. This knowledge is generally utilized for targeted breeding programs to optimize desired traits.

Precision agriculture is an important approach that integrates various technologies that help optimize resource use by tailoring inputs (such as water, fertilizers, and pesticides) based on the specific needs of individual plants or small sections of fields. This improves overall efficiency and minimizes the environmental impact of the agriculture system as well as practices.

Machine learning and artificial intelligence are crucial in analyzing vast datasets generated by monitoring tools. Machine learning algorithms may identify patterns and relationships within data, assisting in predicting optimal plant traits and guiding decision-making in agriculture.

The innovative biosensors, molecular tools, and techniques enable the monitoring of specific biochemical traits in plants. These tools provide insights into nutrient uptake, stress responses, and metabolic pathways, improving resource use efficiency and crop productivity.

Integrating Internet of Things (IoT) devices, including smart sensors and connected platforms, facilitates real-time data collection and decision-making in agriculture. This interconnected system enhances the monitoring and optimization of plant functional traits that determine crop productivity.

6.8 Challenges and Future Directions

Plant functional traits (PFTs) are pivotal in determining resource use efficiency in agroecosystems, influencing various processes such as photosynthesis, nutrient cycling, water use, and crop productivity. However, manipulating PFTs poses multiple challenges that need careful consideration for sustainable and effective outcomes. One major challenge is the complexity of trait interactions between and across plant species. Henceforth, understanding the complex linkage between traits and their responses to changing climatic and environmental conditions is crucial for successful manipulation.

Another significant hurdle is the potential ecological consequences of altering plant functional traits. Changes in traits, such as leaf morphology or root architecture, may have cascading effects on biological diversity, ecosystem structure and function, and nutrient cycling. Hence, balancing enhancing resource use efficiency and maintaining ecological integrity is essential to avoid unintended consequences.

Furthermore, there is a need for a comprehensive understanding of the genetic basis of plant functional traits. Identifying key genes associated with desired traits and developing precise genetic manipulation techniques are critical for targeted trait modifications. It is also noted that ethical considerations, public perception, and regulatory frameworks pose challenges in the genetic manipulation of plants, necessitating transparent communication and stringent oversight.

Looking toward the future, advancements in omics technologies, such as genomics, transcriptomics, and metabolomics, offer promising opportunities for unraveling the complexities of plant functional traits. Further, integrating big data approaches and machine learning algorithms may aid in predicting trait responses to changing environmental conditions, facilitating more informed trait manipulation strategies.

Additionally, collaborative efforts among researchers, policymakers, and stakeholders are crucial to address these challenges and steer research toward sustainable solutions. Future directions in manipulating plant functional traits should prioritize interdisciplinary research, emphasizing the integration of ecological, genetic, and technological perspectives. This holistic approach will enable the development of resilient plant varieties that enhance resource use efficiency while minimizing negative impacts on ecosystems. Besides, a commitment to ethical and sustainable practices will be paramount for successfully manipulating plant functional traits to meet the growing demands for global resource sustainability.

6.9 Environmental and Economic Impacts of PFTs

It is well recognized that PFTs are crucial in shaping the environment and economics, particularly concerning resource use efficiency. These traits directly influence a plant's ability to acquire, utilize, and conserve resources. The environmental impacts of plant functional traits extend beyond individual species, affecting entire ecosystems and their resilience under the influence of climate change and associated environmental challenges.

From an environmental perspective, plant functional traits contribute significantly to biodiversity, ecosystem stability, and climate change mitigation. The PFTs determine resource use efficiency and influence a plant's adaptability to varying environmental conditions. Species with diverse functional traits enhance ecosystem resilience, ensuring the sustainability of crucial ecological processes, such as nutrient cycling, carbon sequestration, and water filtration (Nautiyal et al. 2022). Additionally, specific functional traits, like drought resistance or nitrogen-fixing abilities, determine a plant's capacity to flourish in changing climatic conditions, making them essential for ecosystem adaptation.

The impact of plant functional traits is equally profound on the economic front. Agriculture, forestry, and other resource-dependent industries benefit significantly from PFT-determined productivity (Kumar et al. 2018b). Water and nutrient use efficiency directly influence crop yields, allowing for more sustainable and economically viable agricultural practices. Furthermore, selecting plant species with desirable traits may contribute to the development of bioenergy crops, fostering a renewable and economically feasible alternative to traditional energy sources.

Hence, balancing environmental conservation and economic development requires a nuanced understanding of plant functional traits. Sustainable resource management practices, guided by these traits, are expected to enhance ecosystem services, promote biodiversity, and bolster economic productivity. Moreover, recognizing the interconnectedness of plant functional traits with environmental and economic outcomes becomes imperative to achieve a harmonious coexistence between nature and human society.

6.10 Conclusion

It is concluded that plant function traits, i.e., physiological, morphological, phenological, and biochemical, play crucial roles in determining crop resource use efficiency and productivity. Therefore, the exploration and application of plant functional traits present a promising opportunity to enhance crop resource use efficiency and productivity. Understanding and harnessing the potential role of plant functional traits may optimize agricultural practices to improve water, nutrient, and energy utilization. Integrating the potential role of PFTs into breeding programs and agronomic management strategies holds great significance in developing more resilient and productive crop varieties. As we face the challenges of a growing global population and changing climate, a collaborative and multidisciplinary focus on plant functional traits is required to maximize resource use efficiency and crop productivity and minimize environmental impact. Implementing long-term multidisciplinary and collaborative research on plant functional traits may significantly contribute to a more efficient, sustainable, and resilient future for global food production.

References

Acquaah E (2012) Principles of plant genetics and breeding. John Wiley & Sons

Aerts R, Chapin FS III (2000) The mineral nutrition of wild plants revisited: a re-evaluation of processes and patterns. Adv Ecol Res 30:1–67

Ainsworth EA, Long SP (2005) What have we learned from 15 years of free-air CO_2 enrichment (FACE)? A meta-analytic review of the responses of photosynthesis, canopy properties, and plant production to rising CO_2. New Phytol 165(2):351–372

Ainsworth EA, Rogers A (2007) The response of photosynthesis and stomatal conductance to rising [CO_2]: mechanisms and environmental interactions. Plant Cell Environ 30(3):258–270

Ali H et al (2013) Remediation of heavy metal contaminated ecosystems: potential of indigenous plants in phytoremediation. Environ Sci Pollut Res 20(12):8759–8771

Apurva K, Singh H, Kumar H (2017) Elucidating morphological and growth changes in Arjun tree (*Terminalia arjuna*) grown under elevated temperature condition. Trends Biosci 10(1):332–336

Bardgett RD, van der Putten WH (2014) Belowground biodiversity and ecosystem functioning Nature 515:505–511

Bardgett RD, Mommer L, De Vries FT (2014) Going underground: root traits as drivers of ecosystem processes. Trends Ecol Evol 29(12):692–699

Bartlett MK et al (2014) Global analysis of plasticity in turgor loss point, a key drought tolerance trait. Ecol Lett 17(12):1580–1590

Bengough AG et al (2011) Root responses to soil physical conditions; growth dynamics from field to cell. J Exp Bot 62(1):421–431

Chaves MM et al (2003) Understanding plant responses to drought—from genes to the whole plant. Funct Plant Biol 30(3):239–264

Cobb JN, Biswas PS, Platten JD (2019) Back to the future: revisiting MAS as a tool for modern plant breeding. Theor Appl Genet 132:647–667. https://doi.org/10.1007/s00122-018-3266-4

Collard BCY, Mackill DJ (2008) Marker-assisted selection: an approach for precision plant breeding in the twenty-first century. Philos Trans R Soc B Biol Sci 363(1491):557–572

Collard BCY, Jahufer MZZ, Brouwer JB, Pang ECK (2005) An introduction to markers, quantitative trait loci (QTL) mapping and marker-assisted selection for crop improvement: the basic concepts. Euphytica 142:169–196. https://doi.org/10.1007/s10681-005-1681-5

Comas LH, Callahan HS, Midford PE (2013a) Patterns in root traits of woody species hosting arbuscular and ectomycorrhizas: implications for the evolution of belowground strategies. Ecol Evol 3(4):1101–1111

Comas LH et al (2013b) A method to quantify lateral root branching in field studies: application to the study of maize root response to nitrogen and phosphorus availability. Plant Soil 370(1–2):51–66

Comas LH et al (2013c) Root traits contributing to plant productivity under drought. Front Plant Sci 4:442

De Deyn GB et al (2008) Soil invertebrate fauna enhances grassland succession and diversity. Nature 422(6933):711–713

Dhyani K, Kuniyal HB, Singh H, Sobha (2021) Growth and physiological potential of *Terminalia arjuna* under elevated CO_2 levels in open top chamber condition. J Appl Nat Sci 13(3):1121–1126. https://doi.org/10.31018/jans.v13i3.2490

Díaz S, Kattge J, Cornelissen JH, Wright IJ, Lavorel S, Dray S et al (2016) The global spectrum of plant form and function. Nature 529(7585):167–171

Fierer N, Jackson RB (2006) The diversity and biogeography of soil bacterial communities. Proc Natl Acad Sci 103(3):626–631

Flexas J et al (2016) Mesophyll conductance to CO_2: current knowledge and future prospects. Plant Cell Environ 39(12):2556–2570

Franks SJ et al (2014) The resurrection initiative: storing ancestral genotypes to capture evolution in action. Bioscience 64(3):231–241

Freschet GT, Aerts R, Cornelissen JH (2015) Multiple mechanisms for trait effects on litter decomposition: moving beyond home-field advantage with a new hypothesis. J Ecol 103(4):1096–1105

Gupta SK, Singh H (2017) Observations on sporadic flowering in *Ficus benjamina* L. and *Peltophorum africanum* L. in new forest. Indian Forester 143:290–291

Gupta SK, Ram J, Singh H (2018a) Comparative study of transpiration in cooling effect of tree species in the atmosphere. J Geosci Environ Prot 6:151–166

Gupta SK, Ram J, Singh H, Kumar P (2018b) Comparative study of cooling under shades of some forestry tree species with respect to ambient temperatures. J Geosci Environ Prot 6:51–66

Gupta SK, Ram J, Singh H (2019a) Relationship between leaf area index and atmospheric cooling of tree species. Indian J For 42(3):203–208

Gupta SK, Singh H, Ram J (2019b) Need for a novel paradigm in forestry research: perspectives of comparative atmospheric cooling by different forestry effect tree species. Indian Forester 145(6):509–515

Heffner EL, Sorrells ME, Jannink JL (2009) Genomic selection for crop improvement. Crop Sci 49(1):1–12

Hille F, Charpentier E (2016) CRISPR-Cas: biology, mechanisms and relevance. Philos Trans R Soc Lond Ser B Biol Sci 371(1707):20150496. https://doi.org/10.1098/rstb.2015.0496

Hsu PD et al (2014) Development and applications of CRISPR-Cas9 for genome engineering. Cell 157(6):1262–1278

Jannink JL, Lorenz AJ, Iwata H (2010) Genomic selection in plant breeding: from theory to practice. Brief Funct Genomics 9(2):166–177

Joshi R, Singh H (2020) Carbon sequestration potential of disturbed and non-disturbed forest ecosystem: a tool for mitigating climate change. Afr J Environ Sci Technol 14(11):385–393

Joshi R, Singh H, Chhetri R, Yadav K (2020) Assessment of carbon sequestration potential in degraded and non-degraded community forests in Terai region of Nepal. J For Environ Sci 36(2):13–121. https://doi.org/10.7747/JFES.2020.36.2.113

Joshi R, Singh H, Chhetri R, Poudel SR, Rijal S (2021) Carbon sequestration potential of community forests: a comparative analysis of soil organic carbon stock in community managed forests of far-western Nepal. Eurasian J Soil Sci 10(2):96–104. https://doi.org/10.18393/ejss.825066

Kuiper HA et al (2001) Assessment of the safety of foods derived from genetically modified (GM) crops. Food Chem Toxicol 39(7):663–727

Kumar N, Kumar N, Shukla A, Shankhdhar SC, Shankhdhar D (2015) Impact of terminal heat stress on pollen viability and yield attributes of rice (*Oryza sativa* L.). Cereal Res Commun 43(4):616–626

Kumar N, Shankhdhar SC, Shankhdhar D (2016) Impact of elevated temperature on antioxidant activity and membrane stability in different genotypes of rice (Oryza sativa L.). Indian J Plant Physiol (Now Plant Physiol Rep) 21(1):37–43

Kumar N, Suyal DC, Sharma IP, Verma A, Singh H (2017) Elucidating stress proteins in rice (*Oryza sativa* L.) genotype under elevated temperature: a proteomic approach to understand heat stress response. 3 Biotech 7:205. https://doi.org/10.1007/s13205-017-0856-9. IF: 2.406

Kumar M, Rawat SPS, Hukum S, Ravindranath NH, Kalra N (2018) Dynamic forest vegetation models for predicting impacts of climate change on forests: an Indian perspective. Indian J For 41(1):1–12

Kumar A, Tewari S, Singh H, Singh I, Anand R, Kumar D, Pandey R (2018a) Impact of different agro-forestry systems on growth and yield of turmeric at Tarai region of Uttarakhand, India. J Plant Dev Sci 10(3):157–162

Kumar M, Savita, Singh H, Pandey R, Singh MP, Ravindranath NH, Kalra N (2018b) Assessing vulnerability of forest ecosystem in the Indian Western Himalayan region using trends of net primary productivity. Biodivers Conserv 28:2163–2182. https://doi.org/10.1007/s10531-018-1663-2

Kumar M, Kalra N, Khaiter P, Ravindranath NH, Singh V, Singh H, Sharma S, Rahnamayan (2019a) PhenoPine: a simulation model to trace the phenological changes in *Pinus roxhburghii* in response to ambient temperature rise. Ecol Model 404:12–20

Kumar N, Jeena N, Singh H (2019b) Elevated temperature modulates rice pollen structure: a study from the foothill of the Himalayan agro-ecosystem in India. 3 Biotech 9(5):175. https://doi.org/10.1007/s13205-019-1700-1

Kumar A, Dwivedi GK, Tewari S, Paul J, Anand R, Kumar N, Kumar P, Singh H, Kaushal R (2020a) Carbon mineralization and inorganic nitrogen pools under *Terminalia chebula* Retz.-based agroforestry system in Himalayan foothills, India. For Sci 66(5):634–643. https://doi.org/10.1093/forsci/fxaa012

Kumar A, Dwivedi GK, Tewari S, Jaipaul, Sah VK, Singh H, Kumar P, Kumar N, Kausal R (2020b) Soil organic carbon pools under *Terminalia chebula* Retz. based agroforestry system in Himalayan foothills, India. Curr Sci 118:1098–1103. https://doi.org/10.18520/cs/v118/i7/1098-1103. IF: 1.102. ISSN: 0011-3891

Kumar M, Kalra N, Singh H, Sharma S, Rawat PS, Singh RK, Gupta AK, Kumar P, Ravindranath NH (2021a) Indicator-based vulnerability assessment of forest ecosystem in the Indian Western Himalayas: an analytical hierarchy process integrated approach. Ecol Indic 125:107568. https://doi.org/10.1016/j.ecolind.2021.107568

Kumar M, Phukon SN, Paygude AC, Tyagi K, Singh H (2021b) Mapping phenological functional types (PhFT) in the Indian eastern Himalayas using machine learning algorithm in Google Earth Engine. Comput Geosci 158:104982. https://doi.org/10.1016/j.cageo.2021.104982

Kumar N, Jeena N, Kumar A, Khairakpan R, Singh H (2021c) Comparative response of rice cultivars to elevated air temperature in Bhabar region of Indian Himalaya: status on yield attributes. Heliyon 7(7):e07474. https://doi.org/10.1016/j.heliyon.2021.e07474

Kumar A, Kumar P, Singh H, Kumar N (2021d) Adaptation and mitigation potential of roadside trees with bio-extraction of heavy metals under vehicular emissions and their impact on physiological traits during seasonal regimes. Urban For Urban Green 58:126900. https://doi.org/10.1016/j.ufug.2020.126900

Kumar A, Kumar P, Singh H, Kumar N (2021e) Impact of plant functional traits on infiltration rate under different forests of Kempty watershed in Garhwal Himalaya. Indian J Soil Conserv 49(1):38–44

Kumar A, Kumar P, Singh H, Kumar N (2021f) Modulation of plant functional traits under essential plant nutrients during seasonal regime in natural forests of Garhwal Himalayas. Plant Soil 465:197–212. https://doi.org/10.1007/s11104-021-05003-x

Kumar A, Kumar P, Singh H, Bisht S, Kumar N (2021g) Relationship of physiological plant functional traits with soil carbon stock in the temperate forest of Garhwal Himalaya. Curr Sci 120(8):1368–1373. https://doi.org/10.18520/cs/v120/i8/1368-1373

Kumar A, Tewari S, Singh H, Kumar P, Kumar N, Bisth S, Kushwah S, Nidhi, Kaushal R (2021h) Biomass accumulation and carbon stocks in different agro forestry system prevalent in Himalayan foothills, India. Curr Sci 120(6):1083–1188. https://doi.org/10.18520/cs/v120/i6/1083-1088

Kumar A, Singh H, Kumari G, Bisht S, Malik A, Kumar A, Singh M, Raturi A, Barthwal S, Thakur A, Kaushal R (2022) Adaptive resilience of roadside trees to vehicular emissions via leaf enzymatic, physiological, and anatomical trait modulations. Environ Pollut 313:120191. https://doi.org/10.1016/j.envpol.2022.120191

Kumari A, Singh H (2018) Impact of heat and drought stress on physiological response of *Terminalia arjuna* grown under elevated temperature condition: an adaptive mechanism to climate change. J Pharmacogn Phytochem 7(1):1374–1378

Lopes MS, Araus JL (2019) Integrating genetic, physiological, and environmental effects on flowering time in cereal crops. Curr Opin Plant Biol 49:49–56

Marschner P et al (2011) How relevant is recalcitrance for the stabilization of organic matter in soils? J Plant Nutr Soil Sci 174(1):27–39

Meena D, Singh H, Chaudhari SK (2011) Elucidating strontium response on growth dynamics and biochemical change in *Phaseolus mungo* L. Int J Agric Environ Biotechnol 4:107–113

Mokany K, Raison RJ, Prokushkin AS (2006) Critical analysis of root: shoot ratios in terrestrial biomes. Glob Chang Biol 12(1):84–96

Nautiyal S, Singh H, Kamboj S, Kumar A, Panwar VP, Bala N (2022) Carbon regulating services of a 30-year-old reclaimed limestone mine area in foothills of Himalaya. Indian Forester 148(3):338–343. https://doi.org/10.36808/if/2022/v148i3/168187

Nirmal, Bakshi M, Singh H (2021) Effect of CO_2 elevation on Shisham growth at nursery stage. Indian J Hill Farming 34:272–276

Parmesan C, Yohe G (2003) A globally coherent fingerprint of climate change impacts across natural systems. Nature 421(6918):37–42

Parry MAJ et al (2009) Raising yield potential of wheat. II. Increasing photosynthetic capacity and efficiency. J Exp Bot 60(6):1409–1418

Poorter H et al (2012) Biomass allocation to leaves, stems and roots: meta-analyses of interspecific variation and environmental control. New Phytol 193(1):30–50

Prakash V, Hunney K, Singh H (2022) Effect of elevated carbon dioxide on growth and development of *Santalum album* L. seedlings inoculated with plant growth promoting microorganisms in open top chambers. Int J Environ Agric Biotechnol 7(5):149–159

Puchta H (2017) Applying CRISPR/Cas for genome engineering in plants: the best is yet to come. Curr Opin Plant Biol 36:1–8. https://doi.org/10.1016/j.pbi.2016.11.011

Raturi A, Singh H, Kumar P, Shukla N (2022) Characterizing the post-monsoon CO_2, CH_4, N_2O, and H_2O vapor effluxes from a tropical wetland in the Himalayan foothill. Environ Monit Assess 194:50. https://doi.org/10.1007/s10661-021-09721-8

Rawat AS, Kalra N, Singh H, Kumar M (2020) Application of vegetation models in India for understanding the forest ecosystem processes. Indian Forester 146(2):93–100. https://doi.org/10.36808/if/2020/v146i2/151208

Reich PB, Walters MB, Ellsworth DS (1997) From tropics to tundra: global convergence in plant functioning. Proc Natl Acad Sci 94(25):13730–13734

Reich PB et al (1999) Scaling of respiration to nitrogen in leaves, stems and roots of higher land plants. Ecol Lett 2(3):259–266

Sharma R, Singh H (2021) Alteration in biochemical constituents and nutrients partitioning of *Asparagus racemosus* in response to elevated atmospheric CO_2 concentration. Environ Sci Pollut Res 29:6812–6821. https://doi.org/10.1007/s11356-021-16050-3

Sharma R, Prajapati N, Singh H (2017) Elevated carbon dioxide impacts on bioactive compounds or nutraceuticals properties of medicinal plants. J Pharmacogn Phytochem 8(1):1924–1926
Sharma R, Singh H, Kaushik M, Nautiyal R, Singh O (2018) Adaptive physiological response, carbon partitioning and biomass production of *Withania Somnifera* (L.) Dunal grown under elevated CO_2 regimes. 3 Biotech 8:267. https://doi.org/10.1007/s13205-018-1292-1
Singh H (2021) An integrated approach considering physiological and biophysical based indicators for assessing tolerance of roadside plantations of *Alstonia scholaris* towards urban roadside air pollution: an assessment of adaptation of plantations for mitigating urban roadside air pollution. Trees 37:69. https://doi.org/10.1007/s00468-021-02179-8
Singh H, Verma A (2013a) Physiological responses of rice cultivars to various nitrogen levels. Int J Agric Environ Biotechnol 6(3):383–388
Singh H, Verma A (2013b) Characterization and screening of high nitrogen efficient rice genotype to curtail environmental pollution. Int J Agric Environ Biotechnol 6(1):23–30
Singh H, Verma A, Shukla A (2010a) Comparative study of yield and yield components of hybrid and inbred genotypes of rice (*Oryza Sativa L.*). Int J Agric Environ Biotechnol 3:355–360
Singh H, Verma A, Krishnamoorthy M, Shukla A (2010b) Consequence of diverse nitrogen levels on leaf pigments in five rice genotypes under field emergent circumstance. Int J Bioresour Stress Manag 1:189–193
Singh H, Verma A, Rai SK (2013a) Biochemical evaluation of different rice genotypes grown at various nitrogen levels using SDS-PAGE. Curr Adv Agric Sci 5(1):144–146
Singh H, Verma A, Shukla A (2013b) Guttation fluid as a physiological marker for selection of nitrogen efficient rice (*Oryza sativa* L.) genotypes. Afr J Biotechnol 12(44):6276–6281
Singh (2024) Exploring adaptive modulation in plant functional traits and its impact on the productivity of Acacia auriculiformis under CO_2-enriched environment. Ind Crop Prod 210. https://doi.org/10.1016/j.indcrop.2024.118186
Singh H, Verma A, Ansari MW, Shukla A (2014) Physiological response of rice (*Oryza sativa* L.) genotypes to elevated nitrogen applied under field conditions. Plant Signal Behav 9:e29015. https://doi.org/10.4161/psb.29015
Singh H, Savita, Sharma R, Sinha S, Kumar M, Kumar P, Verma A, Sharma SK (2017a) Physiological functioning of *Lagerstroemia speciosa* L. under heavy roadside traffic: an approach to screen potential species for abatement of urban air pollution. 3 Biotech 7(61):1–10. https://doi.org/10.1007/s13205-017-0690-0
Singh H, Verma A, Kumar M, Sharma R, Gupta R, Kaur M, Negi M, Sharma SK (2017b) Phytoremediation: a green technology to clean up the sites with low and moderate level of heavy metals. Austin Biochem 2(2):1–8
Singh H, Sharma R, Savita, Singh MP, Kumar M, Verma A, Ansari MW, Sharma SK (2018) Adaptive physiological response of *Parthenium hysterophorus* to elevated atmospheric CO_2 concentration. Indian Forester 144:1–14
Singh H, Yadav M, Kumar N, Kumar A, Kumar M (2020) Assessing adaptation and mitigation potential of roadside trees under the influence of vehicular emissions: a case study of *Grevillea robusta* and *Mangifera indica* planted in an urban city of India. PLoS One 15(1):e0227380. https://doi.org/10.1371/journal.pone.0227380
Singh RK, Singh P, Drews M, Kumar P, Singh H, Gupta AK, Govil H, Kaur A, Kumar M (2021a) A machine learning-based classification of LANDSAT images to map land use and land cover of India. Remote Sens Appl Soc Environ 24:100624. https://doi.org/10.1016/j.rsase.2021.100624
Singh S, Singh H, Sharma SK, Nautiyal R (2021b) Seasonal variation in biochemical responses of bamboo clones in the sub-tropical climate of Indian Himalayan foothills. Heliyon 7(4):e06859. https://doi.org/10.1016/j.heliyon.2021.e06859
Singh H, Bandyopadhyaya S, Kumar A, Singh M, Malik A, Kumari G, Raturi A, Yadav SK, Kumar M, Barthwal S, Thakur A, Rawat PS (2022) Understanding the physiological and biophysical response of urban roadside plantations for assessing adaptation and mitigation mechanisms toward vehicular emissions. Urban Clim 44:101183. https://doi.org/10.1016/j.uclim.2022.101183

Singh M, Singh H, Kumar A, Kumar M, Barthwal S, Thakur A (2024) Soil nitrogen availability determines the CO_2 fertilization effect on tree species (*Neolamarckia cadamba*): growth and physiological evidence. Environ Sustain. https://doi.org/10.1007/s42398-023-00300-w
Tardieu F, Tuberosa R (2010) Dissection and modelling of abiotic stress tolerance in plants. Curr Opin Plant Biol 13(2):206–212
Tardieu F, Cabrera-Bosquet L, Pridmore T, Bennett M (2017) Plant phenomics, from sensors to knowledge. Curr Biol 27(15):R770–R783. https://doi.org/10.1016/j.cub.2017.05.055
van der Heijden MG et al (2008) The unseen majority: soil microbes as drivers of plant diversity and productivity in terrestrial ecosystems. Ecol Lett 11(3):296–310
van der Heijden MG, Martin FM, Selosse MA, Sanders IR (2015) Mycorrhizal ecology and evolution: the past, the present, and the future. New Phytol 205(4):1406–1423
Verma S, Kumar N, Verma A, Singh H, Siddique KHM, Singh P (2020) Novel approaches to mitigate heat stress impacts on crop growth and development. Plant Physiol Rep 25:627–644. https://doi.org/10.1007/s40502-020-00550-4
Verma A, Ansari MW, Singh H, Kumar N, Anwar MS, Mudila H, Tuteja N (2021) Proteomics for Brassinosteroid signalling: understanding Brassinosteroids mediated stress responses through advanced proteomics. Plant Gene 26:100282. https://doi.org/10.1016/j.plgene.2021.100282
Wink M (1999) Adaption of plants and arthropods to different biomes and different host plants. Phytochemistry 50(5):639–660
Yadav SK, Singh H, Ginwal HS, Barthwal S (2019a) Elevated CO_2 enhanced growth and physiological process of *Populus deltoides* Bartr. ex Marsh. Indian Forester 145(1):23–27
Yadav SK, Singh H, Nautiyal R, Ginwal HS, Ansari SA, Barthwal S (2019b) Modulation of morpho-physiological responses in *Populus deltoides* by elevated carbon dioxide and temperature. For Sci 66:105–118. https://doi.org/10.1093/forsci/fxz048
York LM (2019) A physiological framework for root phenotypes to improve crop adaptation to low phosphorus availability. Ann Bot 124(6):923–937

Plants Functional Traits and Photosynthetic Efficiency for Enhancing Crop Yield

7

Km. Prachi, Chandra Kanta, Satish Chandra, and Ishwar Prakash Sharma

Abstract

The need for food is growing daily. This demand may be fulfilled using agronomic practices and conventional breeding methods that maximize plant architecture and light absorption to create cultivars with heavier yields. The plants' increased ability to catch light is made possible by their increasing photosynthetic efficiency. The characteristics of the plant have a significant role in controlling the structures and functions of the ecosystem and directly affect the plant's capacity to photosynthesize. Plants fix carbon, transport nutrients, and absorb water through photosynthesis; hence, improvements to plant traits must be necessary for improved photosynthesis. The plant yield that produces the higher-quality grains and biomass is directly attributed to photosynthesis. When comparing the average relative growth rates of C_3 and C_4 plants, the C_4 plants show a higher growth-trait connection since their intercept is bigger.

Keywords

Biomass · Ecosystem · Photosynthesis · Plant · Traits

K. Prachi · I. P. Sharma (✉)
Patanjali Herbal Research Division, Patanjali Research Foundation, Haridwar, Uttarakhand, India

C. Kanta
School of Applied and Life Sciences, Uttaranchal University, Dehradun, Uttarakhand, India

S. Chandra
Department of Botany, Government Degree College Tyuni, Dehradun, Uttarakhand, India

N. Kumar, H. Singh (eds.), *Plant Functional Traits for Improving Productivity*,
https://doi.org/10.1007/978-981-97-1510-7_7

7.1 Introduction

Over the past 50 years, our major crops' agricultural production has risen to consumer demand. The main causes of these gains were advancements in agronomic practices and conventional breeding, which produced higher-yielding cultivars by optimizing plant architecture and light capture. Nevertheless, in many regions of the world, the annual growth in the production of the main crops has plateaued. Higher-yielding cultivars must be developed through novel technical solutions to hold the food supply required to fulfill the demands of the increasing population (Long et al. 2015; Ort et al. 2015). By 2050, there will likely be 9.7 billion people on the planet, up from the current 7.6 billion. To meet the increased demand for plant-based fuels, animal feed, and plant-based proteins, the major food crops' yields will need to increase by 70–100% (Tilman and Clark 2015; Kumar et al. 2016, 2017; Simkin et al. 2019). Estimates of this population growth vary widely. Furthermore, according to Burney et al. (2010), crop output and increasing land clearing for agriculture account for around one-third of global greenhouse gas emissions.

Improving photosynthetic efficiency is an essential first step toward accomplishing these goals. Furthermore, new methods of boosting crop productivity are required, such as modifying pigments related to photosynthesis to enhance light capture and/or photoprotection, since existing breeding strategies are no longer producing appreciable genetically modified gains (Ray et al. 2012, 2013). The maximum yield of plant development in ideal, healthful circumstances is known as the yield potential, commonly known as the biggest yield attainable. The total photosynthetic rate for the growth season is the source of this value. Photosynthetic rate and yield are influenced by increased rates of photosynthetic electron transport (Yadav et al. 2018; Lopez-Calcagno et al. 2020), regulation of Calvin Benson cycle enzymes (Lefebvre et al. 2005; Driever et al. 2017; Lopez-Calcagno et al. 2020; Raines et al. 2022), modification of photorespiration (Timm et al. 2012; Lopez-Calcagno et al. 2018), and increased the rate of non-photochemical extinction dynamics (Kromdijk et al. 2016; Glowacka et al. 2018). Moreover, new research indicates that boosting non-foliar tissue's photosynthetic ability such as that of fruits, stems, and wheat ears may be crucial to raising output (Simkin et al. 2020).

Because leaf and plant functional traits are essential for regulating ecosystem structure, functioning, and resistance to disturbances, it is important to characterize these properties across scales (Reich 2014; Funk et al. 2017; Kumar et al. 2021a, b). Plants are controlled by their structural, biochemical, physiological, and phenological characteristics, which also affect how well they grow, function, and adapt to different environments (Singh et al. 2010a, b, 2013a, b; Singh and Verma 2013a, b). Despite the significance of characterizing these qualities across global biomes, this simple aim is quite difficult due to the flexibility and significant functional variety of leaf and plant functional attributes (Osnas et al. 2018; Kumar et al. 2019; Serbin and Townsend 2020). As a result, only specific biomes have traditionally received worldwide attention (Schimel et al. 2015). The large diversity in leaf characteristics within and across plants is mediated by a wide variety of variables (Serbin et al. 2014; Kumar et al. 2017; Osnas et al. 2018; Neyret et al. 2016; Butler et al. 2017).

Functional traits, for instance, usually vary greatly with average light quality and condition within a canopy (Niinemets 2007; Neyret et al. 2016). Conversely, the lower canopy leaves are typically thinner, have reduced photosynthetic rates, and have modified pigment pools to compensate for the lesser quality of light. Along with local resource gradients, plant characteristics can also vary due to differences in water and fertilizer availability and legacy of disturbance (Singh et al. 2015; Enquist et al. 2017). Since plant traits are functional and influence a plant's performance in many contexts, research on plant functional traits has developed significantly over the past 20 years (Reich 2014). As a result, the link between photosynthesis and the functional features of plants has been summarized in this chapter.

7.2 Plant Functional Traits

The plant functional traits in community and ecosystem ecology are widely used to address how plants respond to environmental modification and contribute to ecosystem functions (Lavorel and Garnier 2002). Any morphological, physiological, or phenological properties of the plant that can be evaluated independently at any level of organization from the cell to the complete organism without considering the environment or other levels of organization are considering its functional qualities (Violle et al. 2007). Plant characteristics are linked to a plant's competitiveness, defense against viruses and aggressors, responses to environmental change and disturbance, and effects on biochemical cycles and disturbance regimes (Lavorel and Garnier 2002; Kumar et al. 2021c, d, 2022).

The primary categories of plant traits are life form, height, leaf, stem, and root characteristics (Cornellissen et al. 2003). Various plant organs have distinct purposes. Through photosynthesis, leaves fix carbon; stems and branches carry nutrients and water, as well as positioning the leaves; roots stabilize the plant and take up nutrients and water (Poorter et al. 2015; Singh et al. 2020). Environmental factors influence how biomass is distributed throughout plant organs; however, functional characteristics may also be useful variables to comprehend biomass distribution (Singh et al. 2014, 2021, 2024; Mensah et al. 2016).

The premise that natural variation represents organisms' adaptation to their surroundings has long guided the study of plant physiological ecology (Raunkiaer 1934). The comparative approach, which correlates changes in organismal function with environmental factors, is predicated on this premise (Caruso et al. 2019). Based on the idea of functional features, significant advancements in ecology have been accomplished during the last 10 years (Diaz et al. 2016). According to Violle et al. (2007), these traits are classified as "morpho-physio-phenological traits, which influence fitness indirectly through their effects on growth, reproduction, and survival." Plant functional traits are thought to significantly regulate crucial ecosystem functions and processes through their impact on soil conditions (Faucon et al. 2017).

Crop productivity depends on the traits of the plants. Bailey-Serres et al. (2019) expounded on a range of features, such as shoot, root, and plasticity traits (left column) and attribute possibilities (right column) for crop improvement (Table 7.1).

7.3 Photosynthesis and Green Revolution

The notable increases in plant yield were observed during the 1960s and 1970s (Furbank et al. 2020). In the decades that followed, crop yields, including those of rice and wheat, initially decreased (Kumar et al. 2016, 2021; Furbank et al. 2019). The breeding program utilized during the Green Revolution, which was mainly driven by the introduction of the main revolutionary dwarfing genes and concentrated on increasing harvest index and grain number, is now widely accepted to have reached a tableland (Furbank et al. 2020). Many elite genotypes of wheat and rice have achieved the theoretical limit of genetic potential for harvest index (Foulkes et al. 2011). The yield equation can take biomass and the harvest index into account, increasing biomass in wheat and rice has recently received a lot of attention without compromising the harvest index (Furbank et al. 2020). This is most easily accomplished by increasing photosynthetic capacity and efficiency through increases in radiation use efficiency.

One of the biological processes in plants that have been researched the most is photosynthesis. Gene suppression is the result of decades of biochemistry and

Table 7.1 Various plants' functional traits and attribute possibilities (Bailey-Serres et al. 2019)

Plant functional traits	Attribute possibilities
1. Shoot traits and plasticity	**1. Natural genetic variation**
(a) Inflorescence architecture and fertility	(a) Stress resilience and recovery mechanisms
(b) Shoot-to-root biomass	(b) Trait pyramiding
(c) Photosynthesis	**2. Gene engineering and editing**
(d) Stomatal movement and density regulation	(a) Spatial, temporal, and inducible control of genes and networks
(e) Assimilate loading and partitioning	(b) Improving protein function, targeting, and turnover
(f) Senescence timing.	(c) Enhancing metabolite pathway and flux
2. Root traits and plasticity	(d) Introducing synthetic traits
(a) Architecture and anatomy	**3. Beneficial soil and leaf microbiome**
(b) Growth dynamics	(a) Seeding and supplementation
(c) Nutrient acquisition and use efficiency	(b) Attraction of beneficial microorganisms
(d) Microbial interactions	**4. Small-molecule delivery**
3. Stress resilience	(a) Response activation
(a) Drought, salinity, flooding, and extreme temperatures (abiotic)	(b) Metabolic regulation
(b) Pests and pathogens (biotic)	**5. Sensor uses for crop management**
(c) Tempered response to minimize growth penalty	(a) Cellular, organ, canopy, and remote

biophysics studying ways and procedures to "titrate" out amounts of key enzymes and find the limits on photosynthetic flux provided by these steps in photosynthesis (Harrison et al. 2001; von Caemmerer et al. 2005). To clarify these restrictions on flow under various environmental circumstances, modeling has been widely employed (von Caemmerer 2000). The investigations have benefited greatly from the leaf-level models developed by Farquhar et al. (1980) as they offer a numerical, mathematical substructure for analyzing the effects of protein level, enzyme activity, and gene expression on the physiology of leaf photosynthetic processes (Furbank et al. 2020).

7.4 Photosynthesis Attributes in Non-foliar Parts

The process of photosynthesis in leaves is well-known and has been extensively studied. There is evidence that photosynthesis can also occur in other parts of the plant, including the petioles and stems (Hibberd and Quick 2002), cereal ears and husks (Maydup et al. 2010; Pengelly et al. 2011), fruits (Carrara et al. 2001; Hiratsuka et al. 2015; Sui et al. 2017), and seeds (Schwender et al. 2004) (Fig. 7.1). These results imply that plants are efficient of fabricating large quantities of photo-assimilates, which are essential for optimal yield. Little information is currently available on how potential modifications to photosynthetic mechanism may affect these tissues that contain chlorophyll, even though chlorophyll fluorescence imaging in non-foliar sections demonstrated the photosystem II operating efficiency and photochemistry, which is suggestive of functional electron transport (Simkin et al. 2020). In non-foliar tissues, photosynthesis has been hypothesized, investigated, and documented.

Stems not only carry out autonomous photosynthesis but also act as short-term repositories for photo-assimilates extracted from leaves (Aschan and Pfanz 2003). In cotton, photosynthesis in the stem has been shown to contribute to yield. It has been observed that seed weight was lowered by 16% when the main stem was maintained in the absence of light (Hu et al. 2012). Simbo et al. (2013) found a reduction in bud dry weight in African baobab and castor beans, providing data in favor of this contention. Certain plants, like *Justicia californica*, only have stems that are capable of photosynthesis; as a result, in the absence of leaves, the flowers and fruits grow (Avila et al. 2014), underscoring the need for stem photosynthesis for successful reproduction. Since deciduous plants do not have leaves in the summer, stem photosynthesis is particularly important to them. *Calicotome villosa* is one example of a plant whose yearly total branch photosynthesis exceeds that of its leaf photosynthesis (Yiotis et al. 2008).

Fruit photosynthesis is especially fascinating since, when ripening, tomatoes change from green photosynthetic metabolism to entirely heterotrophic metabolism (Lytovchenko et al. 2011). Comparing cucumbers to tomatoes and other colored fruits, they vary essentially in that they maintain their green color until they reach full maturity. Additionally, according to Sui et al. (2017), the guard cells on cucumbers have a coating of epicuticular waxes that may diminish their functionality.

Fig. 7.1 Photosynthetic non-foliar plant parts. Stem (**a**, **b**), petiole (**c**, **d**), flower (**e**), grasses inflorescence (**f**, **g**), fruits (**h**–**k**)

Nonetheless, the existence of these pores on the fruit's surface implies that, at least in the case of cucumbers, these fruits can directly absorb some CO_2 from the environment (Simkin et al. 2020). Other non-foliar green tissues besides the fruit pericarp can also do photosynthesis. Significant amounts of chlorophyll are present in the embryos of many species, and these proportions continue until maturity (Smolikova and Medvedev 2016). Although leaf photosynthesis is the primary mechanism for biomass creation, the photosynthetic activity in the ear of the cereals has been reported (Sanchez-Bragado et al. 2014; Simkin et al. 2020).

7.5 Photosynthesis and Plant Yield

The primary process that determines plant yield is photosynthesis; the final yield, which can be either biomass or grain, is mostly determined by how well a plant absorbs light and utilizes it to make biomass throughout the growing season (Long et al. 2006). The maximum yield that is expected from a plant can be produced under ideal growing conditions, free from biotic or abiotic stress, provided that the best-adapted plant variety is produced (Evans and Fischer 1999). Factors influencing yield perspective include light attainability, light collecting, energy conversion, and plant structure. Among these four parameters, energy conversion which depends on photosynthetic effectiveness is the only one that causes yields for major crops like maize, wheat, and rice to be below the maximum that may be achieved (Zhu et al. 2010; Simkin et al. 2019). Since photorespiration can lose up to 50% of fixed carbon under certain circumstances, research on the efficiency of the energy conversion to total biomass is still lacking. The hypothesis that larger yields might be reached by improving photosynthetic CO_2 assimilation is supported by the consistent strong evidence from CO_2 enrichment studies that increased CO_2 absorption may lead to higher yields (Weigel and Manderscheid 2012).

7.6 Transgenic Manipulation of Photosynthesis

Initially, the goal of transgenic modification to increase photosynthetic efficiency was to overexpress a single Calvin-Benson cycle enzyme. From 1992 to 2015, the direct maneuver of this cycle in a range of plants showed that even minute reductions in a few numbers of Calvin-Benson cycle enzymes might impair growth and carbon absorption (Simkin et al. 2019). In *Arabidopsis* (Simkin et al. 2017), tobacco (Lefebvre et al. 2005; Simkin et al. 2015), and tomato (Ding et al. 2016), overexpression of SBPase has demonstrated that elevating the activity of the enzyme enhances the absorption of photosynthetic carbon and biomass output. When tobacco plants were young and expanding, their rates of photosynthetic CO_2 assimilation increased. Additionally, as sucrose and starch accumulated, their biomass increased by 30%. However, when these same plants were fully expanded, there was no discernible increase in photosynthetic rates (Lefebvre et al. 2005).

Subsequent research revealed that tobacco grown in fields with increased CO_2 outside produced much more biomass when SBPase activity was raised (Rosenthal et al. 2011). Additional evidence supporting these findings comes from studies on *Arabidopsis*, where it was found that there was a 42% increase in biomass, increased assimilation of CO_2, and quickened development (Simkin et al. 2017) and in tomatoes, where biomass, starch, and sucrose all accumulated (Ding et al. 2016). According to these findings, SBPase is one of the enzymes that can regulate the carbon flow in a variety of plants' Calvin-Benson cycles. While the simplest explanation for the good impact is a direct result of better CO_2 absorption, a compulsive alternate possibility is that changes in metabolites caused by boosting SBPase activity may work as a signal that modulates plant growth and development. Although the metabolic alterations brought about by varying SBPase activity remain unclear, it is intriguing to observe the developmental shifts in SBPase antisense plants and their reactions to altering other metabolic pathways (Lawson et al. 2006).

Significant improvements in photosynthetic rates in wheat can be attained by increasing SBPase activity. Significantly, the production of grains and biomass was enhanced by these SBPase activity improvements. Plants were cultivated at a high density in one study to inhibit tillering and at a low density in another to promote it. Plants with the lower growth density produced more ears without exhibiting a discernible increase in the number of seeds per ear, in contrast to the higher density when plants had fewer tillers and more seeds per ear. Raising SBPase activity can be advantageous regardless of planting density, as evidenced by the lines with the maximum SBPase activity in both trials showing increases in total seed weight, quantity, and biomass of complete plants (Driever et al. 2017).

The results of overexpressing plant enzymes are corroborated by parallel research that expressed the cyanobacterial enzymes cyFBPase, cySBPase, or the bifunctional fructose-1,6-bisphosphatases/sedoheptulose-1,7-bisphosphatase (cyFBP/SBPase) in higher plants. cySBPase-expressing tobacco plants increased their growth rate and the rate at which photosynthetic CO_2 fixation occurs by more than 20%, resulting in a 1.5-fold increase in ultimate biomass (Tamoi et al. 2006; Giri et al. 2023). In tobacco, overexpression of the Calvin-Benson cycle enzyme FBPA led to increased growth and photosynthesis under conditions of high CO_2, which in turn increased aldolase activity several times over (Uematsu et al. 2012).

7.7 Functional Traits and Their Relationship with Photosynthesis

Photosynthesis impacts the economic characteristics of the leaves and roots. Under hot, high-light circumstances, the C_4 plants exhibit greater average relative growth rates than the C_3 plants because the C_4 plants' growth-trait connection has a larger intercept than the C_3 plants. Because the economic features and the photosynthetic route are unrelated, the structure of the leaf and root does not affect physiological function. Rather, C_3 and C_4 plants gain equally from the growth increase linked to resource-acquiring features (Simpson et al. 2020). Therefore, along the fast-slow trait continuum, C_4 plants have a considerable growth advantage over C_3 plants—that is, assuming the characteristics being studied accurately reflect the ability for resource acquisition under particular development conditions. Based on size-standardized and average plant comparisons, Simpson et al. (2020) argues that leaf functional features are a more reliable predictor of a plant's ecological strategy along the fast-slow continuum than root traits. Specifically, they are correlated with growth rate. Under high temperature and light circumstances, the C_4 plants consistently develop more quickly than the C_3 plants. Therefore, in hot, bright conditions that filter leaf characteristic values to a particular range, C_4 plants can develop faster and have an edge over their C_3 equivalents if all other factors remain similar.

The results indicate that C_4 photosynthesis has a more corresponding role in the development of plants at the "slow" end of the fast-slow continuum. It's interesting to note that there is no relationship between the effects of the photosynthetic pathway and leaf functional attributes on comparative growth rates. This indicates that, regardless of where one falls on the continuum, the overall relative growth rate rise is the same; yet, when the growth rate is already low, the proportionate increase is greater. After the leaf economy is size standardized, the growth rate is particularly dependent on these characteristics. According to Atkinson et al. (2016), C_4 plants have lower leaf dry matter contents and more resource-acquisitive leaves than C_3 plants. The specific leaf area values of C_4 plants are also much higher. Before the C_4 condition in grasses appeared, changes in foliar structure occurred (Christin et al. 2013). Lower densities of leaf tissue than in C_3 plants could have been the outcome of these modifications (Atkinson et al. 2016).

Due in part to their ability to produce more leaf area per unit mass due to higher specific leaf area, C_4 plants have grown more quickly in tropical climates (Simpson et al. 2020). The specific root length and area, as well as the average root diameter, are greater in plants with quicker growth rates (Fort et al. 2013). But there was also a reduced plant biomass in line with these characteristic values. The characteristics of higher plants' roots and leaves, as well as theirs, adjust to various dimensions under common environmental situations (Reich 2014; Simpson et al. 2020). Water availability might be one explanation for the disorganized strategies used here; traits that cause a slow root system can also be useful for extracting water from the deeper layers of the soil (Fort et al. 2013).

7.8 Conclusion

For higher plant biomass and yields, the practical traits of the plant, such as any morphological, physiological, or phenological aspects that are directly related to the plant's capacity for photosynthetic energy, must be altered or enhanced. Photosynthesis immediately affects every component of the plant, including the leaves that fix carbon, the stems and branches that deliver nutrients and water, and the roots that stabilize the plant and absorb nutrients. Different functional properties found in plants, such as the characteristics of their roots and shoots, as well as their flexibility, give rise to varying possibilities for plant yields. The Green Revolution and the advancement of photosynthesis are related; following the Green Revolution's inception, several tactics were created to enhance plants' capacity for photosynthetic processes.

References

Aschan G, Pfanz H (2003) Non-foliar photosynthesis—a strategy of additional carbon acquisition. Flora 198:81–97

Atkinson RRL, Mockford EJ, Bennett C et al (2016) C_4 photosynthesis boosts growth via altered physiology, allocation and size. Nat Plants 2:16038

Avila E, Herrera A, Tezara W (2014) Contribution of stem CO_2 fixation to whole-plant carbon balance in non-succulent species. Photosynthetica 52:3–15

Bailey-Serres J, Parker JE, Ainsworth EA et al (2019) Genetic strategies for improving crop yields. Nature 575(7781):109–118

Burney JA, Davis SJ, Lobell DB (2010) Greenhouse gas mitigation by agricultural intensification. Proc Natl Acad Sci USA 107:12052–12057

Butler EE, Datta A, Flores-Moreno H et al (2017) Mapping local and global variability in plant trait distributions. Proc Natl Acad Sci 114:10937–10946

Carrara S, Pardossi A, Soldatini GF, Tognoni F, Guidi L (2001) Photosynthetic activity of ripening tomato fruit. Photosynthetica 39:75–78

Caruso CM, Eisen KE, Martin RA, Sletvold N (2019) A meta-analysis of the agents of selection on floral traits. Evolution 73(1):4–14

Christin PA, Osborne CP, Chatelet DS et al (2013) Anatomical enablers and the evolution of C_4 photosynthesis in grasses. Proc Natl Acad Sci USA 110:1381–1386

Cornellissen JHC, Lavorel S, Garnier E, Diaz S et al (2003) A handbook of protocols for standardised and easy measurement of plant functional traits worldwide. Aust J Bot 51:334–377

Diaz S, Kattge J, Cornelissen J et al (2016) The global spectrum of plant form and function. Nature 529:167–171

Ding F, Wang M, Zhang S, Ai X (2016) Changes in SBPase activity influence photosynthetic capacity, growth, and tolerance to chilling stress in transgenic tomato plants. Sci Rep 6:32741

Driever SM, Simkin AJ, Alotaibi S et al (2017) Increased SBPase activity improves photosynthesis and grain yield in wheat grown in greenhouse conditions. Philos Trans R Soc Lond Ser B Biol Sci 372:1730

Enquist BJ, Bentley LP, Shenkin A et al (2017) Assessing trait-based scaling theory in tropical forests spanning a broad temperature gradient. Glob Ecol Biogeogr 26:1357–1373

Evans TL, Fischer RA (1999) Yield potential: its definition, measurement, and significance. Crop Sci 39:1544–1551

Farquhar GD, von Caemmerer S, Berry JA (1980) A biochemical model of photosynthetic CO_2 assimilation in leaves of C_3 species. Planta 149:78–90

Faucon MP, Houben D, Lambers H (2017) Plant functional traits: soil and ecosystem services. Trends Plant Sci 22(5):385–394

Fort F, Jouany C, Cruz P (2013) Root and leaf functional trait relations in Poaceae plants: implications of differing resource acquisition strategies. J Plant Ecol 6:211–219

Foulkes MJ, Slafer GA, Davies WJ, Berry PM et al (2011) Raising yield potential of wheat. III. Optimizing partitioning to grain while maintaining lodging resistance. J Exp Bot 62:469–486

Funk JL, Larson JE, Ames GM, Butterfield BJ et al (2017) Revisiting the Holy Grail: using plant functional traits to understand ecological processes. Biol Rev 92:1156–1173

Furbank RT, Jimenez-Berni JA, George-Jaeggli B, Potgieter AB, Deery DM (2019) Field crop phenomics: enabling breeding for radiation use efficiency and biomass in cereal crops. New Phytol 223:1714–1727

Furbank RT, Sharwood R, Estavillo GM, Silva-Perez V, Condon AG (2020) Photons to food: genetic improvement of cereal crop photosynthesis. J Exp Bot 71(7):2226–2238

Giri K, Mishra G, Suyal DC, Kumar N, Doley B, Das N, Baruah RC, Bhattacharyya R, Bora N (2023) Performance evaluation of native plant growth-promoting rhizobacteria for paddy yield enhancement in the jhum fields of Mokokchung, Nagaland, North East India. Heliyon 9:e14588

Glowacka K, Kromdijk J, Kucera K et al (2018) Photosystem II subunit S overexpression increases the efficiency of water use in a field-grown crop. Nat Commun 9(1):868

Harrison EP, Olcer H, Lloyd JC, Long SP, Raines CA (2001) Small decreases in SBPase cause a linear decline in RuBP regeneration rate but do not affect Rubisco carboxylation capacity. J Exp Bot 52:1779–1784

Hibberd JM, Quick WP (2002) Characteristics of C_4 photosynthesis in stems and petioles of C_3 flowering plants. Nature 415(6870):451–454

Hiratsuka S, Suzuki M, Nishimura H, Nada K (2015) Fruit photosynthesis in *Satsuma mandarin*. Plant Sci 241:65–69

Hu YY, Zhang YL, Luo HH, Li W, Oguchi R, Fan DY, Chow WS, Zhang WF (2012) Important photosynthetic contribution from the non-foliar green organs in cotton at the late growth stage. Planta 235:325–336

Kromdijk J, Glowacka K, Leonelli L, Gabilly ST, Iwai M, Niyogi KK, Long SP (2016) Improving photosynthesis and crop productivity by accelerating recovery from photoprotection. Science 354(6314):857–861

Kumar N, Shankhdhar SC, Shankhdhar D (2016) Impact of elevated temperature on antioxidant activity and membrane stability in different genotypes of rice (Oryza sativa L.). Indian J Plant Physiol (Now Plant Physiol Rep) 21(1):37–43

Kumar N, Suyal DC, Sharma IP, Verma A, Singh H (2017) Elucidating stress proteins in rice (*Oryza sativa* L.) genotype under elevated temperature: a proteomic approach to understand heat stress response. 3 Biotech 7:205

Kumar N, Jeena N, Singh H (2019) Elevated temperature modulates rice pollen structure: a study from foothill Himalayan agro-ecosystem in India. 3 Biotech 9:175

Kumar A, Dwivedi GK, Tewari S, Paul J, Anand R, Kumar N, Kumar P, Singh H, Kaushal R (2020a) Carbon mineralization and inorganic nitrogen pools under Terminalia chebula Retz.-based agroforestry system in Himalayan foothills, India. For Sci 66(5):634–643

Kumar A, Kumar P, Singh H, Kumar N (2020b) Adaptation and mitigation potential of roadside trees with bio-extraction of heavy metals under vehicular emissions and their impact on physiological traits during seasonal regimes. Urban For Urban Green 58:126900

Kumar A, Dwivedi GK, Tewari S, Paul J, Sah VK, Singh H, Kumar P, Kumar N, Kaushal R (2020c) Soil organic carbon pools under Terminalia chebula Retz. based agroforestry system in Himalayan foothills, India. Curr Sci 118(7):1098–1103

Kumar N, Jeena N, Kumar A, Khairakpam R, Singh H (2021) Comparative response of rice cultivars to elevated air temperature in Bhabar region of Indian Himalaya: status on yield attributes. Heliyon 7:e07474

Kumar A, Kumar P, Singh H, Bisht S, Kumar N (2021a) Relationship of physiological plant functional traits with soil carbon stock in temperate forest of Garhwal Himalaya. Curr Sci 120(8):1368–1373

Kumar A, Kumar P, Singh H, Kumar N (2021b) Modulation of plant functional traits under essential plant nutrients during seasonal regime in natural forests of Garhwal Himalayas. Plant Soil 465:197–212

Kumar A, Tewari S, Singh H, Kumar P, Kumar N, Bisht S, Kushwaha S, Tamta N, Kaushal R (2021c) Biomass accumulation and carbon stocks in different agroforestry system prevalent in Himalayan foothills, India. Curr Sci 120(6):1083–1088

Kumar A, Kumar P, Singh H, Kumar N (2021d) Impact of plant functional traits on near saturated hydraulic conductivity of soil under different forests of Kempty watershed in Garhwal Himalaya. Indian J Soil Conserv 49(1):38–44. NAAS rating: 5.28. ISSN: 2455-7145

Kumar A, Singh H, Kumari G, Bisht S, Malik A, Kumar N, Singh M, Raturi A, Barthwal S, Thakur A, Kaushal R (2022) Adaptive resilience of roadside trees to vehicular emissions via leaf enzymatic, physiological, and anatomical trait modulations. Environ Pollut 313:120191

Lavorel S, Garnier E (2002) Predicting changes in community composition and ecosystem functioning from traits: revisiting the Holy Grail. Funct Ecol 16:545–556

Lawson T, Bryant B, Lefebvre S, Lloyd JC, Raines CA (2006) Decreased SBPase activity alters growth and development in transgenic tobacco plants. Plant Cell Environ 29:48–58

Lefebvre S, Lawson T, Zakhleniuk OV, Lloyd JC, Raines CA, Fryer M (2005) Increased sedoheptulose-1,7-bisphosphatase activity in transgenic tobacco plants stimulates photosynthesis and growth from an early stage in development. Plant Physiol 138:451–460

Long SP, Zhu XG, Naidu SL, Ort DR (2006) Can improvement in photosynthesis increase crop yields? Plant Cell Environ 29:315–330

Long SP, Marshall-Colon A, Zhu XG (2015) Meeting the global food demand of the future by engineering crop photosynthesis and yield potential. Cell 161:56–66

Lopez-Calcagno PE, Fisk SJ, Brown K et al (2018) Overexpressing the H-protein of the glycine cleavage system increases biomass yield in glasshouse and field-grown transgenic tobacco plants. Plant Biotechnol J 17(1):141–151

Lopez-Calcagno PE, Brown KL, Simkin AJ, Fisk SJ, Lawson T, Raines CA (2020) Simultaneous stimulation of RuBP regeneration and electron transport increases productivity and water use efficiency under field conditions. bioRxiv. https://doi.org/10.1101/2020.01.27.920843

Lytovchenko A, Eickmeier I, Pons C, Osorio S et al (2011) Tomato fruit photosynthesis is seemingly unimportant in primary metabolism and ripening but plays a considerable role in seed development. Plant Physiol 157(4):1650–1663

Maydup ML, Antonietta M, Guiamet JJ, Graciano C, López JR, Tambussi EA (2010) The contribution of ear photosynthesis to grain filling in bread wheat (*Triticum aestivum* L.). Field Crop Res 119(1):48–58

Mensah S, Glèlè Kakaï R, Seifert T (2016) Patterns of biomass allocation between foliage and woody structure: the effects of tree size and specific functional traits. Ann For Res 59:49–60

Neyret M, Bentley LP, Oliveras I, Marimon BS et al (2016) Examining variation in the leaf mass per area of dominant plants across two contrasting tropical gradients in light of community assembly. Ecol Evol 6:5674–5689

Niinemets U (2007) Photosynthesis and resource distribution through plant canopies. Plant Cell Environ 30:1052–1071

Ort DR, Merchant SS, Alric J et al (2015) Redesigning photosynthesis to sustainably meet global food and bioenergy demand. Proc Natl Acad Sci USA 112:8529–9536

Osnas JLD, Katabuchi M, Kitajima K, Wright SJ, Reich PB et al (2018) Divergent drivers of leaf trait variation within plants, among plants, and among functional groups. Proc Natl Acad Sci 115:5480–5485

Pengelly JJ, Kwasny S, Bala S, Evans JR et al (2011) Functional analysis of corn husk photosynthesis. Plant Physiol 156(2):503–513

Poorter H, Jagodzinski AM, Ruiz-Peinado R, Kuyah S et al (2015) How does biomass distribution change with size and differ among plants? An analysis for 1200 plant plants from five continents. New Phytol 208:736–749

Raines CA, Cavanagh AP, Simkin AJ (2022) Improving carbon fixation. In: Ruban A, Murchie E, Foyer C (eds) Photosynthesis in action, 1st edn. Academic Press, New York

Raunkiaer C (1934) The life forms of plants and statistical plant geography. Clarendon, Oxford

Ray DK, Ramankutty N, Mueller ND, West PC, Foley JA (2012) Recent patterns of crop yield growth and stagnation. Nat Commun 3(1):1–7

Ray DK, Mueller ND, West PC, Foley JA (2013) Yield trends are insufficient to double global crop production by 2050. PLoS One 8(6):e66428

Reich PB (2014) The world-wide 'fast–slow' plant economics spectrum: a traits manifesto. J Ecol 102:275–301

Rosenthal DM, Locke AM, Khozaei M, Raines CA, Long SP, Ort DR (2011) Over-expressing the C_3 photosynthesis cycle enzyme sedoheptulose-1-7 bisphosphatase improves photosynthetic carbon gain and yield under fully open air CO_2 fumigation (FACE). BMC Plant Biol 11:123

Sanchez-Bragado R, Molero G, Reynolds MP, Araus JL (2014) Relative contribution of shoot and ear photosynthesis to grain filling in wheat under good agronomical conditions assessed by differential organ δ13C. J Exp Bot 65(18):5401–5413

Schimel D, Pavlick R, Fisher JB, Asner GP, Saatchi S, Townsend P, Miller C, Frankenberg C, Hibbard K, Cox P (2015) Observing terrestrial ecosystems and the carbon cycle from space. Glob Chang Biol 21:1762–1776

Schwender J, Goffman F, Ohlrogge JB, Shachar-Hill Y (2004) Rubisco without the Calvin cycle improves the carbon efficiency of developing green seeds. Nature 432(7018):779–782

Serbin SP, Townsend PA (2020) Scaling functional traits from leaves to canopies. In: Cavender-Bares J, Gamon JA, Townsend FA (eds) Remote sensing of plant biodiversity. Springer Nature, Cham, pp 43–82

Serbin SP, Singh A, McNeil BE, Kingdon CC, Townsend PA (2014) Spectroscopic determination of leaf morphological and biochemical traits for northern temperate and boreal tree plants. Ecol Appl 24(7):1651–1669

Simbo DJ, Van den Bilcke N, Samson R (2013) Contribution of corticular photosynthesis to bud development in African baobab (*Adansonia digitata* L.) and Castor bean (*Ricinus communis* L.) seedlings. Environ Exp Bot 95:1–5

Simkin AJ, McAusland L, Headland LR, Lawson T, Raines CA (2015) Multigene manipulation of photosynthetic carbon assimilation increases CO_2 fixation and biomass yield in tobacco. J Exp Bot 66:4075–4090

Simkin AJ, Lopez-Calcagno PE, Davey PA, Headland LR, Lawson T, Timm S, Bauwe H, Raines CA (2017) Simultaneous stimulation of sedoheptulose 1,7-bisphosphatase, fructose 1,6-bisphophate aldolase and the photorespiratory glycine decarboxylase-H protein increases CO_2 assimilation, vegetative biomass and seed yield in Arabidopsis. Plant Biotechnol J 15:805–816

Simkin AJ, López-Calcagno PE, Raines CA (2019) Feeding the world: improving photosynthetic efficiency for sustainable crop production. J Exp Bot 70(4):1119–1140

Simkin AJ, Faralli M, Ramamoorthy S, Lawson T (2020) Photosynthesis in non-foliar tissues: implications for yield. Plant J 101(4):1001–1015

Simpson KJ, Bennett C, Atkinson RR, Mockford EJ et al (2020) C_4 photosynthesis and the economic spectra of leaf and root traits independently influence growth rates in grasses. J Ecol 108(5):1899–1909

Singh H, Verma A (2013a) Physiological responses of rice cultivars to various nitrogen levels. Int J Agric Environ Biotechnol 6(3):383–388

Singh H, Verma A (2013b) Characterization and screening of high nitrogen efficient rice genotype to curtail environmental pollution. Int J Agric Environ Biotechnol 6(1):23–30

Singh H, Verma A, Shukla A (2010a) Comparative study of yield and yield components of hybrid and inbred genotypes of rice (*Oryza Sativa L.*). Int J Agric Environ Biotechnol 3:355–360

Singh H, Verma A, Krishnamoorthy M, Shukla A (2010b) Consequence of diverse nitrogen levels on leaf pigments in five rice genotypes under field emergent circumstance. Int J Bioresour Stress Manag 1:189–193

Singh H, Verma A, Rai SK (2013a) Biochemical evaluation of different rice genotypes grown at various nitrogen levels using SDS-PAGE. Curr Adv Agric Sci 5(1):144–146

Singh H, Verma A, Shukla A (2013b) Guttation fluid as a physiological marker for selection of nitrogen efficient rice (*Oryza sativa* L.) genotypes. Afr J Biotechnol 12(44):6276–6281

Singh H, Verma A, Ansari MW, Shukla A (2014) Physiological response of rice (*Oryza sativa* L.) genotypes to elevated nitrogen applied under field conditions. Plant Signal Behav 9:e29015. https://doi.org/10.4161/psb.29015

Singh A, Serbin SP, McNeil BE, Kingdon CC, Townsend PA (2015) Imaging spectroscopy algorithms for mapping canopy foliar chemical and morphological traits and their uncertainties. Ecol Appl 25:2180–2197

Singh H, Yadav M, Kumar N, Kumar A, Kumar M (2020) Assessing adaptation and mitigation potential of roadside trees under the influence of vehicular emissions: a case study of Grevillea robusta and Mangifera indica planted in an urban city of India. PLoS One 15(1):e0227380

Singh S, Singh H, Sharma SK, Nautiyal R (2021) Seasonal variation in biochemical responses of bamboo clones in the sub-tropical climate of Indian Himalayan foothills. Heliyon 7(4):e06859. https://doi.org/10.1016/j.heliyon.2021.e06859

Singh M, Singh H, Kumar A, Kumar M, Barthwal S, Thakur A (2024) Soil nitrogen availability determines the CO_2 fertilization effect on tree species (*Neolamarckia cadamba*): growth and physiological evidence. Environ Sustain. https://doi.org/10.1007/s42398-023-00300-w

Smolikova GN, Medvedev SS (2016) Photosynthesis in the seeds of chloroembryophytes. Russ J Plant Physiol 63:1–12

Sui X, Shan N, Hu L, Zhang C, Yu C et al (2017) The complex character of photosynthesis in cucumber fruit. J Exp Bot 68(7):1625–1637

Tamoi M, Nagaoka M, Miyagawa Y, Shigeoka S (2006) Contribution of fructose-1,6-bisphosphatase and sedoheptulose-1,7-bisphosphatase to the photosynthetic rate and carbon flow in the Calvin cycle in transgenic plants. Plant Cell Physiol 47:380–390

Tilman D, Clark M (2015) Food, agriculture and the environment: can we feed the world and save the earth? Daedalus 144:8–23

Timm S, Florian A, Arrivault S, Stitt M, Fernie AR, Bauwe H (2012) Glycine decarboxylase controls photosynthesis and plant growth. FEBS Lett 586(20):3692–3697

Uematsu K, Suzuki N, Iwamae T, Inui M, Yukawa H (2012) Increased fructose 1,6-bisphosphate aldolase in plastids enhances growth and photosynthesis of tobacco plants. J Exp Bot 63:3001–3009

Violle C, Navas ML, Vile D, Kazakou E, Fortunel C, Hummel I, Garnier E (2007) Let the concept of trait be functional. Oikos 116:882–892

von Caemmerer S (2000) Biochemical models of leaf photosynthesis. CSIRO Publishing, Collingwood

von Caemmerer S, Hendrickson L, Quinn V, Vella N, Millgate AG, Furbank RT (2005) Reductions of Rubisco activase by antisense RNA in the C_4 plant *Flaveria bidentis* reduces Rubisco carbamylation and leaf photosynthesis. Plant Physiol 137:747–755

Weigel H, Manderscheid R (2012) Crop growth responses to free air CO_2 enrichment and nitrogen fertilization: rotating barley, ryegrass, sugar beet and wheat. Eur J Agron 43:97–107

Yadav SK, Khatri K, Rathore MS, Jha B (2018) Introgression of UfCyt c6, a thylakoid lumen protein from a green seaweed *Ulva fasciata* Delile enhanced photosynthesis and growth in tobacco. Mol Biol Rep 45(6):1745–1758

Yiotis C, Psaras GK, Manetas Y (2008) Seasonal photosynthetic changes in the green-stemmed Mediterranean shrub *Calicotome villosa*: a comparison with leaves. Photosynthetica 46:262–267

Zhu XG, Long SP, Ort DR (2010) Improving photosynthetic efficiency for greater yield. Ann Rev Plant Biol 61:235–261

Enhancing Water Use Efficiency Through Trait-Based Approaches

8

Siddhant Gupta, Rajeev Ranjan, Amit Bijlwan, Ankita Jha, and Roomesh Kumar Jena

Abstract

Water is critical for environmental sustainability as 60% of agriculture is practiced in semiarid regions, where regular water applications are necessary to complete the crop-growing cycle. Water scarcity and the increasing demands of global agriculture necessitate innovative approaches to enhance water use efficiency (WUE) in crop production. Trait-based approaches offer promising strategies by modifying plant characteristics to optimize water utilization. This comprehensive exploration delves into various trait-based approaches, including deep-rooted systems, stomatal regulation, succulence, altered canopy architecture, improved photosynthetic machinery, and salinity/osmotic stress tolerance, elucidating their mechanisms and advantages. Moreover, it discusses the synergistic potential of crop rotation, intercropping, precision agriculture, and molecular/genetic techniques to pursue improved WUE. Traditional knowledge, sustainable soil management, and educational outreach programs are also integral components, as they bridge the gap between scientific innovation and practical implementation. However, challenges such as genetic complexities, trait stability, environmental concerns, and ethical considerations must be addressed. This chapter concludes with suggestions for advancing research, fostering adoption through policy and incentives, and empowering farmers through education and training. The integration of these strategies promises a more water-efficient, sustainable, and resilient future for global agriculture.

S. Gupta · R. Ranjan (✉) · A. Bijlwan · R. K. Jena
Department of Agrometeorology, G.B. Pant University of Agriculture and Technology, Pantnagar, Uttarakhand, India
e-mail: roomesh.jena@icar.gov.in

A. Jha
ICAR-Indian Institute of Water Management, Bhubaneswar, Odisha, India

N. Kumar, H. Singh (eds.), *Plant Functional Traits for Improving Productivity*,
https://doi.org/10.1007/978-981-97-1510-7_8

Keywords

Water use efficiency · Genetic modification · Stomatal regulation · Environmental impact · Precision agriculture · Traditional knowledge

8.1 Introduction

In semiarid regions, water use efficiency has become an important issue due to increased irrigated areas and increased water requirements for crops. Due to the limited availability of water resources, there is a growing debate about the use of these resources for purposes such as industry, agriculture, and direct human consumption (Bhattacharya 2019). Furthermore, predictions of climate change indicate that temperatures will rise (increasing potential evapotranspiration) and that extreme weather events like heat waves and droughts will occur more frequently. As 60% of agriculture is conducted in semiarid regions, where frequent irrigations are necessary to support the complete growing cycle of crops, the challenge of water availability becomes a critical factor for the sustainability of agricultural practices. In these semiarid areas, where crops predominantly develop during the driest months, precise scheduling of irrigation becomes imperative (Singh et al. 2010a, b, 2013a, b; Singh and Verma 2013a, b). Consequently, a significant portion of scientific research efforts in the realm of crop water use efficiency are directed toward the development and assessment of innovative irrigation techniques. Moreover, there is a focus on evaluating the genetic diversity associated with water use efficiency, particularly concerning rootstocks and cultivars. The emphasis on innovative irrigation approaches seeks to optimize water usage, ensuring that crops receive the necessary moisture without unnecessary waste. Simultaneously, the exploration of genetic diversity in rootstocks and cultivars aims to identify plant varieties that exhibit enhanced water use efficiency, thus contributing to sustainable agricultural practices in water-limited environments. This dual approach, encompassing both technological and biological aspects, underscores the multifaceted nature of addressing water-related challenges in semiarid agricultural regions (Singh et al. 2014, 2021, 2024). The overarching goal is to foster resource-efficient and resilient agricultural systems that can thrive in the face of water scarcity. This is following the social requirement to maximize water usage in viticulture (Bhattacharya 2019).

Water use efficiency (WUE) serves as a metric that captures the balance between two simultaneous physiological processes crucial to plant function: the release of water vapor and the absorption of carbon dioxide. This understanding of WUE is articulated by Fatichi et al. (2023). Essentially, WUE quantifies the efficiency with which plants utilize water to facilitate the essential process of carbon assimilation. This metric is pivotal in assessing the relationship between water consumption and the plant's ability to capture and utilize carbon, providing valuable insights into the plant's overall adaptive strategies and ecological performance. Different soybean cultivars exhibit genetic diversity in WUE (Mian et al. 1996). Because WUE and total biomass output have a positive association, improving WUE in a plant can lead to higher yields (Wright 1996). If genotypes for soybeans are solely selected based on mean yield, the WUE for soybeans may increase (Specht et al. 2001; Hufstetler et al. 2007). Liu et al. (2005) propose that manipulating stomatal partial closure at specific levels of soil water deficit presents an opportunity to enhance water use efficiency (WUE). In other words, by precisely regulating the degree of stomatal closure in response to soil water scarcity, it becomes feasible to optimize the balance between water loss through transpiration and carbon assimilation during photosynthesis. This approach allows for a more efficient utilization of water resources by plants, highlighting the significance of stomatal control as a strategic factor in the enhancement of overall WUE.

8.2 Plant Growth and Yield Concerning Water Use Efficiency

In a comprehensive set of field trials, Cossani et al. (2012) observed significant variations in both water and nitrogen availability, spanning a wide range of water use efficiency (6.3–23 kg grain/ha/mm) and nitrogen use efficiency (2.9–31.8 kg grain/kg available N^{-1} in soil). The study involved three distinct species, i.e., barley, bread wheat, and durum wheat. Nitrogen absorption demonstrated a positive correlation with grain yield and total biomass at the harvest stage for all three species. The field trials further revealed that modifications in water use efficiency resulted in changes in the yield of barley, bread wheat, and durum wheat. Interestingly, while there were some relative differences observed across the species, these differences did not consistently favor barley. Barley did not consistently exhibit more efficient or effective resource utilization compared to bread or durum wheat, especially in conditions with lower yields. The relative disparities in grain yield between wheat and barley, expressed as wheat yield/barley yield, were attributed to differences in their respective nitrogen and water use efficiencies, expressed as wheat resource use efficiency/barley resource use efficiency (Bhattacharya 2019). This underlines the importance of understanding and optimizing resource use efficiency, particularly nitrogen and water, in influencing the crop yields of different species under varying field conditions.

The Passioura equation, extensively discussed in various studies such as those by Blum (2000) and Araus et al. (2002), offers a framework for simulating grain yield in water-limited environments. The equation is expressed as follows:

$$\mathrm{GY} = W \times \mathrm{WUE}_{\mathrm{biomass}} \times \mathrm{HI}$$

where GY represents grain yield, W is the amount of water used by plants (i.e., evapotranspiration), and HI refers to the harvest index.

This model proposes three potential avenues to enhance grain yield:

- **Capturing more water**: Increasing the amount of water absorbed by plants
- **Producing more dry matter per unit of water utilized**: Enhancing water use efficiency in biomass production
- **Adding more assimilates to grains**: Improving the allocation of assimilates to the grain

While this approach seems promising, Passioura (2004) points out that the three elements of the equation are not entirely independent; they cannot interact with each other despite being worth considering separately. The relationships among these elements can be complex. For instance, higher water use efficiency may be linked to lower water consumption, potentially resulting in lower growth and grain yield, particularly under drought conditions. Additionally, crop indices may exhibit drought independence in some cases, but indices dependent on drought are often influenced by post-anthesis water use (Richards et al. 2002). In essence, while the Passioura equation provides a valuable conceptual framework, it is crucial to acknowledge the intricate interactions and dependencies among its components. Understanding these complexities is essential for accurately predicting and optimizing grain yield in water-limited environments.

8.3 Changing Climate

The uptake of water by plants will be governed by four factors of the changing climate. These factors include rising temperatures, rising CO_2 concentrations, erratic precipitation, and shifting humidity. Climate change projections are derived from a series of simulation models that consider various scenarios involving changes in carbon dioxide (CO_2) concentrations and associated forcing functions (Collins et al. 2013). As outlined in the projections by Collins et al. (2013), current CO_2 concentrations were estimated to be around 421 ppm in 2022. However, in the absence of effective mitigation plans, these concentrations are anticipated to rise significantly, reaching a range of 794 to 1142 ppm by the year 2100. These projections highlight the potential future trajectory of CO_2 concentrations and their implications for global climate patterns. Summarizing the findings from Trenberth (2011) and Collins et al. (2013), it is evident that current trends in CO_2 concentration are projected to undergo substantial increases if adequate measures are not implemented to mitigate greenhouse gas emissions. This emphasizes the urgency of adopting effective strategies to curb CO_2 emissions and address the challenges posed by climate change in the coming decades.

- If CO_2 concentrations continue to rise, global average temperatures will rise throughout the twenty-first century. In the scenario with maximum CO_2 emissions, the temperature increase should be between 2.6 and 4.8 °C.
- Although surface warming is greater than ocean warming, the increase in temperature is not uniform across all regions.
- As global temperatures rise, extreme heat will increase, and extreme cold will decrease on daily and seasonal scales.
- As the global mean surface temperature increases, precipitation is expected to rise, potentially at a rate of 1–3%$°C^{-1}$. It's crucial to note, however, that there will be substantial geographical variability in these changes. Different regions may experience different patterns and magnitudes of precipitation changes in response to the warming climate.

- Air's ability to contain water shall rise by 7% $°C^{-1}$. More water may be absorbed by the air, and this shall cause precipitation to be more intense, resulting in more rainfall with each rainfall event.
- Annual surface evaporation increases with increasing temperature, but evaporation over land is correlated with precipitation.

The evolving climate patterns are anticipated to heighten the atmospheric water requirements for crops, potentially leading to constraints in soil water availability. The increased variability in precipitation anticipated during the growing season is a particularly worrisome aspect, and this concern is amplified in soils with limited water-holding capacity. In regions where the ability of the soil to retain water is restricted, the amplified variability in precipitation poses a greater challenge for agricultural activities. Such conditions can lead to fluctuations in soil moisture levels, impacting crop growth and potentially exacerbating water stress during critical stages of the growing season. The increased demand for atmospheric water and the augmented unpredictability of precipitation pose significant challenges, particularly in regions where soils have limited capacity to retain water. This scenario underscores the pressing need for adaptive agricultural practices and innovative water management strategies to navigate the potential limitations imposed by these climate changes. In essence, the evolving precipitation patterns, coupled with soil characteristics, will play a pivotal role in shaping the water dynamics crucial for crop growth. For example, when assessing these fluxes across various latitudes through the application of eddy covariance flux systems, the spatial distribution of carbon and water fluxes is heavily dependent on annual temperatures, precipitation levels, and the duration of the growing season (Xiao et al. 2013). These comparisons identified the variables influencing WUE in various ecosystems, and it was inferred that WUE was influenced by yearly precipitation, gross primary production (GPP), and the length of the growing season (Xiao et al. 2013). When compared, the WUE of forests and coastal wetlands was greater than that of grasslands and croplands. There has been an increase in WUE of wheat, potato, and maize in China, in the previous 50 years which can be linked to a rise in temperature and a decline in precipitation (Guoju et al. 2013).

8.4 Factors Affecting Water Use Efficiency

The soil, climate, crop, and land management techniques all have a substantial impact on crop WUE. According to Ismail and Ozawa (2007), crops cultivated on clayey soils have a 45–64% higher water use efficiency than those grown on sandy soils. Rice cultivated on clayey soils had (25%) greater WUE than rice grown on sandy loam soils, according to Dou et al. (2016). In northern China, warm-dry climates were reported to produce greater WUE for grains such as wheat, oats, potatoes, and maize. According to Wang et al. (2018), applying 60, 120, or 180 kg N ha^{-1} increased winter wheat's water usage efficiency by 5.1%, 13.8%, and 29.3%, respectively. Sinclair (2018) explains how water deficit could enhance WUE. In the

Henan Province of China, adopting conservation tillage methods has demonstrated the potential to enhance WUE in comparison to traditional tillage practices. Research indicates that the application of mulch derived from agricultural residues plays a significant role in reducing evapotranspiration, as evidenced by studies conducted by Xie et al. (2005) and Zheng et al. (2018). Understanding the nuanced interactions between conservation tillage, mulching practices, and their influence on WUE remains a subject of ongoing investigation, with the need for additional research to elucidate the underlying mechanisms governing these observed effects under varied environmental conditions.

8.5 Trait-Based Approaches for Enhancing WUE

8.5.1 Drought-Resistant Traits

8.5.1.1 Deep Root Systems

Plants have many interrelated physiological and molecular mechanisms that are related to drought tolerance, WUE, and yield (Sharma et al. 2023). Although WUE frequently produces smaller plants, there are ways to lessen the growth penalty to crop yield in dry environments (Tardieu 2022). Drought tolerance requires highly conservative survival and recovery strategies. Deep root systems are a key trait that contributes significantly to enhancing WUE in plants. These traits enable plants to take water from deeper soil layers, which can be particularly beneficial in arid or drought-prone regions. Deep-rooted plants possess roots that extend deep into the soil profile, sometimes several meters below the surface. This allows them to tap into groundwater and moisture stored in deeper soil layers, even when surface soils are dry. Deep root systems reduce competition for water with shallow-rooted plants. This can be especially advantageous in intercropping or agroforestry systems, where different crops or trees with varying root depths are planted together. Plants with deep roots are better equipped to withstand periods of drought and water scarcity. They experience less severe water stress during dry spells, allowing them to maintain metabolic processes and photosynthesis. Deep root systems promote sustainable water use by maximizing the utilization of available water resources. This is crucial in regions facing water scarcity.

8.5.1.2 Stomatal Regulation

Stomatal regulation is another vital drought-resistant trait that significantly contributes to enhancing WUE in plants. Stomata are small openings on the leaf surface that control the exchange of gases (carbon dioxide and oxygen) and water vapor. Efficient stomatal regulation helps plants balance the need for CO_2 uptake for photosynthesis while minimizing water lost by transpiration. Under water stress conditions, plants with efficient stomatal regulation mechanisms can partially close their stomata to reduce water loss through transpiration. This conserves water without entirely compromising photosynthesis. Plants can sense environmental factors such as humidity, light intensity, and soil moisture levels. They adjust their stomatal conductance in response to these cues to optimize water use. Efficient stomatal regulation minimizes water loss, especially during periods of water scarcity. This allows the plant to maintain hydration and continue essential physiological processes.

8.5.2 Water Storage and Utilization

Succulence is a trait related to water storage and utilization in plants. Succulent plants have specialized tissues that can store water, allowing them to thrive in arid environments. While succulence is typically associated with certain plant families like Cactaceae and Crassulaceae, understanding and harnessing this trait can offer valuable insights into enhancing water use efficiency (WUE) in a broader range of plants. Succulent plants have specialized tissues, such as parenchyma cells and modified stems or leaves, which can store significant quantities of water. They have reduced surface area for transpiration, which helps minimize water loss. Their thick, fleshy leaves or stems may be covered with a waxy cuticle that further reduces water vapor loss. Succulent tissues serve as water reservoirs, allowing the plant to store surplus water during periods of abundance. This stored water can be utilized during dry spells or droughts, promoting plant survival and growth. Succulent traits have allowed plants to thrive in diverse and challenging environments (Twalla et al. 2022), contributing to the ecological diversity of arid and semiarid regions.

8.5.3 Heat and Drought Tolerance

Heat and drought tolerance are essential traits for enhancing WUE, particularly in regions with high temperatures and water scarcity. These traits involve various mechanisms that allow plants to maintain growth and productivity under stress. Plants with osmotic adjustment mechanisms can regulate the concentration of solutes in their cells, maintaining water uptake even when external water availability is reduced. Heat shock proteins are involved in protecting plant cells from damage caused by high temperatures. They help stabilize cellular structures and enzymes. Heat- and drought-tolerant plants are often more efficient in their water use (Kumar et al. 2015, 2016; Ilyas et al. 2021), allocating water to essential physiological processes even during stress. Some plants adapt to heat and drought stress by developing associations with beneficial soil microbes (Parvathi et al. 2020) that enhance nutrient uptake and water absorption. These traits ensure that limited water resources are allocated efficiently to support essential processes, such as photosynthesis and nutrient uptake. These traits can significantly reduce yield loss during heatwaves and droughts, contributing to food security and economic stability (Ilyas et al. 2021).

8.5.4 Canopy Size

Smaller leaves typically have fewer stomata and a smaller surface area, which leads to reduced transpiration rates. This means that the plant loses less water through evaporation, conserving valuable water resources. They allow the plant to allocate resources more efficiently. The plant can invest in other essential functions, such as root growth, without overcommitting resources to leaf development (Wang et al. 2019). They can improve light penetration into the lower canopy layers, ensuring that lower leaves receive adequate light for photosynthesis. This can lead to more uniform growth and higher overall productivity. Changes in canopy architecture can create a more favorable microclimate within the canopy. Reduced leaf density can improve air circulation and reduce humidity, which can mitigate disease pressure and minimize the risk of fungal infections.

8.5.5 Photosynthetic Efficiency

Enhancing photosynthetic efficiency is a crucial trait-based approach for improving WUE. Improvements in the photosynthetic machinery and carbon assimilation pathways can lead to increased biomass production per unit of water used. Plants with efficient photosystems can capture and utilize light energy more effectively, improving the overall rate of photosynthesis. Some plants, such as C_4 plants, have evolved specialized photosynthetic pathways that allow them to better concentrate CO_2 around the enzyme responsible for carbon fixation (Kumar et al. 2020a, b, c). This reduces water loss through stomata and enhances WUE. Crassulacean acid metabolism (CAM) represents an alternative photosynthetic pathway observed in succulent plants. CAM plants exhibit a distinctive strategy wherein they open their stomata during the night, thereby minimizing water loss throughout the day. This unique approach allows them to efficiently utilize stored carbon dioxide (CO_2), as highlighted in the study by Lim et al. (2019). Plants with enzymes that facilitate efficient carbon fixation and reduce the rate of photorespiration can improve overall WUE (Kumar et al. 2022).

8.5.6 Salinity and Osmotic Stress Tolerance

Some plants can exclude salt ions from their roots, preventing their accumulation in the plant tissues and maintaining normal cellular functions. Certain plants have specialized ion transport mechanisms that allow them to control the uptake and distribution of ions, particularly sodium and potassium. Osmotic adjustment involves the accumulation of organic solutes within the plant cells to balance osmotic pressure and prevent water loss to the surrounding saline or osmotically stressful environment. Halophytes are plants that have evolved specific adaptations to thrive in saline conditions. They possess the ability to secrete salt through specialized glands. Plants with salinity and osmotic stress tolerance can grow in regions where conventional crops struggle due to high soil salinity or water quality issues. Tolerant plants can thrive in conditions where other species cannot, reducing competition for water resources in those environments. In areas affected by salinity and osmotic stress, cultivating tolerant crops can contribute to sustainable agriculture and food security.

8.5.7 Crop Rotation and Intercropping

Crop rotation and intercropping are age-old agricultural practices that remain highly relevant today for enhancing water use efficiency (WUE). These strategies involve deliberate planting of different crops in sequence (crop rotation) or together (intercropping) to optimize water use and resource allocation. Crop rotation allows cultivation of plants with diverse root systems. Some crops may have deep roots, while others have shallow roots. This diversity enables better utilization of soil moisture at various depths. Rotating crops can disrupt the life cycles of pests and diseases that target specific plant species. This can reduce the need for chemical pesticides and minimize the stress that pests and diseases place on crops, indirectly improving WUE. Different crops contribute different nutrients to the soil, and crop rotation can help maintain soil health. Healthy soils retain water more effectively and support better plant growth (Singh et al. 2020; Kumar et al. 2017, 2021a, b).

Intercropping involves planting two or more crop species in close proximity. These species often have complementary water and nutrient requirements, reducing competition for resources and optimizing water use. The presence of multiple crop species can alter the microclimate within the canopy, reducing temperature and humidity extremes. This can mitigate water stress and improve overall crop performance. Intercropping can enhance soil structure, as different root systems improve soil aeration and reduce compaction. Improved soil structure aids in water infiltration and retention. Multiple crop types can help reduce soil erosion, which can be a significant concern in water-limited regions as conserved topsoil retains water more effectively.

8.5.8 Precision Agriculture

Precision agriculture involves the use of technology and data-driven approaches to optimize farming practices, including water management. It allows for precise control and allocation of water resources, reducing wastage and improving WUE. Satellite and drone-based remote sensing technologies play a pivotal role in modern agriculture by offering real-time data on crucial parameters like crop health, soil moisture, and weather conditions. This data serves as a valuable guide for irrigation scheduling, helping farmers avoid both overwatering and underwatering, there by maximizing water use efficiency. Moreover, the integration of automated irrigation systems, like drip irrigation and precision sprinklers, represents a technological leap in water management. The synergy of advanced sensors, satellite imagery, and automated irrigation not only conserves water resources but also contributes to increased crop productivity and sustainable agricultural practices. This technological convergence aligns with the broader goal of achieving precision agriculture, where resource inputs are optimized to enhance efficiency, reduce environmental impact, and ensure long-term agricultural sustainability. These systems can be controlled remotely and programmed to deliver precise amounts of water when needed. Advanced data analytics and modeling help farmers optimize water use by analyzing historical data, weather forecasts, and soil conditions to make irrigation decisions. By reducing water runoff and leaching of nutrients, precision agriculture practices also have environmental benefits, such as minimizing water pollution (Kumar et al. 2019, Kumar et al. 2021a, Kumar et al. 2021b, Kumar et al. 2021c, Kumar et al. 2021d, Kumar et al. 2021e).

8.5.9 Molecular and Genetic Approaches

Molecular and genetic approaches represent cutting-edge methods for enhancing WUE in crops. These approaches leverage biotechnology and genetic engineering to develop plants with specific traits that improve their water use efficiency, using genetic engineering techniques to introduce or modify specific genes into crops. For example, the introduction of genes for drought-responsive proteins can enhance a plant's ability to withstand water stress. Multiple genes can be introduced to create crops with a combination of traits, such as drought resistance, pest resistance, and improved WUE. Techniques like CRISPR-Cas9 enable precise editing of a plant's genome to enhance WUE without introducing foreign genes. Some transgenic crops are designed to use water more efficiently by altering processes like transpiration and photosynthesis. Molecular approaches allow for precise modification of genes related to WUE, resulting in crops with enhanced water use efficiency. Genetic engineering can expedite the breeding process by introducing desired traits directly into crop varieties. Crops with improved WUE require less irrigation, reducing the environmental footprint of agriculture (Kumar et al. 2021c, d).

8.5.10 Traditional Knowledge

Incorporating indigenous practices and traditional knowledge into modern agriculture can be a valuable approach for enhancing water use efficiency (WUE) and promoting sustainable farming systems. Indigenous communities have often developed agricultural practices over generations that are well-suited to local ecosystems and water availability. They often emphasize a diverse range of crops that are adapted to local conditions, including those with natural drought resistance or efficient water use. In regions characterized by arid and semiarid climates, where dryland agriculture is practiced, the cultivation of leguminous fallow crops holds significant importance, as highlighted by Abhiram and Eeswaran (2022). By incorporating leguminous fallow crops into the cropping system, farmers can improve soil health, increase nutrient availability, and contribute to the overall sustainability of agricultural practices in water-limited regions. Additionally, legumes' capacity to reduce soil erosion and their relatively low water requirements make them well-suited for cultivation in areas where water scarcity is a significant concern. This approach aligns with the broader goal of promoting resilient and resource-efficient agricultural systems in arid and semiarid zones. Indigenous communities have developed techniques for harvesting and storing rainwater, such as building cisterns or terraces that slow down water runoff and promote infiltration. Many indigenous practices are suited to arid and semiarid regions, where water conservation is critical. These practices include intercropping, crop rotation, and dry farming techniques. They cultivate and preserve local crop varieties that have evolved to thrive in specific water-limited environments. Some traditional irrigation systems, such as *qanats* or aqueducts, efficiently deliver water to crops without excessive wastage. Incorporating traditional knowledge can help preserve cultural heritage and strengthen community bonds. Many indigenous practices are cost-effective and use locally available materials, making them accessible to resource-constrained farmers (Kumar et al. 2020c).

8.5.11 Soil Management

Soil management is a critical component of enhancing WUE, as soil quality directly impacts water retention, infiltration, and nutrient availability for plants. Adding organic materials such as compost and cover crops improves soil structure, increases water-holding capacity (Ullah et al. 2019), and increases microbial activity. Reduced or no-till practices minimize soil disturbance, reduce evaporation, and help maintain soil structure, which aids in water infiltration. Mulching with organic materials or plastic films conserves soil moisture, reduces evaporation, and moderates soil temperature (Kader et al. 2017). Through regular soil testing, farmers can assess nutrient levels and adjust fertilizer applications to optimize nutrient and water use efficiency. Implementing erosion control measures (Adimassu et al. 2017) such as contour farming and vegetative buffer strips prevents soil erosion and loss of fertile topsoil. Improved soil structure and organic matter content enhance the soil's ability to hold water, reducing the frequency and volume of irrigation needed. Proper soil management guarantees accessibility of vital nutrients to plants, thus promoting to their growth and development and water use efficiency. Sustainable soil management practices promote long-term soil health and fertility, contributing to WUE. Well-managed soils are more resilient to water stress and extreme weather events.

8.6 Challenges and Limitations

Enhancing water use efficiency (WUE) through trait-based approaches faces several technical challenges, one of which is the genetic complexities associated with modifying plant traits. Many traits related to WUE are controlled by multiple genes, making it challenging to pinpoint and modify all the genes responsible for a particular trait. Genes involved in WUE traits often interact with each other and with environmental factors in complex ways, making it difficult to predict the outcomes of genetic modifications. Genetic diversity within plant populations can affect the success of trait modification efforts. Adapting WUE traits across diverse genetic backgrounds can be challenging. Ensuring the stability of modified traits over generations and under changing environmental conditions is another technical challenge. Modified traits may undergo genetic drift over time, leading to loss or reduced expression of the desired trait. Genetic modifications may have unintended consequences, such as impacting other plant functions or interactions with pests and pathogens. The use of genetic engineering and molecular approaches in enhancing WUE raises environmental and ethical concerns. Balancing the enhancement of WUE with sustainable agricultural practices is a crucial challenge. There is a risk of overreliance on technological solutions, such as genetic engineering or precision agriculture, which may not always be accessible to all farmers or may have negative environmental consequences.

8.7 Future Prospects and Recommendations

The future of enhancing water use efficiency through trait-based approaches holds great promise, with ongoing advancements in technology and research. Continued development of genomic tools, such as CRISPR-Cas9 and advanced sequencing technologies, will enable more precise and efficient trait modification. Improved phenotyping technologies will allow for rapid and accurate assessment of plant traits related to WUE, facilitating trait selection in breeding programs. Advances in bioinformatics will aid in identifying candidate genes and their interactions, enhancing our understanding of complex WUE traits. Exploring new research directions is essential for broadening our understanding of WUE and developing effective trait-based approaches. Research efforts should focus on identifying and characterizing novel traits related to WUE, including those influenced by epigenetic factors and microbial interactions (Giri et al. 2023). It should investigate strategies for optimizing multiple traits simultaneously, to create crop varieties that outshine in WUE under diverse environmental conditions. Empowering farmers with such knowledge and skills to implement WUE-enhancing practices is also crucial.

8.8 Conclusion

Water use efficiency is a critical issue in semiarid environments due to increased irrigated areas and crop water requirements. The emphasis is on innovative irrigation approaches and evaluating genetic diversity in relation to water use efficiency. Water use efficiency (WUE) quantifies the efficiency with which plants utilize water to facilitate the essential process of carbon assimilation. Improving WUE in a plant can result in increased yields. The strategy of regulating stomatal partial closure at various levels of soil water deficit offers a chance to improve WUE.

References

Abhiram G, Eeswaran R (2022) Legumes for efficient utilization of summer fallow. In: Advances in legumes for sustainable intensification. Academic Press, London, pp 51–70

Adimassu Z, Langan S, Johnston R, Mekuria W, Amede T (2017) Impacts of soil and water conservation practices on crop yield, run-off, soil loss and nutrient loss in Ethiopia: review and synthesis. Environ Manag 59(1):87–101. https://doi.org/10.1007/s00267-016-0776-1

Araus JL, Slafer GA, Reynolds MP, Royo C (2002) Plant breeding and water stress in C3 cereals: what to breed for? Ann Bot 89:925–940

Bhattacharya A (2019) Water-use efficiency under changing climatic conditions. In: Changing climate and resource use efficiency in plants. Academic Press, London, pp 111–180. https://doi.org/10.1016/b978-0-12-816209-5.00003-9

Blum A (2000) A molecular perspective of crop adaptation and production under drought stress. V. Simposium Hispano-Portugue's de relacioneshı'dricas-Sevilla

Collins M, Knutti R, Arblaster J, Dufresne JL, Fichefet T, Friedlingstein P (2013) Long-term climate change: projections, commitments and irreversibility. In: Stocker TF, Qin D, Plattner G-K, Tignor M, Allen SK (eds) Climate change 2013: the physical science basis. Contribution of Working Group I to the Fifth Assessment Report of the Intergovernmental Panel on Climate Change. Cambridge University Press, Cambridge

Cossani CM, Slafer GA, Savin R (2012) Nitrogen and water use efficiencies of wheat and barley under a Mediterranean environment in Catalonia. Field Crop Res 128:109–118

Dou F, Soriano J, Tabien RE, Chen K (2016) Soil texture and cultivar effects on rice (*Oryza sativa*, L.) grain yield, yield components and water productivity in three water regimes. PLoS One 11(3):e0150549. https://doi.org/10.1371/journal.pone.0150549

Fatichi S, Paschalis A, Bonetti S, Manoli G, Pappas C (2023) Water use efficiency: a review of spatial and temporal variability. In: Encyclopedia of soils in the environment, 2nd edn. Academic Press, London, pp 527–542. https://doi.org/10.1016/B978-0-12-822974-3.00166-X

Giri K, Mishra G, Suyal DC, Kumar N, Doley B, Das N, Baruah RC, Bhattacharyya R, Bora N (2023) Performance evaluation of native plant growth-promoting rhizobacteria for paddy yield enhancement in the jhum fields of Mokokchung, Nagaland, North East India. Heliyon 9:e14588

Guoju X, Fengju Z, Zhengji Q, Yubi Y (2013) Impact of climate change on water use efficiency by wheat, potato and corn in semiarid areas of China. Agric Ecosyst Environ 181:108–114. https://doi.org/10.1016/j.agee.2013.09.019

Hufstetler EV, Boerma HR, Carter TE, Earl HJ (2007) Genotypic variation for three physiological traits affecting drought tolerance in soybean. Crop Sci 47:25–35

Ilyas M, Nisar M, Khan N (2021) Drought tolerance strategies in plants: a mechanistic approach. J Plant Growth Regul 40:926–944. https://doi.org/10.1007/s00344-020-10174-5

Ismail S, Ozawa K (2007) Improvement of crop yield, soil moisture distribution and water use efficiency in sandy soils by clay application. Appl Clay Sci 37:81–89

Kader MA, Senge M, Mojid MA, Nakamura K (2017) Mulching type-induced soil moisture and temperature regimes and water use efficiency of soybean under rain-fed condition in Central Japan. Int Soil Water Conserv Res 5:302–308

Kumar N, Kumar N, Shukla A, Shankhdhar SC, Shankhdhar D (2015) Impact of terminal heat stress on pollen viability and yield attributes of rice (Oryza sativa L.). Cereal Res Commun 43(4):616–626

Kumar N, Shankhdhar SC, Shankhdhar D (2016) Impact of elevated temperature on antioxidant activity and membrane stability in different genotypes of rice (Oryza sativa L.). Indian J Plant Physiol 21(1):37–43

Kumar N, Suyal DC, Sharma IP, Verma A, Singh H (2017) Elucidating stress proteins in rice (Oryza sativa L.) genotype under elevated temperature: a proteomic approach to understand heat stress response. 3 Biotech 7:205

Kumar N, Jeena N, Singh H (2019) Elevated temperature modulates rice pollen structure: a study from foothill Himalayan agro-ecosystem in India. 3 Biotech 9:175

Kumar A, Dwivedi GK, Tewari S, Paul J, Anand R, Kumar N, Kumar P, Singh H, Kaushal R (2020a) Carbon mineralization and inorganic nitrogen pools under Terminalia chebula Retz.-based agroforestry system in Himalayan foothills, India. For Sci 66(5):634–643

Kumar A, Dwivedi GK, Tewari S, Paul J, Sah VK, Singh H, Kumar P, Kumar N, Kaushal R (2020b) Soil organic carbon pools under Terminalia chebula Retz. based agroforestry system in Himalayan foothills, India. Curr Sci 118(7):1098–1103

Kumar A, Kumar P, Singh H, Kumar N (2020c) Adaptation and mitigation potential of roadside trees with bio-extraction of heavy metals under vehicular emissions and their impact on physiological traits during seasonal regimes. Urban For Urban Green 58:126900

Kumar A, Kumar P, Singh H, Bisht S, Kumar N (2021a) Relationship of physiological plant functional traits with soil carbon stock in temperate forest of Garhwal Himalaya. Curr Sci 120(8):1368–1373

Kumar A, Kumar P, Singh H, Kumar N (2021b) Modulation of plant functional traits under essential plant nutrients during seasonal regime in natural forests of Garhwal Himalayas. Plant Soil 465:197–212

Kumar A, Kumar P, Singh H, Kumar N (2021c) Impact of plant functional traits on near saturated hydraulic conductivity of soil under different forests of Kempty watershed in Garhwal Himalaya. Indian J Soil Conserv 49(1):38–44

Kumar A, Tewari S, Singh H, Kumar P, Kumar N, Bisht S, Kushwaha S, Tamta N, Kaushal R (2021d) Biomass accumulation and carbon stocks in different agroforestry system prevalent in Himalayan foothills, India. Curr Sci 120(6):1083–1088

Kumar N, Jeena N, Kumar A, Khairakpam R, Singh H (2021e) Comparative response of rice cultivars to elevated air temperature in Bhabar region of Indian Himalaya: status on yield attributes. Heliyon 7:e07474

Kumar A, Singh H, Kumari G, Bisht S, Malik A, Kumar N, Singh M, Raturi A, Barthwal S, Thakur A, Kaushal R (2022) Adaptive resilience of roadside trees to vehicular emissions via leaf enzymatic, physiological, and anatomical trait modulations. Environ Pollut 313:120191

Lim SD, Lee S, Choi WG, Yim WC, Cushman JC (2019) Laying the foundation for Crassulacean Acid Metabolism (CAM) biodesign: expression of the C4 metabolism cycle genes of CAM in arabidopsis. Front Plant Sci 10:101

Liu Y, Gai JY, Lu H, Wang YJ, Chen SY (2005) Identification of drought tolerant germplasm and inheritance and QTL mapping of related root traits in soybean [Glycine max (L.) Merr.]. Yi Chuan Xue Bao 32:855–863

Mian MAR, Mailey MA, Ashley DA, Wells R, Carter TE, Parrot WA (1996) Molecular markers associated with water use efficiency and leaf ash in soybean. Crop Sci 36:1252–1257

Parvathi MS, Dhanyalakshmi KH, Nataraja KN (2020) Molecular mechanisms associated with drought and heat tolerance in plants and options for crop improvement for combined stress tolerance. In: Hasanuzzaman M (ed) Agronomic crops. Springer, Singapore. https://doi.org/10.1007/978-981-15-0025-1_23

Passioura J (2004) Increasing crop productivity when water is scarce—from breeding to field management. In: New directions for a diverse planet' Proceedings of 4th International Crop Sciences Congress, Brisbane, Australia

Richards RA, Rebetzke GJ, Condon AG, van Herwaarden AF (2002) Breeding opportunities for increasing the efficiency of water use and crop yield in temperature cereals. Crop Sci 42:111–121

Sharma N, Raman H, Wheeler D, Kalenahalli Y, Sharma R (2023) Data-driven approaches to improve water-use efficiency and drought resistance in crop plants. Plant Sci 336:111852
Sinclair TR (2018) Effective water use required for improving crop growth rather than transpiration efficiency. Front Plant Sci 9:14–42
Singh H, Verma A (2013a) Physiological responses of rice cultivars to various nitrogen levels. Int J Agric Environ Biotechnol 6(3):383–388
Singh H, Verma A (2013b) Characterization and screening of high nitrogen efficient rice genotype to curtail environmental pollution. Int J Agric Environ Biotechnol 6(1):23–30
Singh H, Verma A, Shukla A (2010a) Comparative study of yield and yield components of hybrid and inbred genotypes of rice (Oryza Sativa L.). Int J Agric Environ Biotechnol 3:355–360
Singh H, Verma A, Krishnamoorthy Mand Shukla A (2010b) Consequence of diverse nitrogen levels on leaf pigments in five rice genotypes under field emergent circumstance. Int J Bioresour Stress Manage 1:189–193
Singh H, Verma A, Rai SK (2013a) Biochemical evaluation of different rice genotypes grown at various nitrogen levels using SDS-PAGE. Curr Adv Agric Sci 5(1):144–146
Singh H, Verma A, Shukla A (2013b) Guttation fluid as a physiological marker for selection of nitrogen efficient rice (Oryza sativa L.) genotypes. Afr J Biotechnol 12(44):6276–6281
Singh H, Verma A, Ansari MW, Shukla A (2014) Physiological response of rice (Oryza sativa L.) genotypes to elevated nitrogen applied under field conditions. Plant Signal Behav 9:e29015. https://doi.org/10.4161/psb.29015
Singh H, Yadav M, Kumar N, Kumar A, Kumar M (2020) Assessing adaptation and mitigation potential of roadside trees under the influence of vehicular emissions: a case study of Grevillea robusta and Mangifera indica planted in an urban city of India. PLoS One 15(1):e0227380
Singh S, Singh H, Sharma SK, Nautiyal R (2021) Seasonal variation in biochemical responses of bamboo clones in the sub-tropical climate of Indian Himalayan foothills. Heliyon 7(4):e06859. https://doi.org/10.1016/j.heliyon.2021.e06859
Singh M, Singh H, Kumar A, Kumar M, Barthwal S, Thakur A (2024) Soil nitrogen availability determines the CO_2 fertilization effect on tree species (Neolamarckia cadamba): growth and physiological evidence. Environ Sustain. https://doi.org/10.1007/s42398-023-00300-w
Specht J, Chase K, Macrander M, Graef G, Chung J, Markwell J, Germann M, Orf J, Lark K (2001) Soybean response to water. Crop Sci 41:493–509
Tardieu F (2022) Different avenues for progress apply to drought tolerance, water use efficiency and yield in dry areas. Curr Opin Biotech 73:128–134
Trenberth KE (2011) Changes in precipitation with climate change. Clim Res 47:123–138
Twalla JT, Ding B, Cao G (2022) Roles of stomata in gramineous crops growth and biomass production. Cereal Res Commun 50:603–616. https://doi.org/10.1007/s42976-021-00216-3
Ullah H, Arenas RS, Ferdous Z, Attia A, Dutta A (2019) Improving water use efficiency, nitrogen use efficiency, and radiation use efficiency in field crops under drought stress: a review. Adv Agron 156:109–157
Wang L, Palta JA, Chen W, Chen Y, Deng X (2018) Nitrogen fertilization improved water-use efficiency of winter wheat through increasing water use during vegetative rather than grain filling. Agric Water Manag 197:41–53
Wang C, He J, Zhao T-H, Cao Y, Wang G, Sun B, Yan X, Guo W, Li M-H (2019) The smaller the leaf is, the faster the leaf water loses in a temperate forest. Front Plant Sci 10:58
Wright G (1996) Review of ACIAR selection for water use efficiency in legumes project recommends further research. ACIAR Food Legume Newslett:2–3
Xiao J, Sun G, Chen J, Chen H, Chen S, Dong G et al (2013) Carbon fluxes, evapotranspiration, and water use efficiency of terrestrial ecosystems in China. Agric For Meteorol 18:76–90. https://doi.org/10.1016/j.agrformet.2013.08.007
Xie ZK, Wang YJ, Li FM (2005) Effect of plastic mulching on soil water use and spring wheat yield in arid region of Northwest China. Agric Water Manag 75:71–83
Zheng J, Fan J, Zhang F, Yan S, Guo J, Chen D, Li Z (2018) Mulching mode and planting density affect canopy interception loss of rainfall and water use efficiency of dryland maize on the Loess Plateau of China. J Arid Land 10:794–808

Trait-Based Approaches to Improve Nutrient Uptake Efficiency in Crops

9

Amit Joshi and Jharna Maiti

Abstract

In the face of global population growth and changing climatic conditions, ensuring food security remains a paramount challenge. Meeting the rising demand for food while minimizing the environmental footprint of agriculture calls for innovative strategies to enhance crop productivity. One promising approach is to improve nutrient uptake efficiency in crops through trait-based interventions. Trait-based approaches leverage the genetic diversity within crop species to identify and harness specific traits that enhance nutrient uptake efficiency. These traits may include root architecture, nutrient transporters, and symbiotic associations with beneficial microbes. By understanding the genetic basis of these traits, scientists and breeders can develop crop varieties with improved nutrient-acquisition capabilities. One key trait that has garnered significant attention is root architecture. Root traits such as root length, density, and branching patterns play a critical role in nutrient acquisition from the soil. Genetic modification or breeding for desirable root traits can lead to crops that are better equipped to explore larger soil volumes, access nutrients efficiently, and tolerate nutrient-stressed conditions. Additionally, nutrient transporters are vital components of nutrient uptake efficiency. Research has identified specific genes responsible for the uptake of essential nutrients like nitrogen, phosphorus, and micronutrients. Modifying or selecting for crop varieties with enhanced transporter activity can improve nutrient uptake and utilization, reducing the need for excessive fertilization. Moreover, harnessing the power of beneficial microorganisms, such as mycorrhizal fungi and nitrogen-fixing bacteria, presents another avenue for trait-based enhancement of nutrient uptake. These symbiotic associations can enhance nutrient acquisition, particularly in nutrient-poor soils, and reduce the environ-

A. Joshi (✉) · J. Maiti
Department of Biochemistry, Kalinga University, Naya-Raipur, Chhattisgarh, India
e-mail: amit.joshi@kalingauniversity.ac.in; jharna.maiti@kalingauniversity.ac.in

N. Kumar, H. Singh (eds.), *Plant Functional Traits for Improving Productivity*,
https://doi.org/10.1007/978-981-97-1510-7_9

mental impact of chemical fertilizers. While trait-based approaches hold immense promise, they are not without challenges. By harnessing the genetic diversity of crops and targeting specific traits related to nutrient acquisition, researchers and breeders can develop crop varieties that thrive in nutrient-limited environments. However, successful implementation of these approaches requires collaboration among scientists, breeders, policymakers, and farmers to address the challenges and ensure equitable access to the benefits of trait-based crop improvement.

Keywords

Trait-based approaches · Nutrient uptake efficiency · Crop productivity · Root architecture · Nutrient transporters · Beneficial microorganisms · Sustainable agriculture · Food security

9.1 Introduction

A key factor in determining crop output is nutrient uptake efficiency, and agricultural productivity is necessary to meet the world's food demand. In order to maintain sustainable food supply, there is a growing need to optimize agricultural methods as the world's population grows. Utilizing trait-based methods to improve crop nitrogen uptake efficiency is one viable way to accomplish this goal (Kumar et al. 2021; Cheng et al. 2023). Traditional agricultural operations often require the use of fertilizers to deliver necessary nutrients to crops. But a large amount of these nutrients are lost by volatilization, runoff, or leaching, which causes problems for the environment and inefficiencies in the economy (Fu and Niu 2023). In order to overcome these obstacles, scientists are investigating novel approaches that concentrate on the genetic characteristics of crops to enhance their capacity to absorb, apply, and preserve nutrients. In trait-based techniques, particular genetic features that affect a crop's ability to absorb and use nutrients are identified and modified. These characteristics could include nutrient transporters, root architecture, and physiological mechanisms involved in nutrient absorption. Scientists want to create crops with better nutrient usage efficiency by comprehending the genetic foundation of these features, which would ultimately support resource-efficient and sustainable agriculture. Root architecture is a crucial characteristic that scholars are focusing on (Kumar et al. 2019; En et al. 2023). By searching the soil for minerals and water, roots are essential to the uptake of nutrients. A plant's capacity to efficiently absorb nutrients can be greatly impacted by characteristics including branching patterns, root length, and density. Optimizing these features through genetic manipulation has the potential to improve nitrogen uptake and decrease the amount of fertilizer required (Singh et al. 2013a, b). Furthermore, the identification and manipulation of genes linked to nutrient transporters is a focus of research (Yang et al. 2023a). The plant's intake and distribution of vital nutrients are carried out by these transporters. Scientists hope to decrease nutrient losses and environmental effect by increasing crop absorption and utilization efficiency through modifying the expression of these genes. Furthermore, precise and targeted changes of particular genes linked to nutrition intake are now possible, thanks to developments in molecular breeding techniques like CRISPR-Cas9. With the use of this technology, crops with improved nutrient uptake efficiency can be developed without causing unfavorable side effects. Sustainable agriculture has a bright future if trait-based methods are developed to increase crop nutrient uptake efficiency. Through the use of contemporary molecular techniques and genetic insights, scientists want to create crops that are better able to absorb, use, and hold onto nutrients (Hull et al. 2023). This will ultimately lead to higher agricultural yields, less negative environmental effects, and increased global food security. In order to guarantee the appropriate and secure application of these cutting-edge technologies, it is crucial that we traverse ethical issues and regulatory frameworks as we go deeper into the field of trait-based crop development.

9.2 Nutrient Uptake in Crop Plants

The vital component of agricultural sustainability and production is nutrient absorption efficiency. The increasing demands of a growing global population on crop output make it imperative to improve the way in which crops utilize nutrients. The several trait-based methods used to increase crop nutrient uptake efficiency are examined in this chapter, with an emphasis on agronomic, physiological, and genetic techniques (Kumar et al. 2015; Riseh et al. 2023). The utilization of trait-based methods to increase crop nutrient uptake efficiency is essential for raising agricultural output while reducing environmental effects. These tactics center on locating and utilizing particular genetic features in crop plants that promote effective uptake and application of vital nutrients. A crucial characteristic is root design, which affects the plant's capacity to search a greater area of the soil for nutrients. Deep and wide root systems give varieties an edge in terms of nutrient intake since they can reach deeper soil layers for nutrients. Furthermore, increased membrane transporters and nutrient storage capacities two features linked to improved nutrient transport within the plant are being worked on (Junedi et al. 2023).

9.2.1 The Importance of Nutrient Efficiency in Agriculture

The foundation of modern agriculture is excellent nutrient management, which is required to produce crops that are both high yielding and sustainable. The need for food is expected to rise as the world's population expands, necessitating increased agricultural production. On the other side, improper fertilizer application can lead to both cost inefficiencies and environmental damage. It is consequently critical to establish trait-based techniques to boost crop nutrient uptake efficiency (Pathak 2023).

9.2.2 Trait-Based Approaches in Crop Improvement

Trait-Based Approaches to Improve Nutrient Uptake Efficiency

- Crops that are bred to have strong root systems will be better able to search the soil for nutrients. Improved nutrient uptake can be achieved via characteristics including deeper roots, more root branching, and root hair proliferation.
- Through gene manipulation, the ability of the plant to receive and transfer vital nutrients from the soil to the aboveground tissues can be enhanced. These genes encode nutrient transporter proteins (Mamrutha et al. 2023).
- Enhancing nutrient uptake can be achieved by utilizing advantageous connections with microbes like as mycorrhizal fungus. These fungi improve the surface area accessible for nutrient absorption and lengthen the root system.
- Better nutrient absorption and utilization can result from identifying and adding genes that control nutrient utilization within the plant. This will guarantee that the ingested nutrients are effectively used for growth and development.
- By using GPS-guided variable rate fertilizer application, among other precision agricultural techniques, nutrient delivery may be optimized based on the unique requirements of the crop during various phases of growth (Fasusi et al. 2023).

9.2.3 Existing Studies on Enhancing Nutrient Uptake Efficiency

In agricultural research, trait-based strategies to improve crop nutrient uptake efficiency have drawn a lot of interest. These tactics concentrate on locating and exploiting particular genetic features in crop varieties that maximize uptake and application of nutrients. Numerous studies have already been conducted that have looked into different ways to enhance this crop performance factor. A well-known field of study is the morphology and architecture of roots. Studies have looked into characteristics that affect a plant's capacity to absorb nutrients from the soil, such as branching patterns, surface area, and root length. Scientists are trying to improve nitrogen uptake by choosing and breeding crops with more effective root systems (Kumar et al. 2016; Saudy et al. 2023).

9.3 Genetic and Molecular Basis of Nutrient Uptake Traits

One important component affecting agricultural production and sustainability is crop nutrient uptake efficiency. With the use of genetic and molecular interventions, trait-based techniques seek to improve the attributes of crop nutrient uptake. These methods offer a sustainable and environmentally friendly way to address the worldwide problem of feeding an expanding population with the least amount of chemical fertilizer consumption (Coskun and White 2023).

9.3.1 Genetic Basis of Nutrient Uptake Traits

9.3.1.1 QTL Mapping

The process of quantitative trait locus (QTL) mapping facilitates the identification of genomic areas linked to nutrient absorption features in crops. With the use of this technique, scientists may identify and target the precise genes that govern the efficiency of nutrient absorption (Chen et al. 2023).

9.3.1.2 Candidate Gene Approach

Scientists can create crops with enhanced nutrient absorption features by discovering and targeting particular genes involved in nutrient transport and metabolism. For instance, it is possible to alter the genes that code for transporters and nutrient uptake-related enzymes to increase their effectiveness.

9.3.1.3 Genome-Wide Association Studies (GWAS)

Across a wide range of crops, GWAS can reveal relationships between genetic variants and nutrient absorption features. These correlations shed light on the genetic variety that underlies nutritional efficiency (Waszczuk et al. 2023).

9.3.2 Molecular Basis of Nutrient Uptake Traits

Plants mostly rely on transporter proteins anchored in the cell membranes of their roots to absorb nutrients. Crops with improved nutrient absorption capacities may be developed as a result of our growing understanding of the structure and regulation of these proteins (Table 9.1):

1. **Root Architecture**: For the intake of nutrients, the molecular underpinnings of root architecture and development are essential. The ability of a crop to search the soil for nutrients is influenced by genes that affect architecture, branching, and root growth.
2. **Nutrient Sensing and Signaling**: Nutrient absorption is significantly regulated by molecular pathways involved in signaling and detecting nutrients. By focusing on these mechanisms, crops may be able to adapt to settings with reduced nutrient availability.
3. **Nutrient Use Efficiency (NUE) Genes**: Important molecular targets are genes linked to efficient utilization of nutrients. Crops can be engineered to use absorbed nutrients more effectively, minimizing waste, by modifying these genes (Günzel 2023).

9.3.3 Nutrient Transporters and Their Role

The identification and selection of particular genetic features linked to nutrient transporters is a key component of trait-based strategies to increase crop nutrient

Table 9.1 Genes enlisted below to indicate uptake

SI. No.	Ion	Transporter gene	Function
1	Nitrate	NRT1, NRT2	Uptake of nitrate from the soil into plant roots
2	Ammonium	AMT1, AMT2	Uptake of ammonium ions from the soil
3	Iron	IRT1, IRT2	Uptake of iron, especially under low-iron conditions
4	Phosphate	PHT1, PHT2	Uptake of phosphate from the soil into roots
5	Potassium	HAK, KUP	Uptake of potassium ions
6	Zinc	ZIP	Transport of zinc ions into plant roots
7	Water	PIP (plasma membrane intrinsic proteins)	Facilitate water transport across membranes
8	Various (small solutes: ammonium, boron)	NIP (nodulin 26-like intrinsic proteins)	Transport of various solutes including nutrients
9	Nitrate (high affinity)	HATS	High-affinity nitrate transport

uptake efficiency. The transportation of vital nutrients from the soil into the plant's cells is greatly aided by nutrient transporters. Specialized proteins called transporters are found in the roots of plants and are in charge of absorbing, moving, and distributing nutrients like potassium, phosphorus, and nitrogen, among other necessary elements (Krishna et al. 2023).

9.3.3.1 Nitrate Transporters (NRTs)

The uptake of nitrate ions (NO^{3-}) from the soil is carried out by nitrate transporters. They play a critical role in the plant's capacity to absorb and use nitrogen, a nutrient that is necessary for growth and development (Singh and Verma 2013a, b; Xia et al. 2023).

9.3.3.2 Phosphate Transporters (PHTs)

Phosphate transporters facilitate the uptake of phosphate ions ($H_2PO_4^-$ and HPO_4^{2-}) from the soil. Phosphorus is a key element in processes like photosynthesis, energy transfer, and nucleic acid synthesis (Rui et al. 2023).

9.3.3.3 Potassium Transporters (HAKs)

The intake of potassium ions (K^+), which is necessary for preserving cellular osmotic potential, enzyme activation, and general plant metabolism, is facilitated by potassium transporters (Shen et al. 2023).

9.3.3.4 Iron Transporters (IRTs)

Iron ions (Fe^{2+} and Fe^{3+}) are taken up from the soil by iron transporters. For functions like photosynthesis and respiration, iron is essential (Pandey and Abid 2023).

9.3.3.5 Zinc Transporters (ZIPs) and Iron-Regulated Transporters (IRT1)

Zinc ions (Zn^{2+}), a micronutrient that is crucial for plant growth and development, are absorbed in part through ZIP transporters. As was already established, IRT1 is also involved in the intake of iron (Ahmad et al. 2023).

9.3.4 Symbiotic Relationships with Beneficial Microbes

Under the soil lies a hidden but important alliance in the complex web of plant life: symbiotic connections between plants and helpful microorganisms. This dynamic partnership has become a pillar of sustainable agriculture, shaped over millions of years of development. Understanding and utilizing these symbiotic relationships offers creative ways to improve plant health, nutrient uptake, and overall crop output as mankind grapples with issues of food security and environmental sustainability. Mutualism stands out as a crucial component in the spectrum of interactions that occur in symbiotic partnerships between plants and helpful bacteria. Mutualistic relationships produce benefits for both the plant and the microbe, bringing harmony and balance to the complex dance of nature (Eriksen and Ballard 2020; Giri et al. 2023).

The mycorrhizal association, in which certain fungi create a symbiotic relationship with plant roots, is one of the best-known instances. By expanding the reach of plant roots and improving nutrient and water absorption in return for the carbohydrates the plant photosynthesizes, these mycorrhizal fungi benefit plants. This mutualistic dance strengthens the structure and resilience of the soil in addition to enhancing the plant's ability to absorb nutrients. This is advantageous to the plant as well as the health of the soil and the cycling of nutrients (Fan et al. 2023).

On the other hand, by transforming atmospheric nitrogen into a form that plants can use, nitrogen-fixing bacteria allow plants to absorb this essential nutrient (Singh et al. 2010a, b). These bacteria live in nodules on the roots of the plant, which is how this process happens. Establishing symbiotic partnerships with these bacteria can help plants absorb nitrogen more efficiently and use less artificial fertilizers. There is a wealth of unrealized potential for sustainable agriculture when symbiotic connections between plants and helpful bacteria are investigated. Understanding the complexities of these collaborations opens us new avenues for innovation that help us maintain ecological balance, grow healthier crops, and conserve resources (Kumar et al. 2019; Ejsing 2023). The exploration of plant-microbe symbiosis has the potential to lead to more resilient and environmentally friendly agriculture in the future.

9.3.5 Genes and Pathways Involved

The identification and modification of particular genes and pathways involved in nutrient absorption and utilization are key components of trait-based strategies to improve crop nutrient uptake efficiency. The design of the root system, which is essential for nutrient absorption, is one important target. Since they have a direct effect on surface area and the ability to absorb nutrients, genes controlling the development of root hair and lateral root growth are particularly interesting. Furthermore, the uptake of nutrients depends on channels and transporters found in the membranes of the root cells. Proton pumps and nitrate transporter genes, for example, are examples of genes that may be modified. Furthermore, the total efficiency of nutrient usage is greatly influenced by genes involved in nutrient remobilization inside the plant, especially during senescence (Siddiqui et al. 2023).

9.4 Trait-Based Strategies for Nutrient Uptake Enhancement

The utilization of trait-based techniques to boost nutrient uptake in crops entails the identification and exploitation of particular genetic features that facilitate the effective absorption of crucial nutrients from the soil. In order to maximize agricultural output while limiting resource inputs, these qualities are essential. Root architecture, which controls the amount and dispersion of roots in the soil, is one important characteristic. Deep, broad root systems enable varieties to draw nutrients from lower soil levels, minimizing reliance on fertilizers supplied on the surface. Furthermore, characteristics linked to mutualistic associations between crops and advantageous microbes, such as mycorrhizal fungi, improve the assimilation of nutrients. These organisms associate with plant roots in mutually beneficial ways that facilitate the exchange of nutrients for carbohydrates (Kumar et al. 2022; Jin 2023).

9.4.1 Genetic Modification Techniques

Agriculture has undergone a revolution because of genetic modification techniques, which have allowed scientists to increase crops' efficiency in absorbing nutrients. Manipulating the genes in charge of metabolic pathways and nutrient transporters is one such strategy. Through targeted gene introduction or modification, scientists can enhance a plant's capacity to take in and utilize vital nutrients from the soil. Furthermore, trait-based strategies make use of cutting-edge breeding methods such as quantitative trait locus (QTL) mapping and marker-assisted selection (MAS) (Table 9.2; Fig. 9.1). These techniques help find and choose plants that have favorable characteristics associated with efficient nutrient uptake. Breeders can expedite the evolution of crops with higher nutrient-utilization capacities by concentrating on particular genetic markers linked to effective nutrient absorption (Bhaskar et al. 2023).

9.4.2 Breeding Programs and Trait Selection

Trait-based techniques are used in breeding programs to improve crop nutrient uptake efficiency, which is an important aspect of sustainable agriculture. Using this method, plants with desirable nutrient-acquisition and nutrient-utilization features are chosen and propagated. Important characteristics include root architecture,

Table 9.2 QTL and marker-assisted selection technique for genetic modification in plants

Technique	Description
QTL Mapping	Identifies Quantitative Trait Loci (QTLs) associated with specific traits by analyzing genetic markers
Marker-Assisted Selection (MAS)	Uses molecular markers linked to desired traits for selective breeding
Marker-Assisted Introgression (MAI)	Transfers specific genomic regions containing desired traits from one plant to another using markers
Marker-Assisted Backcrossing (MAB)	Repeatedly backcrosses a plant with desired traits to a recurrent parent using markers
Genomic Selection	Predicts an individual's genetic value using dense marker information throughout the genome
Gene Editing (e.g., CRISPR-Cas9)	Precisely modifies plant DNA sequences to introduce or alter specific traits

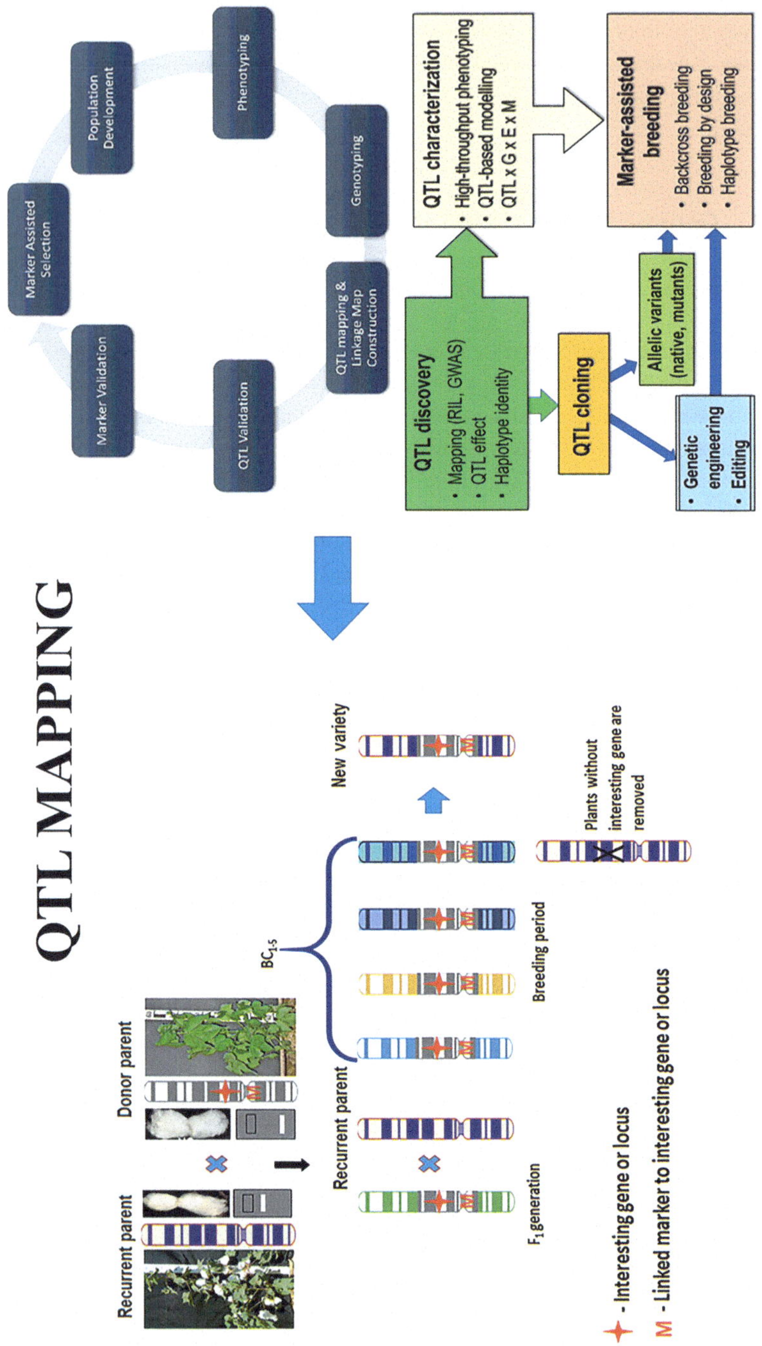

Fig. 9.1 QTL mapping and marker-assisted selection

mycorrhizal relationships, and nutrient transporter efficiency. One important characteristic affecting nutrient uptake is root architecture. Plants that have well-branched, deep roots are less dependent on surface nutrients because they may obtain nutrients from deeper soil layers. By increasing the effective reach of the root system, mycorrhizal association symbiotic connections between plants and soil fungi improve nutrient absorption (Liu et al. 2023a, 2023b).

It is critical that nutrients are transported through plants efficiently. Uptake rates are influenced by features that control the expression and function of nutrient transporters in roots; selective breeding focuses on these qualities to improve the plants' capacity to absorb and use vital nutrients.

9.4.3 Microbiome Manipulation for Enhanced Nutrient Uptake

The use of trait-based strategies to manipulate the microbiome to improve the efficiency of nutrient uptake in crops has gained popularity recently. This novel approach maximizes nutrient uptake by taking use of the symbiotic link between plants and the bacteria they are associated with, hence enhancing crop quality and yield. The intentional selection or enhancement of advantageous microbial populations in the rhizosphere the soil around plant roots is known as microbiome management. These microorganisms are essential to the processes of mineralization, solubilization, and nutrient cycling. Researchers want to improve crops' availability and accessibility of vital nutrients including micronutrients, phosphorus, and nitrogen by utilizing this natural mechanism (Philippot et al. 2023).

Research on the identification and characterization of particular plant characteristics that facilitate the formation and recruitment of advantageous microbial consortia is one potential field of study. Plants that have exudation patterns that are high in organic acids and sugars, for example, may support a more functional and diversified microbiome, which is well-known for its role at mobilizing nutrients in plants (Fig. 9.2; Table 9.3).

Furthermore, plants can have specific features tailored for improved nutrient uptake through the use of genetic engineering techniques. Through the insertion of genes corresponding to distinct exudates or root architecture, scientists can produce crops whose microbiome is optimized for the best possible supply of nutrients.

9.5 Case Studies of Successful Trait-Based Approaches

The application of trait-based techniques has demonstrated its significance in augmenting crop nutrient uptake efficiency, consequently advancing sustainable agricultural methodologies. Using this method, particular genetic characteristics that help crops absorb and use nutrients more efficiently are identified and selected. The effectiveness of these strategies in transforming agricultural productivity is demonstrated by a number of successful case studies.

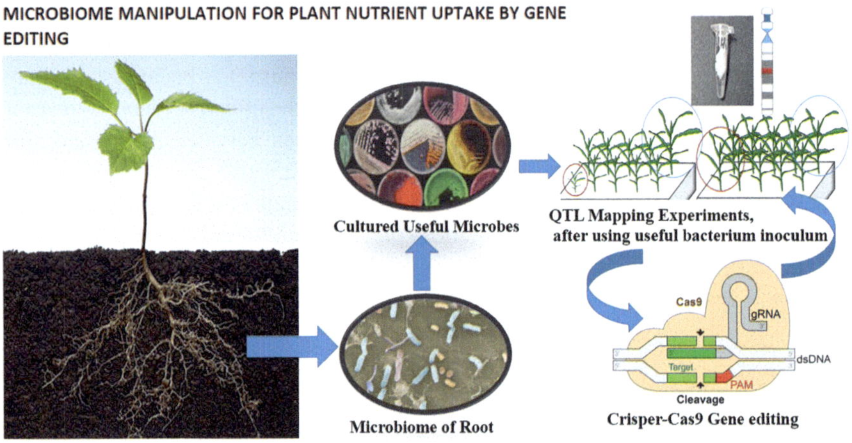

Fig. 9.2 Microbiome manipulation for promoting plant nutrient uptake

Table 9.3 Microbiome manipulation techniques for plant nutrient uptake

Technique	Description
Rhizosphere Engineering	Manipulation of the rhizosphere microbial community to enhance nutrient availability and uptake by the plant
Symbiotic Microbe Inoculation	Introducing beneficial microbes (e.g., mycorrhizal fungi, nitrogen-fixing bacteria) to improve nutrient uptake
Biofertilizers	Application of microbial formulations or organic materials to enrich soil fertility and aid in nutrient absorption
Plant-Microbe Interaction Studies	Understanding the communication and interaction between plants and their associated microbes for better nutrient uptake
Genetic Modification for Root Traits	Modifying plant genetics to enhance root architecture for improved nutrient absorption and water uptake
Precision Agriculture Techniques	Implementing technology-driven strategies to deliver nutrients precisely to plants based on their specific requirements

9.5.1 Case Study 1: Drought-Resistant Maize Varieties

A prestigious agricultural institute's researchers developed drought-resistant maize cultivars using trait-based methods. Through an emphasis on features associated with root design and efficiency of nutrient uptake, scientists were able to develop deeper and more effective root systems in maize plants. This led to significant output increases by enabling the crops to receive nutrients and water even in dry conditions (Singh et al. 2023).

9.5.2 Case Study 2: Nitrogen-Use Efficient Rice Strains

In a noteworthy instance, researchers utilized trait-based breeding methodologies to augment rice's nitrogen-use effectiveness. They were successful in creating rice strains that required less nitrogen fertilizer while retaining high yields by finding and integrating genes linked to increased nitrogen absorption and use. This lessened the negative effects that excessive nitrogen application had on the ecosystem while also saving farmers money on inputs (Zhang et al. 2023).

9.5.3 Challenges and Considerations

Enhancing crop nutrient absorption efficiency through trait-based methods has enormous promise for tackling issues related to global food security. Nonetheless, there are a number of important issues and factors to take into mind. First and foremost, a significant obstacle is the genetic complexity of nutrition absorption features. A single feature is frequently influenced by several genes, and these interactions are complex and poorly understood. Because of this intricacy, identifying and successfully manipulating the essential genes takes advanced genetic techniques and substantial research resources. Second, nutrient intake is significantly influenced by environmental variability. The temperature, management techniques, and state of the soil can all affect a crop's capacity to take up and use nutrients. Creating features that function well in a variety of settings is a significant task (Bacha and Iqbal 2023).

9.5.4 Potential Trade-Offs and Unintended Consequences

Using trait-based strategies to improve crop nutrient uptake efficiency has potential to address issues with global food security. They do, however, have possible trade-offs and unforeseen repercussions that need to be properly thought through.

The possibility of decreased overall plant vigor is one possible trade-off. The plant's resistance to biotic and abiotic stressors may be compromised if other crucial agronomic traits are neglected in favor of nutrient absorption qualities. Furthermore, genetic alterations meant to enhance nutrition absorption might unintentionally interfere with other biochemical pathways, with unexpected physiological repercussions (Singh et al. 2014, 2021, 2024). The possible effect on soil microbial populations is a further cause for concern. Modifications in the characteristics of nutrient intake could have an impact on the rhizosphere and possibly upset the symbiotic connections with helpful bacteria. The dynamics of the nitrogen cycle and soil health may be impacted in a cascade manner by this (Ruhanen et al. 2023).

9.5.5 Regulatory and Ethical Concerns

The safety and environmental effects of introducing genetically modified characteristics into agricultural systems need to be thoroughly assessed by regulatory organizations. Thorough evaluations are necessary to guarantee that these changes don't unintentionally damage organisms that aren't the intended target or disturb ecosystems. Additionally, to give customers clarity regarding the origin and nature of genetically modified crops, traceability protocols and explicit labeling should be implemented.

A key factor in the acceptance of trait-based techniques is ethical issues. The ideas of equity and accessibility should guide farmers and seed firms in order to ensure that everyone may benefit from increased nutrient absorption efficiency, not just large-scale commercial enterprises. In addition, it is important to exercise caution when managing intellectual property rights associated with genetically modified features in order to strike a balance between innovators' interests and the general welfare (Rafeeq et al. 2023). The ethical and regulatory considerations must be balanced with the potential advantages of trait-based techniques in order to responsibly progress agricultural biotechnology. To create strong frameworks that support the equitable and sustainable adoption of these technologies in agriculture, players from the public and private sectors must work together.

9.5.6 Adoption and Accessibility by Farmers

Farmers are essential to agricultural innovation because they are open to adopting trait-based strategies, which can result in significant gains in nutrient utilization. By improving particular plant characteristics linked to nutrient uptake, transport, and utilization, these methods help crops flourish even in soils lacking in nutrients. The adoption of trait-based techniques is greatly aided by the availability of resources and information. Farmers can be informed about these practices through extension services, agricultural training programs, and knowledge-sharing platforms. Partnerships between governmental organizations, agricultural groups, and research institutes can also help to advance accessibility and offer essential assistance (Coleman et al. 2023). Furthermore, it's critical to support the cost-effectiveness and accessibility of trait-enhanced seeds and technologies. Incentives, grants, and subsidies can all help lower the financial obstacles to implementing these strategies. Additionally, demonstrating the advantages of trait-based strategies through the establishment of demonstration plots and field trials can boost farmer confidence and encourage broader adoption.

9.5.7 Sustainability and Environmental Implications

Sustainable agriculture has brought much focus to trait-based methods for improving crop nutrient uptake efficiency. By creating crop types with certain characteristics that maximize nutrient utilization, these methods help to reduce the demand for synthetic fertilizers and lessen the negative environmental effects of overuse. A crucial characteristic is root architecture, in which crops are designed to have larger, deeper root systems that can reach deeper soil layers for nutrients. This lessens the possibility that nutrients may leak into bodies of water, causing pollution. Decreased greenhouse gas emissions from fertilizer application and production are another benefit of increased nutrient usage efficiency (Karmaker et al. 2023). The creation of crops with improved nutrient transporters, which can absorb nutrients more efficiently, is another encouraging feature. Because these crops use less fertilizer, agriculture's carbon footprint is smaller.

9.6 Future Directions and Research Gaps

The development of high-throughput phenotyping techniques to precisely and effectively screen sizable populations for nutrient absorption features should also be a primary focus of research. This will facilitate the selection of superior genotypes with increased nutrient usage efficiency and speed up the breeding process.

Field-based validation of trait-based techniques across a variety of agroecosystems and environmental situations is desperately needed. This will make sure that these methods are resilient and flexible in actual farming situations (Tang et al. 2023).

9.6.1 Emerging Technologies and Tools

These cutting-edge instruments and technologies have the potential to address issues with global food security by increasing agricultural output while reducing negative environmental effects. This field is progressing due to a number of significant breakthroughs:

1. **Genomic Selection**: Researchers can find and produce crops with enhanced nutrient absorption properties by using genetic profiling and selection. By using targeted breeding, nitrogen runoff is decreased and overfertilization is not necessary (Song et al. 2023).
2. **Precision Agriculture**: Precision agriculture makes use of sensor technologies, drones, and data analytics to supply nutrients precisely to crops according to their needs. This maximizes nutrient use and reduces waste (Su et al. 2023).
3. **Biotechnology**: It is possible to create crops through genetic modification to express genes that improve their capacity to absorb nutrients, resulting in more effective nutrient consumption in difficult settings (Kumar et al. 2017).
4. **Microbiome Engineering**: Plants absorb nutrients more effectively when beneficial soil microorganisms are controlled to optimize nutrient cycling and root-microbe interactions (Vimal et al. 2023).
5. **Nanotechnology**: In order to minimize losses and enhance plant uptake, controlled and effective nutrient release is made possible by nanoscale fertilizers and nutrient delivery systems (Dey et al. 2023).

Together, these cutting-edge technologies help to create an agriculture that is more resilient and sustainable, enabling crops to flourish with fewer fertilizer inputs while satisfying the expanding needs of the world's population.

9.6.2 Integrating Trait-Based Approaches with Sustainable Agriculture

Trait-based techniques have great potential to improve crop nutrient uptake efficiency when combined with sustainable agriculture. Farmers can limit their impact on the environment, optimize resource allocation, and lower input costs by utilizing genetic characteristics that affect nutrient absorption and usage. A crucial characteristic is root architecture, which is essential for nutrient uptake. Crops that are bred to have longer, deeper root systems can reach deeper into the soil and absorb nutrients that might otherwise go unnoticed. Furthermore, characteristics associated with the movement of nutrients within plants might improve the effectiveness of the transfer of nutrients from roots to shoots, guaranteeing that essential components get to the locations where they are most required (Lu et al. 2023). Moreover, characteristics that affect nutrient use efficiency (NUE) are critical. Higher NUE crops can flourish in environments with restricted nutrient availability, which makes them ideal for sustainable farming methods. These characteristics lessen the requirement for overfertilization by allowing plants to utilize the nutrients that are readily available.

9.6.3 Unexplored Aspects of Nutrient Uptake Efficiency

In recent years, there has been a noticeable increase in interest in trait-based methods for increasing crop nutrient uptake efficiency. Trait-based approaches seek to improve plants' capacity to effectively absorb and use nutrients from the soil, whereas traditional breeding strategies have concentrated on yield enhancement. This may result in farming methods that are more resource-efficient and sustainable (Yang et al. 2023b). The possibility of using the inherent genetic variety found in crops is one undiscovered component of this strategy. Breeders can create crop varieties that flourish in a variety of soil situations by recognizing and selecting for particular features associated with nutrient uptake efficiency. Certain plant types, for instance, can have unique root structures that improve nutrient absorption, while other plant kinds might have systems in place to maximize nutrient use inside the plant.

One area that needs more research is the impact of microbial populations on the efficiency of nitrogen uptake. Mycorrhizal fungi, among other beneficial soil microbes, work in symbiotic partnerships with plant roots to facilitate nutrient intake. Gaining insight into and control of these interactions may present a viable path toward increasing crop nutrient uptake efficiency (Riseh et al. 2023).

9.7 Conclusion

In conclusion, trait-based strategies have great potential for improving crop nutrient uptake efficiency and provide a long-term answer to the world's food security problems. Researchers and agronomists can create crop varieties with enhanced nutrient uptake capacities by identifying and modifying important genetic features linked to nutrient acquisition and use. By doing this, crop output and quality are increased while also lessening the negative environmental effects of excessive fertilizer application. Moreover, the use of trait-based methods can be crucial in reducing the adverse impacts of nitrogen runoff on ecosystems. A crop's ability to efficiently absorb nutrients allows it to make better use of the resources at hand, reducing waste and potential environmental damage. These methods can also aid in the production of crops that can flourish in a greater variety of situations by making them more resistant to nutritional deficiencies.

References

Ahmad I, Rawoof A, Islam K, Momo J, Anju T, Kumar A, Ramchiary N (2023) Diversity and expression analysis of ZIP transporters and associated metabolites under zinc and iron stress in capsicum. Plant Physiol Biochem 196:415–430

Bacha SAS, Iqbal B (2023) Advancing agro-ecological sustainability through emerging genetic approaches in crop improvement for plants. Funct Integr Genomics 23(2):145

Bhaskar R, Pandey SP, Kim H, Jayakodi SK, Gupta MK, Han SS (2023) Nanobionics for sustainable crop production: recent development to regulate plant growth and protection strategies from pests. OpenNano 15:100198

Chen KX, Dai DY, ling W, Yang LM, Li DD, Chao W et al (2023) SLAF marker based QTL mapping of fruit-related traits revealed a major-effect candidate locus ff2. 1 for flesh firmness in melon. J Integr Agric 22:3331–3345

Cheng Y, Zhang T, Hu X, Liu Z, Liang Q, Yan S et al (2023) Drip fertigation triggered by soil matric potential reduces residual soil nitrate content and improves maize nitrogen uptake and yield stability in an arid area. Eur J Agron 150:126932

Coleman D, Gallagher RV, Falster D, Sauquet H, Wenk E (2023) A workflow to create trait databases from collections of textual taxonomic descriptions. Eco Inform 78:102312

Coskun D, White PJ (2023) Ion-uptake mechanisms of individual cells and roots: short-distance transport. In: Marschner's mineral nutrition of plants. Academic Press, London, pp 11–71

Dey D, Chowdhury S, Sen R (2023) Insight into recent advances on nanotechnology-mediated removal of antibiotic resistant bacteria and genes. J Water Process Eng 52:103535

Ejsing M (2023) Living with others: on multispecies resurgence in the altered forest landscapes of the Anthropocene. J Polit Ecol 30(1):316–334

En Y, Yamaji N, Ma JF (2023) Linking root morphology and anatomy with transporters for mineral element uptake in plants. Plant Soil 484(1–2):1–12

Eriksen C, Ballard S (2020) Alliances in the Anthropocene: fire, plants, and people. Springer Nature, Singapore

Fan W, Chen Y, Yuan B, Chen X, Wang Q, Zhao S, Jia W (2023) Enhancing anammox nitrogen removal by static magnetic field exposure: performance, microbial community and symbiotic relationship analysis. J Water Process Eng 53:103709

Fasusi OA, Babalola OO, Adejumo TO (2023) Harnessing of plant growth-promoting rhizobacteria and arbuscular mycorrhizal fungi in agroecosystem sustainability. CABI Agric Biosci 4(1):26

Fu X, Niu H (2023) Key technologies and applications of agricultural energy internet for agricultural planting and fisheries industry. Inf Process Agric 10(3):416–437

Giri K, Mishra G, Suyal DC, Kumar N, Doley B, Das N, Baruah RC, Bhattacharyya R, Bora N (2023) Performance evaluation of native plant growth-promoting rhizobacteria for paddy yield enhancement in the jhum fields of Mokokchung, Nagaland, North East India. Heliyon 9:e14588

Günzel D (2023) Is there a molecular basis for solvent drag in the renal proximal tubule? Pflügers Arch 475(2):277–281

Hull SC, Charles J, Caplan AL (2023) Are we what we eat? The moral imperative of the medical profession to promote plant-based nutrition. Am J Cardiol 188:15–21

Jin Q (2023) Building microbial kinetic models for environmental application: a theoretical perspective. Appl Geochem 158:105782

Junedi MA, Mukhopadhyay R, Manjari KS (2023) Alleviating salinity stress in crop plants using new engineered nanoparticles (ENPs). Plant Stress 9:100184

Karmaker CL, Al Aziz R, Palit T, Bari AM (2023) Analyzing supply chain risk factors in the small and medium enterprises under fuzzy environment: implications towards sustainability for emerging economies. Sustain Technol Entrep 2(1):100032

Krishna TA, Maharajan T, Ceasar SA (2023) The role of membrane transporters in the biofortification of zinc and iron in plants. Biol Trace Elem Res 201(1):464–478

Kumar N, Kumar N, Shukla A, Shankhdhar SC, Shankhdhar D (2015) Impact of terminal heat stress on pollen viability and yield attributes of rice (Oryza sativa L.). Cereal Res Commun 43(4):616–626

Kumar N, Shankhdhar SC, Shankhdhar D (2016) Impact of elevated temperature on antioxidant activity and membrane stability in different genotypes of rice (Oryza sativa L.). Indian J Plant Physiol 21(1):37–43

Kumar N, Suyal DC, Sharma IP, Verma A, Singh H (2017) Elucidating stress proteins in rice (Oryza sativa L.) genotype under elevated temperature: a proteomic approach to understand heat stress response. 3 Biotech 7:205

Kumar N, Jeena N, Singh H (2019) Elevated temperature modulates rice pollen structure: a study from foothill Himalayan agro-ecosystem in India. 3 Biotech 9:175

Kumar N, Jeena N, Kumar A, Khairakpam R, Singh H (2021) Comparative response of rice cultivars to elevated air temperature in Bhabar region of Indian Himalaya: status on yield attributes. Heliyon 7:e07474

Kumar A, Singh H, Kumari G, Bisht S, Malik A, Kumar N, Singh M, Raturi A, Barthwal S, Thakur A, Kaushal R (2022) Adaptive resilience of roadside trees to vehicular emissions via leaf enzymatic, physiological, and anatomical trait modulations. Environ Pollut 313:120191

Liu CY, Hao Y, Wu XL, Dai FJ, Abd-Allah EF, Wu QS, Liu SR (2023a) Arbuscular mycorrhizal fungi improve drought tolerance of tea plants via modulating root architecture and hormones. Plant Growth Regul 102:13–22

Liu Y, Hu B, Chu C (2023b) Toward improving nitrogen use efficiency in rice: utilization, coordination, and availability. Curr Opin Plant Biol 71:102327

Lu K, Jähnig SC, Wu H, Xie Z, Chen X, He F (2023) Trait-based approach of aquatic insects to track recovery of wetland ecosystems in Northeast China. Ecol Indic 155:111012

Mamrutha HM, Khobra R, Krishnappa G, Wadhwa Z, Tiwari R, Singh G, Pratap Singh G (2023) Importance of integrating physiological breeding to augment crop breeding. In: Translating physiological tools to augment crop breeding. Springer Nature, Singapore, pp 1–6

Pandey SN, Abid M (2023) Zinc biofortification: role of ZIP family transporters in the uptake of zinc from the soil up to the grains. In: Mineral biofortification in crop plants for ensuring food security. Springer Nature, Singapore, pp 105–120

Pathak H (2023) Impact, adaptation, and mitigation of climate change in Indian agriculture. Environ Monit Assess 195(1):52

Philippot L, Chenu C, Kappler A, Rillig MC, Fierer N (2023) The interplay between microbial communities and soil properties. Nat Rev Microbiol. https://doi.org/10.1038/s41579-023-00980-5

Rafeeq H, Afsheen N, Rafique S, Arshad A, Intisar M, Hussain A et al (2023) Genetically engineered microorganisms for environmental remediation. Chemosphere 310:136751

Riseh RS, Vazvani MG, Kennedy JF (2023) The application of chitosan as a carrier for fertilizer: a review. Int J Biol Macromol 252:126483

Ruhanen HM, Mofikoya AO, Vesterbacka A, Kivimäenpää M, Blande JD (2023) Trait-based cropping of brassicaceous plants: effects on ecosystem services and crop yield. Biol Control 187:105389

Rui W, Ma J, Wei N, Zhu X, Li Z (2023) Genome-wide analysis of the PHT gene family and its response to mycorrhizal symbiosis in tomatoes under phosphate starvation conditions. Int J Mol Sci 24(12):10246

Saudy HS, Salem EM, Abd El-Momen WR (2023) Effect of potassium silicate and irrigation on grain nutrient uptake and water use efficiency of wheat under calcareous soils. Gesunde Pflanzen 75(3):647–654

Shen L, Fan W, Li N, Wu Q, Chen D, Luan J et al (2023) Rice potassium transporter OsHAK18 mediates phloem K+ loading and redistribution. Plant J 116(1):201–216

Siddiqui N, Gabi MT, Kamruzzaman M, Ambaw AM, Teferi TJ, Dadshani S et al (2023) Genetic dissection of root architectural plasticity and identification of candidate loci in response to drought stress in bread wheat. BMC Genomic Data 24(1):38

Singh H, Verma A (2013a) Physiological responses of rice cultivars to various nitrogen levels. Int J Agric Environ Biotechnol 6(3):383–388

Singh H, Verma A (2013b) Characterization and screening of high nitrogen efficient rice genotype to curtail environmental pollution. Int J Agric Environ Biotechnol 6(1):23–30

Singh H, Verma A, Shukla A (2010a) Comparative study of yield and yield components of hybrid and inbred genotypes of rice (*Oryza Sativa L.*). Int J Agric Environ Biotechnol 3:355–360

Singh H, Verma A, Krishnamoorthy Mand Shukla A (2010b) Consequence of diverse nitrogen levels on leaf pigments in five rice genotypes under field emergent circumstance. Int J Bioresour Stress Manage 1:189–193

Singh H, Verma A, Rai SK (2013a) Biochemical evaluation of different rice genotypes grown at various nitrogen levels using SDS-PAGE. Curr Adv Agric Sci 5(1):144–146

Singh H, Verma A, Shukla A (2013b) Guttation fluid as a physiological marker for selection of nitrogen efficient rice (*Oryza sativa* L.) genotypes. Afr J Biotechnol 12(44):6276–6281

Singh H, Verma A, Ansari MW, Shukla A (2014) Physiological response of rice (*Oryza sativa* L.) genotypes to elevated nitrogen applied under field conditions. Plant Signal Behav 9:e29015. https://doi.org/10.4161/psb.29015

Singh S, Singh H, Sharma SK, Nautiyal R (2021) Seasonal variation in biochemical responses of bamboo clones in the sub-tropical climate of Indian Himalayan foothills. Heliyon 7(4):e06859. https://doi.org/10.1016/j.heliyon.2021.e06859

Singh A, Pandey H, Pandey S, Lal D, Chauhan D, Aparna et al (2023) Drought stress in maize: stress perception to molecular response and strategies for its improvement. Funct Integr Genomics 23(4):296

Singh M, Singh H, Kumar A, Kumar M, Barthwal S, Thakur A (2024) Soil nitrogen availability determines the CO_2 fertilization effect on tree species (*Neolamarckia cadamba*): growth and physiological evidence. Environ Sustain. https://doi.org/10.1007/s42398-023-00300-w

Song H, Dong T, Yan X, Wang W, Tian Z, Sun A et al (2023) Genomic selection and its research progress in aquaculture breeding. Rev Aquac 15(1):274–291

Su J, Zhu X, Li S, Chen WH (2023) AI meets UAVs: a survey on AI empowered UAV perception systems for precision agriculture. Neurocomputing 518:242–270

Tang C, Jiang B, Ejaz I, Ameen A, Zhang R, Mo X, Wang Z (2023) High-throughput phenotyping of nutritional quality components in sweet potato roots by near-infrared spectroscopy and chemometrics methods. Food Chem X 20:100916

Vimal SR, Singh JS, Kumar A, Prasad SM (2023) Plant genotype-microbiome engineering as nature-based solution (NbS) for regeneration of stressed agriculture: a review. Sci Hortic 321:112258

Waszczuk MA, Jonas KG, Bornovalova M, Breen G, Bulik CM, Docherty AR et al (2023) Dimensional and transdiagnostic phenotypes in psychiatric genome-wide association studies. Mol Psychiatry. https://doi.org/10.1038/s41380-023-02142-8

Xia X, Wei Q, Xiao C, Ye Y, Li Z, Marivingt-Mounir C et al (2023) Genomic survey of NPF and NRT2 transporter gene families in five inbred maize lines and their responses to pathogens infection. Genomics 115(2):110555

Yang H, Li J, Wu G, Huang X, Fan G (2023a) Maize straw mulching with no-tillage increases fertile spike and grain yield of dryland wheat by regulating root-soil interaction and nitrogen nutrition. Soil Tillage Res 228:105652

Yang H, Zhang WP, Xu HS, Yu RP, Su Y, Surigaoge S et al (2023b) Trade-offs and synergies of plant traits co-drive efficient nitrogen use in intercropping systems. Field Crop Res 302:109093

Zhang H, Zhang J, Yang J (2023) Improving nitrogen use efficiency of rice crop through an optimized root system and agronomic practices. Crop Environ 2:192–201

Plant Functional Traits in Crop Breeding: Advancement and Challenges

10

Divya Chaudhary, Neeraj Pal, Arushi Arora, Bhamare Deepak Prashant, and Sreya Venadan

Abstract

Plant functional traits have emerged as crucial factors in crop breeding, offering a pathway to address the multifaceted challenges facing agriculture today. These traits, encompassing features like leaf morphology, root architecture, and physiological attributes, govern how plants interact with their environment and influence their performance in various ecological contexts. One of the paramount advantages of integrating plant functional traits into crop breeding lies in optimizing resource utilization. Recent advancements have facilitated the incorporation of plant functional traits into breeding programs. High-throughput phenotyping technologies enable efficient measurement of these traits on a large scale, enhancing selection processes. Trait stacking, combining multiple functional traits within a single variety, optimizes performance across diverse environments, enhancing adaptability and resilience. Plant functional traits are

D. Chaudhary (✉)
Department of Genetics and Plant Breeding, G.B. Pant University of Agriculture and Technology, Pantnagar, Uttarakhand, India

N. Pal
Department of Molecular Biology and Genetic Engineering, G.B. Pant University of Agriculture and Technology, Pantnagar, Uttarakhand, India

A. Arora
Department of Plant Breeding and Genetics, Punjab Agricultural University, Ludhiana, Punjab, India

B. D. Prashant
School of Agricultural Biotechnology, Punjab Agricultural University, Ludhiana, Punjab, India

S. Venadan
Department of Biotechnology, ICAR Directorate of Onion and Garlic Research, Pune, Maharashtra, India

N. Kumar, H. Singh (eds.), *Plant Functional Traits for Improving Productivity*,
https://doi.org/10.1007/978-981-97-1510-7_10

increasingly integrated into precision agriculture practices, enabling farmers to tailor crop management strategies based on trait data, thereby optimizing resource allocation and improving yields. While the promise of plant functional traits is substantial, challenges persist. Complex interactions between traits and environmental variability make accurate trait predictions challenging. Novel crop varieties developed with modified traits may face consumer resistance, necessitating effective communication and education. Smaller breeding programs with limited resources may struggle to access advanced technologies. Addressing these challenges demands collaboration among scientists, breeders, policymakers, and stakeholders. This chapter explores the significance of plant functional traits in crop breeding, recent advancements, and the challenges that researchers and breeders face in integrating these traits into breeding programs.

Keywords

Plant functional traits · Crop breeding · Adaptation · Precision agriculture · Abiotic stress

10.1 Introduction

In the realm of crop breeding, the exploration and manipulation of plant functional traits have emerged as pivotal endeavors, shaping the trajectory of agricultural advancement. The chapter titled "Plant Functional Traits in Crop Breeding: Advancement and Challenges" delves into the intricacies of harnessing plant characteristics to optimize crop performance, resilience, and yield. Understanding and manipulating functional traits, which encompass various physiological, morphological, and biochemical features, hold the key to developing crops that can thrive in diverse environmental conditions, resist pests and diseases, and contribute to sustainable agricultural practices (Wani et al. 2016).

This chapter offers a comprehensive overview of recent advancements in incorporating plant functional traits into crop breeding strategies, exploring their multifaceted roles in influencing crop adaptation to changing climates, enhancing resource use efficiency, and improving overall agricultural sustainability. Alongside these advancements, the chapter delves into the challenges associated with integrating functional traits into breeding programs, addressing issues ranging from complex trait interactions to the imperative need for interdisciplinary collaboration. Integral to this discourse is the exploration of enhancing plant functional traits through high-throughput phenotyping (HTP) and advanced breeding technologies, elucidating how marker-assisted selection (MAS) and genomic tools contribute to the targeted improvement of plant functional traits (Fig. 10.1). Automated and non-invasive HTP is a platform that empowers breeders to systematically assess plant functional traits on a significant scale, facilitating the identification and selection of plants exhibiting desired characteristics (Zhang et al. 2023). Concurrently, Perez-de-Castro et al. (2012) add to this technological storyline by investigating the utilization of genomic selection. This method involves choosing individuals based on

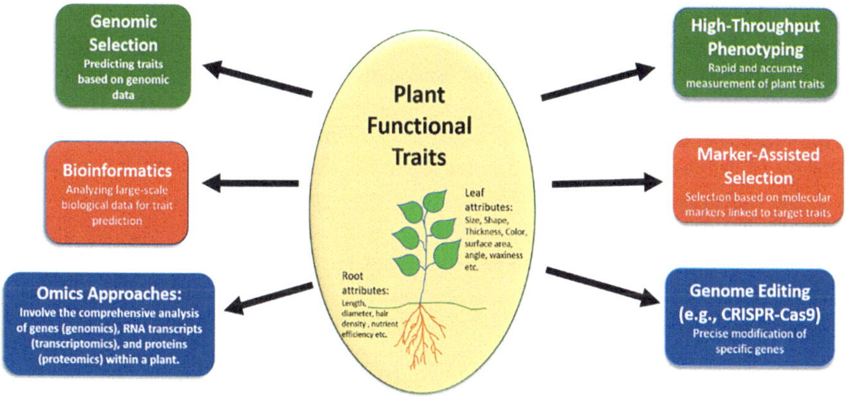

Fig. 10.1 Enhancing plant functional traits through advanced plant breeding technologies

their complete genetic profile rather than specific traits. This approach accelerates the breeding process by predicting an individual's performance based on its genomic profile. These technologies play a pivotal role in accelerating the breeding process and expanding the scope of trait manipulation, offering new dimensions to crop improvement strategies (Chen et al. 2019).

The significance of plant functional traits in crop breeding is underscored by their direct influence on a crop's ability to adapt to changing environmental conditions, especially in the face of the considerable threat posed by climate change. This challenge becomes evident through changes in temperature, precipitation trends, and the rising occurrence of extreme weather events. Therefore, it is crucial to prioritize the comprehension and selection of plant functional traits that provide resilience. Various researchers emphasize the pivotal role of functional traits in enhancing crop resilience, with traits such as tolerance to drought and heat, and efficient nutrient use emerging as crucial components of a crop's adaptive capacity (Gonzalez Guzman et al. 2022). The chapter delves into how breeding programs can strategically target these traits, thereby contributing to the development of crops capable of withstanding the challenges posed by the rapidly changing climate, ultimately ensuring global food security.

While the potential benefits of manipulating plant functional traits are substantial, their integration into breeding programs is not without challenges. Wani et al. (2016) delve into the complexities associated with trait interactions and underscore the need for a holistic understanding of plant biology. Functional traits often interact with each other, making it challenging to predict the combined effects of multiple traits on crop performance accurately. The chapter critically evaluates these challenges, offering insights into the limitations and pitfalls of current approaches. Emphasizing the importance of interdisciplinary collaboration, the discussion highlights the necessity of bringing together experts from diverse fields such as genetics, physiology, and agronomy. Through collaborative initiatives of this nature, a thorough comprehension of the genetic and environmental factors shaping functional traits can be realized. This, in turn, establishes pathways for the development of more effective and sustainable breeding strategies.

This chapter navigates the dynamic landscape of plant breeding, offering a critical examination of the current state of knowledge while pinpointing research gaps and suggesting avenues for future exploration. Through these insights, it actively contributes to the ongoing scientific dialogue, fostering a deeper understanding of both the potential and limitations of harnessing plant functional traits for crop improvement. The promise of plant functional traits in crop breeding lies in their capacity to enable the development of varieties better suited to changing environmental conditions, more resource-efficient, and environmentally sustainable. However, overcoming the challenges associated with integrating these traits necessitates collaboration among breeders, ecologists, geneticists, and other stakeholders. Looking forward, as advancements persist, plant functional traits are poised to play an increasingly vital role in shaping the future landscape of crop improvement and ensuring global food security.

10.2 Plant Functional Traits: Importance in Crop Breeding and Agriculture

Plant functional traits play a pivotal role in shaping crop breeding and advancing sustainable agriculture, offering a nuanced understanding of characteristics crucial for crop performance and resilience. Recent research highlights the relevance of specific traits in addressing contemporary agricultural challenges. For instance, research emphasizes the importance of traits like stomatal conductance and root architecture in conferring drought tolerance, addressing the increasing relevance of water scarcity in agriculture. Additionally, advancements in understanding nutrient use efficiency underscore the significance of traits influencing nutrient uptake and allocation, contributing to optimized nutrient utilization and reduced environmental impacts (Lynch 2019).

The emphasis has turned to the chemical attributes, particularly secondary metabolites, as they play a crucial role in the development of crops that can resist pests and diseases, aligning with sustainable pest management practices. As pests and diseases evolve, staying informed about recent findings in plant chemical ecology is crucial for designing resilient crops. Reproductive traits remain central to yield potential, guiding breeders in developing crops with increased productivity. Recent studies emphasize the importance of traits related to flowering time, seed size, and production, enabling the selection of varieties that optimize reproductive success (Höft et al. 2018). Furthermore, characters associated with temperature tolerance are gaining significance in the face of climate change challenges. Ongoing research on the genetic basis of temperature-related traits aids in developing crops capable of withstanding fluctuating environmental conditions.

Technological innovations, such as HTP, have significantly advanced breeding programs by enabling quick and effective measurement of functional traits (Zhang et al. 2023). This accelerates the breeding process, allowing for the identification and selection of individuals with desired traits more efficiently. Recent studies focusing on root traits highlight their importance not only in plant-soil interactions but also in contributing to soil health and ecosystem services, including carbon sequestration (Tracy et al. 2020). Emphasizing stress-resistant traits, such as resistance to salinity or extreme temperatures, aligns with sustainable agriculture goals and promotes the development of crops resilient to diverse environmental challenges.

Staying informed about recent findings in stress physiology aids in incorporating the latest knowledge into breeding programs. The continuous integration of recent scientific findings and technological advancements ensures that these breeding programs stay at the forefront of innovation, contributing to a more sustainable and resilient agricultural future. The dynamic nature of ongoing research underscores the need for breeders to stay engaged with the latest developments in plant functional traits for informed decision-making in crop improvement.

10.3 Resource Utilization and Efficiency

10.3.1 Water Use Efficiency Traits

Selecting water-efficient cultivars is crucial for minimizing water consumption in regions facing water scarcity. Excessive groundwater extraction (covering 64% of irrigated areas) has led to rapid depletion of groundwater tables, underscoring the importance of enhancing water use efficiency (WUE) to mitigate stress on this critical resource. The challenge lies in identifying and integrating various traits that significantly increase WUE while proving effective in real-world conditions. The composite phenotype, enabling a plant to use less water while maintaining carbon assimilation, encompasses traits like transpiration efficiency (TE), stomatal development, root development, cuticle composition, etc. in response to environmental conditions.

Water use efficiency (WUE) in plants is intricately linked to factors such as TE, which, in turn, is influenced by photosystem activity, CO_2 concentration in substomatal cavities, and stomatal activity. Specific genetic adaptations, exemplified by the SDD1 (Stomatal Density and Distribution) gene, play a crucial role in governing the differentiation of meristemoids and guard cells. This genetic control results in a reduction in stomatal density, ultimately enhancing drought tolerance without compromising leaf area or plant size, thereby boosting overall WUE (Büssis et al. 2006). The EPF (Epidermal Patterning Factor) and EPFL (Epidermal Patterning Factor-like) gene families contribute to the regulation of stomatal density during the development of Arabidopsis leaves. Overexpression of EPF1 or EPF2 leads to a decrease in stomatal numbers without affecting photosynthetic capacity, thus increasing WUE (Franks et al. 2015). Conversely, EPFL9 promotes stomatal differentiation, resulting in an increase in stomatal numbers (Torii 2012). The ERECTA family, specifically the TMM (Too Many Mouths) receptor, modulates the activity of EPF, thereby influencing stomatal development. YODA, positioned downstream of TMM, regulates the fate of guard cells, impacting stomatal numbers.

ANGUSTIFOLIA3 (AN3), identified as a key regulator, enhances WUE by repressing YODA, leading to lower stomatal density and transpiration without compromising biomass accumulation (Meng and Yao 2015).

Deep root systems, characterized by a high density at depth, play a crucial role in improving yield in water-limited environments. Plants adapted to drought, such as sorghum, naturally display vertically oriented root structures. The genetic components that shape root traits provide promising avenues for enhancing water use efficiency (WUE) and addressing crop loss in challenging conditions. Studies emphasize the significance of the *Deeper Rooting 1* (*DRO1*) gene in rice, contributing to steeper root angles and bolstering drought tolerance through a deeper root system. In Arabidopsis, *ANGUSTIFOLIA3* (*AN3*) plays a role in regulating drought tolerance by influencing WUE through enhanced root traits. Conversely, in rice, *OsNAC9* modifies root architecture, leading to improved drought resistance and grain yield (Meng and Yao 2015).

Enhanced wax deposition within the cuticle emerges as a strategic approach to minimize water loss in plants. The thin cuticle layer enveloping plant tissues, consisting of cutin and waxes, plays a crucial role as a protective barrier against water loss. Insights from mutant studies, such as tomato eceriferum6 (cer6) and lacs1, lacs2, bodyguard, and glossyhead1, underscore the significance of both waxes and cutin in preserving water status (Lü et al. 2011). Key genes involved in cuticle biosynthesis, including transcription factors (ERFs, MYBs, HD-Zip IV), play pivotal roles in this process. Examples include *CsCER1* in cucumber and the *SHN* clade genes in Arabidopsis, influencing alkane biosynthesis, cuticle permeability, and drought resistance (Wang et al. 2015). In rice, *OsGL1-6* affects cuticular wax accumulation, enhancing drought resistance. The *ECERIFERUM9* (*CER9*) gene in Arabidopsis, identified as a negative regulator of cuticle lipid synthesis, not only improves the response to water deficit but also enhances water use efficiency (WUE). This suggests its potential as a molecular breeding target for enhancing drought tolerance and WUE in crops (Zhou et al. 2013).

Overall, plant architecture encompassing traits related to roots, stomata, and cuticles plays a pivotal role in shaping water use efficiency (WUE). The three-dimensional organization of plants, guided by endogenous mechanisms, holds the promise of improving drought tolerance.

10.3.2 Nutrient Use Efficiency Traits

Nitrogen use efficiency (NUE), a key metric calculated by dividing the grain dry yield by the available nitrogen supply, carries significant implications for both profitability and the reduction of greenhouse gas emissions. The components influencing NUE encompass nitrogen uptake efficiency (NUpE) and nitrogen utilization efficiency (NUtE). Various traits contribute to these efficiencies, including chlorophyll content, plant height, biomass, N accumulation indices, and agronomic traits. A critical factor in achieving high NUpE is root system architecture (RSA), where desirable traits involve an increase in root dry weight, length, density, number, diameter, and the length of lateral roots/root hairs (Saini et al. 2021).

In 1989, Stamp explored the interplay between photosynthesis, enzyme activities, and leaf morphological characteristics, examining their impact on growth and dry matter accumulation in maize. To address variations in physiological measurements, Stamp conducted a controlled greenhouse experiment (Stamp 1989). The findings revealed a robust correlation between nitrogen concentration in leaf tissue and the structural/physiological parameters of the photosynthetic apparatus. Favorable shoot development and growth were observed under medium and high nitrogen conditions, evident in a low root-to-shoot dry matter ratio, aligning with similar observations elsewhere. In a study by Cirilo et al. (2009), morphophysiological traits related to nitrogen capture, utilization, and partitioning, along with considerations of plant architecture and leaf senescence during the post-silking period, were identified as desirable for broad adaptation in maize. The research highlighted that high nitrogen partitioning to grains after silking correlated with strong hybrid performance under high nitrogen availability and warm climates. Conversely, superior grain yield performance under limited nitrogen was strongly associated with an efficient canopy sustaining resource capture until maturity. Root architecture significantly influences grain yield under both high and low nitrogen conditions (Coque and Gallais 2006), underscoring the pivotal role of root traits in nitrogen use efficiency (NUE).

Studies focusing on nitrogen-related root traits, such as uptake, metabolism, distribution, and storage, reveal an increasingly clear connection between NUE and these traits. The plant's ability to quickly develop its root system, rather than simply maintaining a larger root-to-shoot ratio over the lifecycle, can be beneficial, particularly in sandy soils with high leaching potential (Liao et al. 2006). Root length density, representing the root length per volume, is crucial for nutrient acquisition, as greater density increases the root surface area without a proportional increase in carbon allocation. Deep root systems offer access to a greater soil profile for enhanced water and nutrient absorption, but their effectiveness depends on soil conditions. Root hairs play a key role in nutrient uptake, particularly under nutrient stress conditions, as they increase the root surface area. While root diameter does not significantly affect nitrate or ammonium uptake capacity, the plasticity of the root system allows for considerable flexibility in responding to nutrient supply and demand. Root exudates, including proteolytic enzymes that digest proteins, can contribute to nitrogen capture, although their significance in high inorganic nitrogen availability remains uncertain. Efforts are also made to enhance NUE by increasing nitrate reductase (NR) activity, which has shown limited success. Despite attempts to overexpress NR for improved NUE, the impact has been minimal. *Alanine aminotransferase* (*AlaAT*) overexpression in roots has demonstrated greater success in increasing NUE, leading to higher yields in transgenic canola, particularly at low nitrogen levels (Good et al. 2007). This success highlights the potential of modifying downstream nitrogen assimilatory processes to enhance NUE. By storing nitrogen in forms not regulated by uptake processes, there is potential to increase uptake under conditions where it would typically be downregulated.

10.3.3 Traits for Energy Capture and Conversion

Photosynthesis, a vital physiological function of leaves, plays a crucial role in whole-plant energy capture. However, its efficiency depends not only on the photosynthetic response of individual leaves but also on their integration into an effective canopy, considering the costs of producing and maintaining their photosynthetic capacity. This intricate process is pivotal for providing green plants with the majority of their chemical energy, influencing their ability to compete and reproduce.

Light significantly influences plant photosynthetic capacities, especially with an abundance of far-red light in the canopy environment. This alters leaf dimensions, enhances radiation use efficiency, and affects leaf area, playing a pivotal role in radiation use efficiency (Yang et al. 2020). The angle of the leaf directly affects light interception, and stomata, primary channels for water and gas exchange on the leaf epidermis, respond uniquely to far-red light across various plant species. High far-red light can either increase or reduce stomata numbers, influencing stomatal conductance and density.

Studies emphasize the potential of supplementing with far-red light, particularly in a low red/far-red (R/Fr) ratio, to enhance plant biomass. Red, blue, and green light demonstrated no significant effect on lettuce, but far-red light supplementation significantly increased both dry and fresh weight. Tomatoes, soybean, and lettuce biomass experienced notable boosts under high far-red light conditions (Yang et al. 2020). Moreover, reducing the R/Fr ratio promotes the distribution of plant dry matter to the stem. The leaf, as the primary organ of photosynthesis, regulates light transmission through its mesophyll tissue, adapting its structure to optimize light reception.

Consistent findings from various studies highlight that the addition of far-red light increases the net photosynthetic rate and leaf area while reducing leaf thickness. This reduction is attributed to low R/Fr conditions, decreasing palisade tissue thickness by reducing cell size and number, leading to an overall reduction in leaf thickness. The interplay between light and leaf physiology underscores the complexity of optimizing photosynthetic efficiency in plants, considering their traits for energy capture and conversion (Zhang et al. 2019).

10.4 Adaptation and Resilience

Utilizing technological innovations for crop adaptation and resilience holds significant implications for sustaining productivity in the face of changing climatic conditions. This involves addressing the challenges posed by more variable rainfall and temperatures, including extreme weather events, compared to current conditions. When establishing selection criteria, it is crucial to take into account the distribution of pathogens, given the increased likelihood of major pest and disease outbreaks, as well as the emergence of novel pathogens (Langridge et al. 2021). Continuous breeding emerges as a crucial strategy to enhance the adaptive capacity of major crops in diverse environments (Fig. 10.2). It's important to acknowledge that selection processes have effectively eliminated minor negative effects, resulting in incremental gains through the fixation of major loci controlling adaptation in elite germplasm. Recent significant advancements in high-throughput phenotyping, crop modeling, data processing and analysis, and genome-wide analysis hold great promise for tailoring various traits to counterpart the demands of unstable environments.

10.4.1 Traits for Abiotic Stress Tolerance

Harnessing technological advancements for crop resilience against abiotic stresses (drought, heat, salinity, and cold) is crucial for global agriculture, particularly during the reproductive phase. Water scarcity affects cereals' flower initiation, and elevated temperatures induce sterility in cereals. Salinity challenges high-yield rice cultivation on salt-affected soils. Addressing nutrient deficiencies globally enhances crop yield potential and protects against various stresses. Research on genetics and molecular mechanisms of stress tolerance is pivotal. Analyzing genetic data reveals the complexity of phenotypic responses, requiring exploration at genetic and molecular levels. Identifying regulatory hubs and gene manipulation strategies is essential for accelerating crop genetic improvements.

Manipulating root architecture is a strategic approach to address stress conditions, including water and nutrient deficiencies. The DRO1 locus in rice, altering root architecture for vertical and deeper roots, enhances drought tolerance and nitrogen acquisition without penalizing rice growth (Arai-Sanoh et al. 2014). Cytokinin catabolic enzyme CKX overexpression leads to root-specific cytokinin degradation,

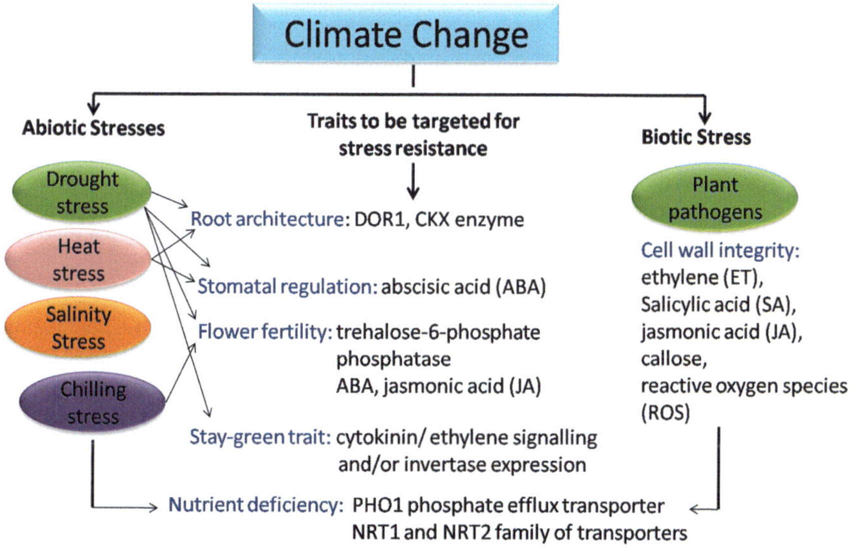

Fig. 10.2 Traits regulated under various stresses in response to climate change

resulting in higher root biomass, increased drought tolerance, and enhanced nutrient accumulation (Werner et al. 2010). Single gene manipulations can induce scalable phenotypic alterations relevant to plant performance under various abiotic stress conditions.

Stomatal regulation, crucial for plant response to abiotic stresses, exhibits varying patterns. Under heat stress, stomata open to cool leaves through transpiration, while under drought, stomata close to minimize water loss. When facing simultaneous drought and heat stress, stomata collectively remain closed, prioritizing the regulatory mechanisms induced by drought over heat stress. Engineering ABA receptors to accept non-ABA chemicals as agonists can contribute to field-scale manipulation of stomatal functions. Investigating the ABA signaling machinery elucidates its mechanism of operation, comprising SnRK2/OST1, PP2C, and PYR/PYL/RCAR proteins (Takahashi et al. 2018).

Maintaining flower fertility and sustained seed/fruit set is crucial for yield under stress. In maize, ear-specific expression of trehalose-6-phosphate phosphatase increases kernel set and yield under drought without growth penalties (Nuccio et al. 2015). Rice flower fertility under drought is influenced by stress adaptation hormones like ABA and JA. Anther-specific transgenic enhancement of ABA catabolism mitigates sterility under cold stress, emphasizing the critical role of spatiotemporal regulation in stress-adaptive pathways.

The stay-green trait, indicating senescence and resource allocation regulation, is crucial for yield under optimal and stress conditions. Natural allelic variants or transgenic alteration of cytokinin/ethylene signaling and invertase expression effectively manipulate this trait. Single gene manipulations, like senescence-induced expression of the IPT gene, increase yield performance under stress (Albacete et al. 2015).

10.4.2 Traits for Biotic Stress Resistance

Biotic stress encompasses plant pathogens like bacteria, fungi, viruses, nematodes, and insects, often causing physiological changes, biomass loss, altered flowering, reduced seed set, and the accumulation of protective metabolites. Pathogen resistance is often governed by single genes, requiring stability testing under diverse stress scenarios.

Preserving cell wall integrity is crucial for threat recognition and response. Mutations in cellulose synthaseA3 (CESA3) trigger lignin production, substituting missing cell wall components. Inhibiting cellulose biosynthesis induces stress responses, activating ethylene, salicylic acid, jasmonic acid, callose, and reactive oxygen species, and altering cell wall composition (Hamann 2012). These findings highlight the dynamic nature of plant defense mechanisms, underscoring the importance of intact cell walls in effectively combating environmental challenges.

10.5 Advancements in Trait Selection

10.5.1 High-Throughput Phenotyping (HTP)Technologies

Phenotyping of functional traits is essential in the improvement of crops with high productivity and adaptability. However, the process of phenotyping several plant traits is a tedious process that requires repeated multilocation trials under diverse environments for efficient trait selection. Additionally, identification and phenotyping of functional traits in plants pose a major obstacle in the field of plant breeding (Zhang et al. 2023). The recording of phenotypic traits is also labor-intensive and time-taking. The HTP technologies offer an automated, efficient, large-scale, and precise measurement of plant traits. Moreover, these technologies can screen traits at their initial stages eliminating the need for waiting up to their maturity stages to record the data. HTP technologies involve several imaging techniques. The frequently used imaging systems used in HTP with their description and functional traits studied using each technique are presented in Table 10.1.

10.5.2 Genomic Tools and Marker-Assisted Selection

10.5.2.1 Genomic Tools

The progress in genome breeding methods, particularly the development of high-throughput next-generation sequencing (NGS) tools and similar techniques, has transformed plant breeding systems. These technologies have helped in generating several tools that help deal with the complex traits in plants. The common and widely used genomic tools are discussed as follows:

(a) Whole Genome Sequencing (WGS): Whole genome sequencing is the complete sequencing of the entire genome. It provides comprehensive information on the genetic variation of an organism (Ahmetov et al. 2022). For WGS, two massive parallel sequencing platforms, viz., Roche Genome Sequencer FLX (GS-FLX) and Illumina Genome Analyzer (Illumina-GA), are widely used.
(b) Genotyping by Sequencing (GBS): With the progression in the NGS systems, excellent, low-cost, and high-throughput markers called single nucleotide polymorphism (SNP) markers were developed. GBS is an enzyme-based complexity reduction method that uses restriction endonucleases to target smaller portions of the genome that fairly represent the organism. It is coupled with DNA-barcoded adapters for producing multiplex libraries of a plant sample for sequencing.
(c) Transcriptome Sequencing: A transcriptome is the initial result of genome expression and represents a key link between DNA and the phenotype of the trait. Transcriptome sequencing enables researchers to characterize coding and noncoding regions responsible for the trait of interest at various growth stages. It focuses on a subset of relevant target genes and transcripts affecting the trait of study. This method can profile thousands of genes in one go and create an

Table 10.1 Commonly used high-throughput phenotyping techniques with respect to functional traits analyzed in plants

S. No.	HTP techniques	Description	Functional trait analyzed	References
1.	Visible light imaging	This method is based on capturing the digital images of plants. The sensors utilized in this imaging process are capable of detecting light within the wavelength range of 400–700 nm, yielding reflected values resultant to the red, green, and blue components	Drought stress	Campbell et al. (2020)
			Drought stress	Kim et al. (2020)
			Biomass yield	Jiang and Yang (2022)
			Plant growth, leaf water content, and water use measurement	Ge et al. (2016)
2.	Thermal imaging	This technique acquires temperature data from the surface of plants and allows the study of several complex traits. Thermal imaging cameras are quite affordable and robust	Canopy temperature	Deery et al. (2016)
			Vegetation stress	Sagan et al. (2019)
			Water stress tolerance	Banerjee et al. (2020)
3.	Hyperspectral imaging	It combines spectroscopy and imaging techniques. Hyperspectral cameras capture images in the wavelength between 350 and 2500 nm	Chlorophyll index	Capolupo et al. (2015)
			Leaf water content, macronutrient concentrations in leaves	Pandey et al. (2017)
			Photosynthetic plant traits	Suarez et al. (2021)
			Salt stress tolerance	Sytar et al. (2017)
			Drought stress	Asaari et al. (2022)
4.	Fluorescence imaging	Fluorescence refers to the emission of light by a plant upon absorbing radiation, specifically of shorter wavelength	Dynamic photosynthesis	McAusland et al. (2019)
			Biotic stress	Pérez-Bueno et al. (2019)
			Drought stress	Yao et al. (2018)
			Leaf chlorophyll content	Zhang et al. (2022)

5.	Tomographic imaging	The procedure is conducted using Magnetic Resonance Imaging (MRI), Positron Emission Tomography (PET), or X-ray Computed Tomography (X-ray CT). MRI employs nuclear magnetic resonance techniques to produce images	Photoassimilate transport dynamics (PET)	Karve et al. (2015)
			Morphological and anatomical stem properties (X-ray CT)	Gomez et al. (2018)
			Root traits (MRI)	Pflugfelder et al. (2017)
			Grain characteristics (X-ray CT)	Hu et al. (2020)
			Microstructure of plant leaves (X-ray CT)	Mathers et al. (2018)

overall picture of cell function. It uses a NGS platform to know the presence and quantity of desired RNA in a biological sample and to analyze continuously changing cellular transcriptome (Anamika et al. 2016).

(d) TILLING and EcoTILLING: TILLING (Targeting Induced Local Lesions in Genomes) represents a non-transgenic reverse genetics strategy. This method requires prior knowledge of the DNA sequence and utilizes a mismatch endonuclease to identify induced mutations. It can identify various mutation types in a gene, such as missense, nonsense, silent, and splice-site mutations. On the other hand, EcoTILLING, a derivative of TILLING, investigates natural genetic variations in populations to identify SNPs, including rare ones. This approach is both practical and efficient for functional genomic studies related to various plant traits.

(e) High-Density Genetic Maps: These are powerful tools that allow identification of genomic regions responsible for functional and complex traits in plants. These maps are widely used in comparative genomic studies, QTL mapping, and MAS, and also provide important tools for analyzing the genome structure and studies related to evolution in plants (Gong et al. 2022).

(f) Genome-Wide Association Studies (GWAS): GWAS is a powerful and ubiquitous tool that helps in dissecting the genetic variations associated with complex traits. Besides its usefulness in identifying genes or loci affecting several traits from diverse natural populations, it reveals the genetic architecture underlying these complex traits.

10.5.2.2 Marker-Assisted Selection (MAS)

MAS is the indirect selection of such genomic regions in plants that are associated with the expression of traits of interest using molecular markers. MAS is widely used for manipulating qualitative traits. However, in the case of complex or polygenic traits, this method is quite complicated. The MAS technique is being routinely used to transfer several traits (Table 10.2). Several crop varieties for commercially important traits have been successfully developed through MAS and released for cultivation.

10.5.3 Trait Stacking and Multi-Trait Selection

The amalgamation of various traits or genes within plants is referred to as gene or trait stacking. This strategy proves advantageous for plant breeders who seek to incorporate more than one trait into crops. In transgenic plants, it is possible to combine different genes or traits in a single plant, either in a single step or through several steps. To increase yields, developing crops with multiple traits affecting yield and other commercially important traits is required. The stacking of traits increases crop performance and thereby results in the expression of multiple genes in a single product (Ainley et al. 2013). The transgenic technology has enabled a rapid increase in the production of crops with stacked traits. Stacked trait crops like herbicide-tolerant and insect-resistant cotton and maize are developed. These

Table 10.2 Some functional traits transferred by MAS

S. No.	Trait transferred using MAS	References
1	Bacterial blight and blast of rice	Mehta et al. (2019)
2	Leaf curl virus, tobacco mosaic virus, bacterial wilt, late blight, and gray leaf spot in tomato	Hanson et al. (2016)
3	Resistance to fusarium wilt in chickpea	Bharadwaj et al. (2022)
6	Drought tolerance in rice	Singh et al. (2022)
7	Drought tolerance in pearl millet	Rani et al. (2022)
8	Heat tolerance in rice	Ye et al. (2022)
9	Heat tolerance in wheat	Bellundagi et al. (2022)

transgenic crops have immense potential to revolutionize agriculture. Plant breeders often select multiple traits like plant height, grain size, early maturity, drought tolerance, disease resistance, etc. Few examples of functional trait stacking in plants are presented in Table 10.3.

A common approach used for this multi-trait selection is termed index selection. In this selection method, the approach is to assign different weights to individual traits based on their relative economic importance. In the present era, most crop improvement programs target at the selection of multiple traits (Beavis et al. 2023). The types of selections for multi-traits are presented in Table 10.4.

It is well evident that the efficiency of selection would be much higher if multiple traits were selected during plant breeding programs. However, the identification of these genotypes that efficiently combine with the high-performing traits poses a challenge for the breeders. Nevertheless, novel techniques like multi-trait genotype-ideotype distance index (MGIDI) for genotype selection are efficient for successful multi-trait selections in plants (Olivoto and Nardino 2020).

10.6 Precision Agriculture and Trait Integration

10.6.1 Precision Agriculture: Concepts and Applications

The utilization of precision agriculture started around the middle of the 1980s. It is a revolutionary technique that can enhance agricultural production and is capable of optimizing management practices and inputs, thereby preserving the environment (Rabia et al. 2022). Precision agriculture is defined as a farm management system based on information and technology for identification, analysis, and management of variability within fields for optimization of the land resources in terms of sustainability, profitability, and protection. In this technique, novel information technologies can be used in order to make better decisions in crop fields. Precision agriculture increases the efficiency of crop production by dealing with the natural variability within a field (Banu 2015). The benefits of precision agriculture can be achieved through the reduction in water usage, herbicides, fertilizers, and pesticides. The concept relies on the management of the entire field based on the recognition of site-specific differences within fields and subsequent management of actions to be undertaken accordingly. Precision farming is considered as an agricultural revolution by scientists and researchers involved in maximizing benefits from crop plants (Fig. 10.3).

10.6.2 Integration of Functional Traits into Precision Agriculture

10.6.2.1 Nutrient Use Efficiency

Advanced technologies like geographic information systems, global positioning systems, sophisticated software, and the utilization of sensors and specialized equipment in precision agriculture have elevated the application of this technology

Table 10.3 Trait stacking of functional traits in plants

S. No.	Traits stacked	References
1.	Blast resistance, gall midge resistance, bacterial blight resistance, submergence and salinity tolerance in rice	Das and Rao (2015)
2.	Three resistant genes in potato against *Phytophthora infestans*	Zhu et al. (2012)
3.	Drought tolerance and salinity tolerance in rice	Shailani et al. (2021)
4.	Weed-, insect-, and drought-tolerant maize	Malenica et al. (2021)
5.	Herbicide- and insecticide-resistant soybean	Zanatta et al. (2020)
6.	Brown streak disease resistance, mosaic disease resistance, and enhanced iron and zinc in cassava	Narayanan et al. (2021)

Table 10.4 Types of selections involved in multi-trait selection

S. No.	Type of selection	Description
1	Tandem selection	This selection method involves selection for a single trait until that particular trait is improved, followed by a second trait, and this criterion is followed until the trait has been improved to a desirable level
2	Independent culling	In this method of selection for each trait, a certain level of merit is established. All the individuals falling below the selected merit level are discarded
3	Index selection	This criterion of selection involves the selection of all traits simultaneously by utilization of an index of net merit
4	Multistage selection	In this type of selection, different traits at several stages are selected during the development of the crop

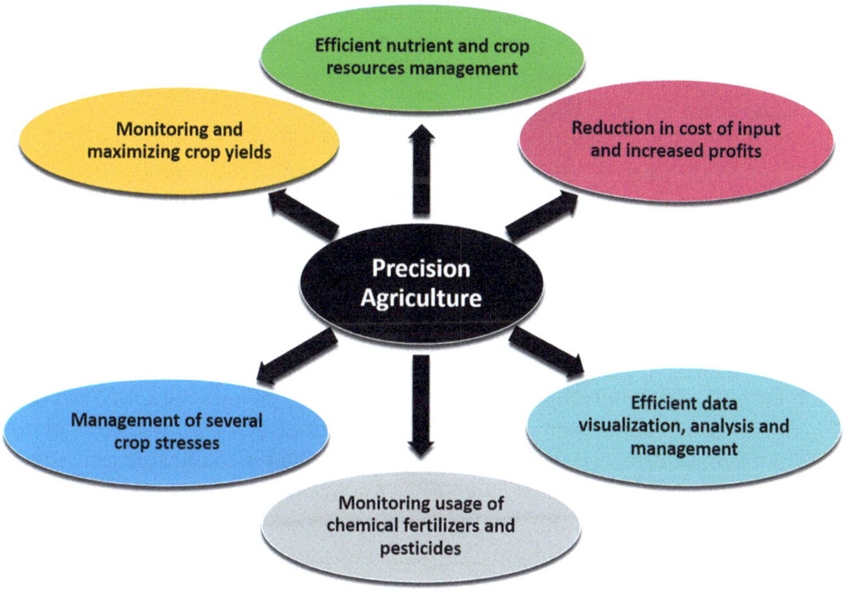

Fig. 10.3 Benefits of precision agriculture

to enhance nutrient use efficiency. Precision agriculture matches the fertilizer inputs to the site-specific field conditions by accurate measurement of crop nutrient status, spatial variability of soil, and their relation to the crop responses (Hedley 2015).

10.6.2.2 Disease Management

Plant diseases are major challenges to food security globally, resulting in heavy loss of crop yields, thereby impacting the affected regions socioeconomically. The integration of precision agriculture for managing diseases in fields has the potential to reduce the usage of pesticides by their application in the amount as much they are needed. This will reduce the adverse effects on the environment and reduce the excessive use of chemicals in fields. Utilizing sensors for disease detection and implementing precision agriculture methods for targeted pesticide application are effective means of integrated disease management (Singh et al. 2023).

10.6.2.3 Drought Stress Management

Around 80–95% of plant biomass consists of water, playing a vital role in essential plant processes such as growth, development, and metabolism. Consequently, drought is recognized as a significant environmental stress, particularly in regions prone to drought. The emergence of certain high-throughput techniques integrated into precision agriculture such as digital image processing, computer vision, and spectral sensors has led to nondestructive, rapid, feasible, accurate, and automated processing of massive data analysis for drought-related traits (Kim et al. 2023).

10.6.2.4 Estimation of Plant Biomasses

The development of advanced technology facilities for HTP has brought attention to the growing significance of estimating the biomass of individual plants from their two-dimensional images. The conventional approaches to biomass detection typically rely on measurements that involve destructive methods. Therefore, accurate, rapid, and nondestructive methods for monitoring the biomass make a way toward smart and precision agriculture techniques. According to researchers, the combination of RGB data and the application of advanced algorithms techniques are capable of obtaining higher accuracy in data that too at a lower cost (Wang et al. 2021).

10.6.2.5 Predicting Crop Yields

Maximizing the crop yield with low inputs and investment remains an ultimate goal of the farming community for increasing the efficiency of agricultural production. Moreover, researchers utilizing precision agriculture involved in research can utilize a hierarchical categorization of specific temporal and spatial that in turn affects these crop yields. The traditional approaches for crop yield prediction include field surveys that are time-consuming along with the use of several types of expensive sensors. However, advancements in electronic technologies and other efficient sensing systems have proved as excellent technologies for predicting yields in fields. Precision agriculture techniques can administer and monitor the soil and crop health, thereby helping in accessing crop productivity. Additionally, several specialized models utilized in precision agriculture are capable of accurately predicting these crop yields (Mythili and Rangaraj 2021).

10.6.3 Case Studies of Precision Agriculture with Functional Traits

Research show that precision agriculture has often shown partial efficiency as promised as these technologies are mostly focused on increasing resolutions of spatial variations yields and soils but lower efforts toward incorporating physiological principles of the crop responses to environmental variations. Toward this direction, Monzon et al. (2018) followed the whole-farm precision agriculture approach focused on physiological processes to bridge gaps in yields, increase profits, and reduce risk to farmers in the dryland farms in the region of southern Pampas. Crops such as wheat and barley, maize, soybean, and sunflower were mainly grown in these fields. Different conditions with respect to zones considering depth of soils, water table, and frost risks were considered. An in-depth analysis of risk assessment and improvements in terms of farm profit and outputs was performed. Following the research, in comparison with neighboring farms, San Lorenzo produced 46% higher gross margins and 54% higher farm outputs, with a higher net income. Likewise, Bucci et al. (2020) conducted a study in Italy where they assessed the economic viability of a farm through the application of precision agriculture technologies. The results demonstrated that the adoption of these technologies led to a substantial increase in wheat yield (both soft and durum) and a noteworthy reduction in production costs.

The sustainable improvement of agricultural efficiency is essential for supporting a continuously growing global population amid limited resources. To specifically tackle challenges like water scarcity and variations in crop water status across fields, precision irrigation methods depend on precise data regarding crop water status, soil moisture levels, and weather conditions. Several studies have shown that the canopy temperature parameters in plants can access indicators influencing plant water stress, thereby proving it as an efficient tool for irrigation water management. Nevertheless, several criteria such as cost-effective and efficient measurement techniques are still a constraint. In this context, Martínez et al. (2017) evaluated the effectiveness of thermal imaging and infrared thermometry in monitoring water stress levels in sugar beet crops. They carried out multiple experiments with the aim of creating a cost-effective infrared temperature system suitable for installation on an unmanned aerial vehicle (UAV) to acquire thermal data. The study also emphasized the comparison of canopy temperature measurements collected using low-cost platforms using conventional thermal cameras. Additionally, the investigation delved into the components of a cost-effective system designed to derive temperature-based water stress indices. To accomplish these aims, plots subjected to water deficit and plots with sufficient irrigation were established. Their results proved that lightweight canopy temperature systems were reliable and efficient. However, weather conditions and area covered by the infrared sensors were major constraints during the study. On the other hand, Costa et al. (2022) focused on the application of UAV-based hyperspectral imaging to forecast grain yields and associated traits in wheat under heat stress conditions with the aim of stabilizing genotypes. Preliminary findings suggest that the models for each studied trait exhibited satisfactory accuracy.

Although precision agriculture has been utilized since times, novel techniques and their efficient use in maximizing the efficiency of plant functional traits have recently been practiced. However, with respect to the current scenarios, utilization of novel and automated precision agriculture techniques and models is capable of maximizing yields through manipulating these plant functional traits.

10.7 Challenges in Integrating Plant Functional Traits

Plant functional traits play a pivotal role in crop breeding, facilitating the creation of crop varieties that can better withstand changing environmental conditions and are capable of meeting the needs of a growing global population. However, there are several challenges associated with incorporating plant functional traits into crop breeding programs:

10.7.1 Trait Complexity

Identifying and manipulating the genetic components responsible for various functional traits pose a challenge due to the influence of multiple genes and environmental factors on these traits. Crop yield is a complex trait governed by the interactions of multiple genes, each contributing to various components like grain number, biomass, grain weight, etc. Similarly, drought tolerance involves the coordination of genes controlling water use efficiency, osmotic adjustment, and root architecture. Further, the root traits are again controlled by several genes with some extent of interaction and environmental influence (Comas et al. 2013). The quantitative nature of polygenic traits makes them complex, and difficult to analyze, control, or predict with accuracy. They exhibit a continuous range of phenotypic expression and are extremely sensitive to environmental fluctuations making it difficult to discriminate and distinguish individuals for the desired traits.

10.7.2 Constraints in Phenotyping

Limited use of functional trait analysis in developing improved crop lines is largely due to technical challenges in phenotyping plant traits (Ghanem et al. 2015). Although declining sequencing costs provide abundant genomic data, linking it to phenotypes for agricultural applications remains a challenge. A crucial approach involves large-scale, high-throughput screening for candidate germplasm carrying genes controlling important traits, complemented by detailed examination of subtle phenotypes in smaller subsets. Challenges include laborious and inconsistent measurement of functional traits, as well as the need for replicated trials in diverse environments. Automation and high-throughput methods like spectroscopy, noninvasive imaging, robotics, and high-performance computing can overcome these bottlenecks in phenotyping.

10.7.3 Trait-Environment Interactions

The trait-environment interactions pose significant hurdles in the integration of plant functional traits into breeding programs. Breeding for specific traits may lack consistency across regions and changing climates due to the dynamic interaction between genotype and environment. Therefore, the breeding programs are implemented in genotype-environment systems that are dynamic and continually shifting, presenting an ever-evolving target for plant breeders.

10.7.4 Trade-Offs Between Traits

Enhancing one functional trait may come at the expense of another reflecting divergent selection pressures. Correlations between traits can pose challenges in the independent improvement of each trait, and any change in one trait existing within complex networks may result in affecting multiple others (Poorter et al. 2009). Further, trade-offs may exhibit environmental and temporal sensitivity wherein the balance between traits changes with time and environmental conditions. Breeding programs, hence, need to strike a balance between different traits to achieve optimal crop performance.

10.7.5 Genetic Diversity and Adaptability

Maintaining productivity while enhancing genetic diversity poses a challenge in modern crop breeding. Emphasis on a few high-yielding traits results in genetic uniformity, reducing diversity and narrowing the genetic base. This focus makes crops less adaptable and vulnerable to diseases, pests, and environmental changes. Traits for adaptation to diverse environments may conflict with those optimizing yield under specific conditions. The introduction of new genetic material to improve traits can pose challenges to maintaining the existing diversity. Further, breeders need to ensure that the introduced traits for enhancing diversity must be adaptive to the local environmental conditions for sustaining productivity (Kahane et al. 2013).

10.7.6 Ethical and Societal Considerations

Integrating plant functional traits into breeding programs raises ethical and societal considerations. Potential ecological impacts on local ecosystems and biodiversity, coupled with uncertainties in technologies like genome editing, demand attention. The precautionary principle, advocating caution in the face of scientific uncertainty, guides ethical and regulatory measures. Ensuring fair access to genetic resources for all stakeholders, especially farmers and local communities, is crucial. Breeders and scientists must provide transparent information on integrated traits and address concerns related to food safety. Long-term health impacts of trait-integrated crops should be assessed (Simberloff et al. 2013). It is essential to address these concerns and ensure responsible crop breeding practices.

10.7.7 Regulatory Hurdles

Developing crops with modified traits faces complex regulatory challenges, notably with consumer acceptance of genetically modified organisms (GMOs). The use of CRISPR/Cas9 generates diverse risks and benefits, requiring extensive interdisciplinary discussions involving politicians, industry, the public, and academics. Regulatory standards vary across countries, creating trade barriers and necessitating compliance with multiple frameworks. Uncertainty in defining GMOs complicates regulatory oversight for breeding technologies (Sprink et al. 2016). Post-release monitoring for agronomic and environmental performance is essential. Enforcing effective liability and redress mechanisms for unforeseen negative impacts is crucial. A clear understanding of regulatory environments, stakeholder involvement, and efforts to harmonize international standards are vital for navigating regulatory obstacles.

10.7.8 Data Analysis and Integration

Effective use of functional trait data in breeding requires advanced data integration and analysis. Standardizing data across platforms is crucial, including integration of omics data. Another limitation is in the handling of missing data which is common in large-scale phenotyping studies which require robust strategies (Little and Rubin 2019). Experimental approaches to studying the effect of a trait on crop yield can be expensive and sometimes even impossible in multisite, multi-season trials. Overcoming these issues requires a combination of advanced statistical methods, bioinformatics, machine learning, modeling, and interdisciplinary collaborations for identifying trait-gene associations and predicting crop performance.

10.7.9 Long Breeding Cycles

Long breeding cycles, often spanning several years, hinder the efficiency and effectiveness of trait improvement in crop breeding (Lobell and Gourdji 2012). This delay results in a loss of time and resources and hampers adaptability to rapidly changing environmental conditions. For traits involving resistance mechanisms, extended breeding cycles increase vulnerability to emerging pests and diseases. Additionally, cultivars may lose market relevance as they struggle to align with changing demands and preferences. Speeding up the breeding process while ensuring trait improvements requires a strategic approach integrating advanced breeding technologies, efficient phenotyping, and collaborative efforts to expedite the development and delivery of enhanced cultivars to farmers.

10.7.10 Accessibility to Advanced Technologies

Many breeding programs may lack access to state-of-the-art equipment and technologies, especially in developing regions, thus posing a major bottleneck in trait measurement and genetic analysis (Lybbert and Sumner 2012). Access to proprietary technologies is restricted by intellectual property laws, thus limiting their use in breeding programs. Ethical and regulatory issues may also create barriers in the use of modern technology. Further, access to shared resources and expertise can be seriously restricted in the face of limited collaborations and networking. Coordinated efforts in infrastructure development, organizing training programs, formation of supportive policies, and collaborations can ensure equitable access to breeding for efficient trait incorporation. Despite these challenges, the integration of plant functional traits in crop breeding holds great promise for developing more resilient, productive, and sustainable crop varieties. Collaboration among scientists, breeders, and policymakers is crucial to overcome these obstacles and align crop breeding with global population demands and changing environments.

10.8 Crop Varieties and Sustainability: Traits and Impact

10.8.1 Notable Crop Varieties Developed with Functional Traits

Plant breeding based on functional traits has been followed from a long time, utilizing conventional and molecular approaches (Akhtar et al. 2023). Phenotypic selection was the conventional method before the advent of molecular breeding. Molecular techniques, such as QTL mapping and GWAS, have identified genes for specific traits, enabling the development of functional markers for faster and cost-effective varietal improvement. Notable examples include improving the Indian rice variety Swarna through marker-assisted backcross breeding (MABB), incorporating Sub-1 from an exotic variety. Table 10.5 provides detailed information on crop varieties developed using different breeding approaches.

10.8.2 Impact of Functional Traits on Crop Productivity and Sustainability

Plant functional trait significantly contributes to crop productivity. The constraint of macronutrient nitrogen has a great impact on most of the world's agriculture areas. Identification of traits and mapping of quantitative trait loci have revealed NUE's established significant association with the productivity of crops particularly in maize, rice, and wheat (Sandhu et al. 2021). The association between photosynthetic gain and yield is well established, and improvement in the rate of photosynthesis using genetic enhancement has significantly improved the yield of crop plants (Theeuwen et al. 2022). The stay-green trait also imparts significantly toward net photosynthetic gain; moreover, identification of the new genetic loci and genes involved in stay-green characters exhibits great potential to be further utilized for improvement in photosynthetic gain. Leaf length is another parameter contributing significantly toward increased photosynthesis. Plant height-associated QTLs in soybean revealed genes underlying this complex mechanism and its potential impact on yield (Yu et al. 2022). The stem bark is an integral part of photosynthesis transport and storage, and genomic regions associated with it have been investigated using genomic mapping technique. The negative impact of lignin present in stems and barks of trees on postharvest process application is well known. Genome-wide association studies in sorghum have facilitated the identification of regions associated with lignin formation (Niu et al. 2020) which can be further eliminated using different breeding strategies. Being a crucial parameter impacting the yield of crop plants, the plant root-related traits are also important. Root proliferation to different

Table 10.5 Notable plant varieties improved using functional traits

S. No.	Crops	Variety	Functional trait	Effect of trait	Method
1.	Rice	Swarna Sub-1	Plant height	Flooding tolerance	MABB
		SambhaMahsuri Sub-1	Plant height	Flooding tolerance	MABB
		DRR Dhan 43	Root characteristics	Drought tolerance	MABB
		CR DHAN 801	Root characteristics	Drought tolerance	MABB
2.	Wheat	HI 8802	Root characteristics	Drought tolerance	Modified pedigree method
		PBW396	Root characteristics	Drought tolerance	Conventional breeding
3.	Maize	HM-5	Plant height	Flooding tolerance	Conventional breeding
		PMH-2	Plant height	Flooding tolerance	Conventional breeding
4.	Sugarcane	Co 98014 (Karan-1)	Plant height	Flooding tolerance	Conventional breeding
5.	Pearl millet	CZP-9802	Root characteristics	Drought tolerance	Heterosis
6.	Barley	RD 2660	Root characteristics	Drought tolerance	Conventional breeding
7.	Soybean	NRC 7	Root characteristics	Drought tolerance	Conventional breeding

environmental as well as physiological stimuli is of great potential for sustainable crop yield. Another trait, root growth angle, has increased the yield in rice plants by improving lodging resistance and enhancing nutrient uptake from soil (Zhu et al. 2022).

The role of plant functional traits toward stress tolerance and thereby maintaining significant yield is well established. In the case of drought conditions, plants with improved root architecture exhibited good response and without significant yield loss. Leaf rolling is another mechanism to avoid drought stress in plants, and genomic regions associated with it are successfully mapped (Verma et al. 2020). Stomatal conductance mediates salt-related stress in barley (Liu et al. 2017). The plant-microbe association is significantly explained using plant functional traits. Genomic-assisted disease resistance breeding has facilitated the understanding of complex genomes like brassica (Amas et al. 2023). Fungal disease advancement in genomics has shaped the research in the understanding of functional genomics traits.

10.9 Emerging Trends in Functional Trait-Based Crop Breeding and Its Potential

Genomics tools such as quantitative trait loci (QTL) mapping and genome-wide association mapping are important aspects in the process of identification of gene (s) or loci associated with particular traits. Next-generation sequencing methodologies, advances in data analytics, and generation of precise mutations using genome editing tools like CRISPR/Cas9 have enhanced the accuracy as well as fastened the process of characterizing different traits. Analysis of functional traits using genomic mapping is a well-established method for characterizing different functional traits. Moreover, candidate gene identification using genome-wide association mapping has been promising in case of functional traits such as nitrogen use efficiency, chlorophyll content, root characteristics, and seedling biomass (Xiong et al. 2023). Advanced sequencing technologies have enabled researchers to understand complex regulation of various genes in diverse environment or physiological conditions. Whole genome RNA-seq analysis has enlightened the difference in two drought-adapted wheat varieties. The mRNA-Sequencing-based analysis revealed the presence of 2406 DEGs associated with nitrogen use efficiency in wheat (Kaur et al. 2022). Transcriptome analysis of primary roots in rice revealed 1243 DEGs and further explained the role of flavonoids in development of roots in rice.

Application of bacterial CRISPR/Cas9 system in plant has revolutionized the research in plant domain. The potential CRISPR/Cas9 system has been utilized for enhancing photosynthesis in rice by mutating light-harvesting complex genes and thereby increasing photosynthesis per photon absorbed (Caddell et al. 2023). Similarly, knocking out of hexokinase gene possessed positive impact on enhancing photosynthesis efficiency in rice. Mutation of HvARE1 gene in barley using CRISPR/Cas9 demonstrated the potential of same gene in nitrogen use efficiency. The application of CRISPR/Cas9 is also well documented in case of drought tolerance in various crop plants. Leaf rolling is another functional trait efficiently managed by this bacterial editor, thereby impacting the photosynthesis (Han et al. 2023). This impactful technique can further be utilized for improving various functional traits and thereby generating sustainable plants for diverse ecosystem.

Plant functional traits are essential aspects for measuring ecosystem services which helps to determine performance of plants in a given ecosystem. Advancement in molecular biology techniques has facilitated the identification, validation, and transfer process of these important plant functional traits that can significantly uplift the capability of plant to live in a certain ecosystem.

10.10 Conclusions

Plant functional traits stand as the cornerstone of agricultural progress, providing key insights and avenues for shaping the future of crop cultivation. In the context of our earlier discussion, the exploration of plant functional traits underscores their pivotal role in determining a plant's adaptability and overall performance. The integration of these traits into breeding programs, coupled with advanced technologies like HTP and genomic selection, opens unprecedented avenues for precision breeding, accelerating the development of crops with desired characteristics. Cutting-edge technologies, such as CRISPR/Cas9 gene editing, precision breeding, and omics approaches, offer unprecedented precision and efficiency in trait manipulation. These tools enable scientists to delve into the intricate molecular mechanisms governing plant responses to environmental stressors, leading to the development of crops with enhanced adaptability and performance. For example, the mutation of the *HvARE1* gene in barley using CRISPR/Cas9 has been shown to improve nitrogen use efficiency. MAS has played a key role in speeding up the breeding cycle, and various cultivars with improved plant functional traits, such as the flooding-tolerant rice cultivar "Swarna Sub-1," "SambhaMahsuri," and the drought-tolerant wheat cultivars "PBW396" and "HI 8802," have been developed.

Furthermore, the acknowledgment of challenges inherent in integrating plant functional traits reveals the intricate landscape that characterizes trait manipulation. These challenges, including trait complexity, constraints in phenotyping, trait-environment interactions, trade-offs between traits, genetic diversity and adaptability, and accessibility to advanced technologies, underscore the necessity for a collaborative and interdisciplinary approach. This recognition serves as a pivotal factor in refining breeding strategies and advancing our understanding of the intricate relationships between genetics, environment, and functional traits. As we chart the path forward, sustainability, resilience, and innovation emerge as guiding principles. The integration of traditional breeding methods with advanced technologies presents a synergistic approach that maximizes the benefits of both. This approach not only accelerates the breeding process but also ensures that the resulting crops exhibit a balance between productivity and environmental responsibility.

References

Ahmetov II, Hall EC, Semenova EA, Pranckevičienė E, Ginevičienė V (2022) Advances in sports genomics. Adv Clin Chem 107:215–263. https://doi.org/10.1016/bs.acc.2021.07.004

Ainley WM, Sastry-Dent L, Welter ME, Murray MG, Zeitler B, Amora R, Corbin DR, Miles RR, Arnold NL, Strange TL, Simpson MA (2013) Trait stacking via targeted genome editing. Plant Biotechnol J 11(9):1126–1134. https://doi.org/10.1111/pbi.12107

Akhtar S, Rao E, Uike A, Saatu M (2023) Plant breeding strategies: traditional and modern approaches. In: Jhangek DK et al (eds) Genetic revolution in agriculture: unleashing the power of plant genetics. Elite Publishing House, New Delhi

Albacete A, Cantero-Navarro E, Großkinsky DK, Arias CL, Balibrea ME, Bru R, Fragner L, Ghanem ME, González MDLC, Hernández JA, Martínez-Andújar C (2015) Ectopic overexpression of the cell wall invertase gene CIN1 leads to dehydration avoidance in tomato. J Exp Bot 66(3):863–878

Amas JC, Thomas WJ, Zhang Y, Edwards D, Batley J (2023) Key advances in the new era of genomics-assisted disease resistance improvement of brassica species. Phytopathology 113(5):771–785. https://doi.org/10.1094/PHYTO-08-22-0289-FI

Anamika K, Verma S, Jere A, Desai A (2016) Transcriptomic profiling using next generation sequencing-advances, advantages, and challenges. In: Next generation sequencing-advances, applications and challenges, vol 9. IntechOpen, London, pp 7355–7365

Arai-Sanoh Y, Takai T, Yoshinaga S, Nakano H, Kojima M, Sakakibara H, Kondo M, Uga Y (2014) Deep rooting conferred by DEEPER ROOTING 1 enhances rice yield in paddy fields. Sci Rep 4(1):5563

Asaari MSM, Mertens S, Verbraeken L, Dhondt S, Inzé D, Bikram K, Scheunders P (2022) Non-destructive analysis of plant physiological traits using hyperspectral imaging: a case study on drought stress. Comput Electron Agric 195:106806

Banerjee K, Krishnan P, Das B (2020) Thermal imaging and multivariate techniques for characterizing and screening wheat genotypes under water stress condition. Ecol Indic 119:106829

Banu S (2015) Precision agriculture: tomorrow's technology for today's farmer. J Food Process Technol 6(8):1. https://doi.org/10.4172/2157-7110.1000468

Beavis W, Lamkey K, Mahama AA (2023) Multiple trait selection. In: Quantitative genetics for plant breeding. Iowa State University Digital Press, Ames

Bellundagi A, Ramya KT, Krishna H, Jain N, Shashikumara P, Singh PK, Singh GP, Prabhu KV (2022) Marker-assisted backcross breeding for heat tolerance in bread wheat (*Triticumaestivum* L.). Front Genet 13:1056783. https://doi.org/10.3389/fgene.2022.1056783

Bharadwaj C, Jorben J, Rao A, Roorkiwal M, Patil BS, Ahammed SK, Saxena DR, Yasin M, Jahagirdar JE, Sontakke PL, Pithia MS (2022) Development of high yielding Fusarium wilt resistant cultivar by pyramiding of "genes" through marker-assisted backcrossing in chickpea (*Cicer arietinum* L.). Front Genet 13:1747. https://doi.org/10.3389/fgene.2022.924287

Bucci G, Bentivoglio D, Belletti M, Finco A (2020) Measuring a farm's profitability after adopting precision agriculture technologies: a case study from Italy. Acta IMEKO 9(3):65–74. https://doi.org/10.21014/acta_imeko.v9i3.799

Büssis D, von Groll U, Fisahn J, Altmann T (2006) Stomatal aperture can compensate altered stomatal density in Arabidopsis thaliana at growth light conditions. Funct Plant Biol 33(11):1037–1043

Caddell D, Langenfeld NJ, Eckels MJ, Zhen S, Klaras R, Mishra L, Bugbee B, Coleman-Derr D (2023) Photosynthesis in rice is increased by CRISPR/Cas9-mediated transformation of two truncated light-harvesting antenna. Front Plant Sci 14:1050483. https://doi.org/10.3389/fpls.2023.1050483

Campbell MT, Grondin A, Walia H, Morota G (2020) Leveraging genome-enabled growth models to study shoot growth responses to water deficit in rice. J Exp Bot 71(18):5669–5679. https://doi.org/10.1093/jxb/eraa280

Capolupo A, Kooistra L, Berendonk C, Boccia L, Suomalainen J (2015) Estimating plant traits of grasslands from UAV-acquired hyperspectral images: a comparison of statistical approaches. ISPRS Int J Geo Inf 4(4):2792–2820

Chen K, Wang Y, Zhang R, Zhang H, Gao C (2019) CRISPR/Cas genome editing and precision plant breeding in agriculture. Annu Rev Plant Biol 70:667–697. https://doi.org/10.1146/annurev-arplant-050718-100049

Cirilo AG, Dardanelli J, Balzarini M, Andrade FH, Cantarero M, Luque S, Pedrol HM (2009) Morpho-physiological traits associated with maize crop adaptations to environments differing in nitrogen availability. Field Crop Res 113(2):116–124

Comas LH, Becker SR, Cruz VMV, Byrne PF, Dierig DA (2013) Root traits contributing to plant productivity under drought. Front Plant Sci 4:442

Coque M, Gallais A (2006) Genomic regions involved in response to grain yield selection at high and low nitrogen fertilization in maize. Theor Appl Genet 112(7):1205–1220

Costa L, McBreen J, Ampatzidis Y, Guo J, Gahrooei MR, Babar MA (2022) Using UAV-based hyperspectral imaging and functional regression to assist in predicting grain yield and related traits in wheat under heat-related stress environments for the purpose of stable yielding genotypes. Precision Agric 23:622–642. https://doi.org/10.1007/s11119-021-09852-5

Das G, Rao GJN (2015) Molecular marker assisted gene stacking for biotic and abiotic stress resistance genes in an elite rice cultivar. Front Plant Sci 6:698. https://doi.org/10.3389/fpls.2015.00698

Deery DM, Rebetzke GJ, Jimenez-Berni JA, James RA, Condon AG, Bovill WD, Hutchinson P, Scarrow J, Davy R, Furbank RT (2016) Methodology for high-throughput field phenotyping of canopy temperature using airborne thermography. Front Plant Sci 7:1808. https://doi.org/10.3389/fpls.2016.01808

Franks PJ, Doheny-Adams W, Britton-Harper ZJ, Gray JE (2015) Increasing water-use efficiency directly through genetic manipulation of stomatal density. New Phytol 207(1):188–195

Ge Y, Bai G, Stoerger V, Schnable JC (2016) Temporal dynamics of maize plant growth, water use, and leaf water content using automated high throughput RGB and hyperspectral imaging. Comput Electron Agric 127:625–632. https://doi.org/10.1016/j.compag.2016.07.028

Ghanem ME, Marrou H, Sinclair TR (2015) Physiological phenotyping of plants for crop improvement. Trends Plant Sci 20(3):139–144

Gomez FE, Carvalho G, Shi F, Muliana AH, Rooney WL (2018) High throughput phenotyping of morpho-anatomical stem properties using X-ray computed tomography in sorghum. Plant Methods 14:1–13. https://doi.org/10.1186/s13007-018-0326-3

Gong W, Ma L, Gao Q, Wei B, Zhang J, Liu X, Gong P, Wang Z, Zhao G (2022) Construction of a high-density genetic linkage map and identification of flowering-related QTL in erect milkvetch (Astragalusadsurgens). Crop J 10(4):1141–1150. https://doi.org/10.1016/j.cj.2022.01.008

Gonzalez Guzman M, Cellini F, Fotopoulos V, Balestrini R, Arbona V (2022) New approaches to improve crop tolerance to biotic and abiotic stresses. Physiol Plant 174(1):e13547. https://doi.org/10.1111/ppl.13547

Good AG, Johnson SJ, De Pauw M, Carroll RT, Savidov N, Vidmar J, Lu Z, Taylor G, Stroeher V (2007) Engineering nitrogen use efficiency with alanine aminotransferase. Botany 85(3):252–262

Hamann T (2012) Plant cell wall integrity maintenance as an essential component of biotic stress response mechanisms. Front Plant Sci 3:77

Han Y, Yang J, Wu H, Liu F, Qin B, Li R (2023) Improving rice leaf shape using CRISPR/Cas9-mediated genome editing of SRL1 and characterizing its regulatory network involved in leaf rolling through transcriptome analysis. Int J Mol Sci 24(13):11087. https://doi.org/10.3390/ijms241311087

Hanson P, Lu SF, Wang JF, Chen W, Kenyon L, Tan CW, Tee KL, Wang YY, Hsu YC, Schafleitner R, Ledesma D (2016) Conventional and molecular marker-assisted selection and pyramiding of genes for multiple disease resistance in tomato. Sci Hortic 201:346–354. https://doi.org/10.1016/j.scienta.2016.02.020

Hedley C (2015) The role of precision agriculture for improved nutrient management on farms. J Sci Food Agric 95(1):12–19. https://doi.org/10.1002/jsfa.6734

Höft N, Dally N, Hasler M, Jung C (2018) Haplotype variation of flowering time genes of sugar beet and its wild relatives and the impact on life cycle regimes. Front Plant Sci 8:2211. https://doi.org/10.3389/fpls.2017.02211

Hu W, Zhang C, Jiang Y, Huang C, Liu Q, Xiong L, Yang W, Chen F (2020) Nondestructive 3D image analysis pipeline to extract rice grain traits using X-ray computed tomography. Plant Phenomics 2020:3414926. https://doi.org/10.34133/2020/3414926

Jiang Y, Yang Y (2022) High-throughput phenotyping for plant growth and biomass yield of switchgrass under a controlled environment. Grass Res 2(1):1–7

Kahane R, Hodgkin T, Jaenicke H, Hoogendoorn C, Hermann M, Keatinge JDH, d'Arros Hughes J, Padulosi S, Looney N (2013) Agrobiodiversity for food security, health and income. Agron Sustain Dev 33:671–693

Karve AA, Alexoff D, Kim D, Schueller MJ, Ferrieri RA, Babst BA (2015) In vivo quantitative imaging of photoassimilate transport dynamics and allocation in large plants using a commercial positron emission tomography (PET) scanner. BMC Plant Biol 15(1):1–11. https://doi.org/10.1186/s12870-015-0658-3

Kaur S, Shamshad M, Jindal S, Kaur A, Singh S, Kaur S (2022) RNA-seq-based transcriptomics study to investigate the genes governing nitrogen use efficiency in Indian wheat cultivars. Front Genet 13:85391. https://doi.org/10.3389/fgene.2022.853910

Kim SL, Kim N, Lee H, Lee E, Cheon KS, Kim M, Baek J, Choi I, Ji H, Yoon IS, Jung KH (2020) High-throughput phenotyping platform for analyzing drought tolerance in rice. Planta 252(3):38

Kim J, Lee C, Park J, Kim N, Kim SL, Baek J, Chung YS, Kim K (2023) Comparison of various drought resistance traits in soybean (Glycine max L.) based on image analysis for precision agriculture. Plants 12(12):2331. https://doi.org/10.3390/plants12122331

Langridge P, Braun H, Hulke B, Ober E, Prasanna BM (2021) Breeding crops for climate resilience. Theor Appl Genet 134(6):1607–1611

Liao M, Palta JA, Fillery IR (2006) Root characteristics of vigorous wheat improve early nitrogen uptake. Aust J Agric Res 57(10):1097–1107

Little RJ, Rubin DB (2019) Statistical analysis with missing data, vol 793. Wiley, Hoboken

Liu X, Fan Y, Mak M, Babla M, Holford P, Wang F, Chen G, Scott G, Wang G, Shabala S, Zhou M (2017) QTLs for stomatal and photosynthetic traits related to salinity tolerance in barley. BMC Genomics 18(1):1–13. https://doi.org/10.1186/s12864-016-3380-0

Lobell DB, Gourdji SM (2012) The influence of climate change on global crop productivity. Plant Physiol 160(4):1686–1697

Lü S, Zhao H, Parsons EP, Xu C, Kosma DK, Xu X, Chao D, Lohrey G, Bangarusamy DK, Wang G, Bressan RA (2011) The glossyhead1 allele of ACC1 reveals a principal role for multidomain acetyl-coenzyme A carboxylase in the biosynthesis of cuticular waxes by Arabidopsis. Plant Physiol 157(3):1079–1092

Lybbert TJ, Sumner DA (2012) Agricultural technologies for climate change in developing countries: policy options for innovation and technology diffusion. Food Policy 37(1):114–123

Lynch JP (2019) Root phenotypes for improved nutrient capture: an underexploited opportunity for global agriculture. New Phytol 223(2):548–564. https://doi.org/10.1111/nph.15738

Malenica N, Dunić JA, Vukadinović L, Cesar V, Šimić D (2021) Genetic approaches to enhance multiple stress tolerance in maize. Genes 12(11):1760. https://doi.org/10.3390/genes12111760

Mathers AW, Hepworth C, Baillie AL, Sloan J, Jones H, Lundgren M, Fleming AJ, Mooney SJ, Sturrock CJ (2018) Investigating the microstructure of plant leaves in 3D with lab-based X-ray computed tomography. Plant Methods 14(1):1–12. https://doi.org/10.1186/s13007-018-0367-7

Martínez J, Egea G, Agüera J, Pérez-Ruiz M (2017) A cost-effective canopy temperature measurement system for precision agriculture: a case study on sugar beet. Precis Agric 18:95–110. https://doi.org/10.1007/s11119-016-9470-9

McAusland L, Atkinson JA, Lawson T, Murchie EH (2019) High throughput procedure utilising chlorophyll fluorescence imaging to phenotype dynamic photosynthesis and photopro-

tection in leaves under controlled gaseous conditions. Plant Methods 15(1):1–15. https://doi.org/10.1186/s13007-019-0485-x
Mehta S, Singh B, Dhakate P, Rahman M, Islam MA (2019) Rice, marker-assisted breeding, and disease resistance. In: Disease resistance in crop plants: molecular, genetic and genomic perspectives. Springer, Cham, pp 83–111
Meng LS, Yao SQ (2015) Transcription co-activator Arabidopsis ANGUSTIFOLIA 3 (AN 3) regulates water-use efficiency and drought tolerance by modulating stomatal density and improving root architecture by the transrepression of YODA (YDA). Plant Biotechnol J 13(7):893–902
Monzon JP, Calviño PA, Sadras VO, Zubiaurre JB, Andrade FH (2018) Precision agriculture based on crop physiological principles improves whole-farm yield and profit: a case study. Eur J Agron 99:62–71. https://doi.org/10.1016/j.eja.2018.06.011
Mythili K, Rangaraj R (2021) Crop recommendation for better crop yield for precision agriculture using ant colony optimization with deep learning method. Ann Romanian Soc Cell Biol 25:4783–4794
Narayanan N, Beyene G, Chauhan RD, Grusak MA, Taylor NJ (2021) Stacking disease resistance and mineral biofortification in cassava varieties to enhance yields and consumer health. Plant Biotechnol J 19(4):844–854. https://doi.org/10.1111/pbi.13511
Niu H, Ping J, Wang Y, Lv X, Li H, Zhang F, Chu J, Han Y (2020) Population genomic and genome-wide association analysis of lignin content in a global collection of 206 forage sorghum accessions. Mol Breeding 40:1–13. https://doi.org/10.1007/s11032-020-01151-7
Nuccio ML, Wu J, Mowers R, Zhou HP, Meghji M, Primavesi LF, Paul MJ, Chen X, Gao Y, Haque E, Basu SS (2015) Expression of trehalose-6-phosphate phosphatase in maize ears improves yield in well-watered and drought conditions. Nat Biotechnol 33(8):862–869
Olivoto T, Nardino M (2020) MGIDI: a novel multi-trait index for genotype selection in plant breeding. BioRxiv. https://doi.org/10.1101/2020.07.23.217778
Pandey P, Ge Y, Stoerger V, Schnable JC (2017) High throughput in vivo analysis of plant leaf chemical properties using hyperspectral imaging. Front Plant Sci 8:1348
Pérez-Bueno ML, Pineda M, Barón M (2019) Phenotyping plant responses to biotic stress by chlorophyll fluorescence imaging. Front Plant Sci 10:1135. https://doi.org/10.3389/fpls.2019.01135
Perez-de-Castro AM, Vilanova S, Cañizares J, Pascual L, Blanca M, Diez M, Prohens J, Picó B (2012) Application of genomic tools in plant breeding. Curr Genomics 13(3):179–195
Pflugfelder D, Metzner R, van Dusschoten D, Reichel R, Jahnke S, Koller R (2017) Non-invasive imaging of plant roots in different soils using magnetic resonance imaging (MRI). Plant Methods 13(1):1–9. https://doi.org/10.1186/s13007-017-0252-9
Poorter H, Niinemets Ü, Poorter L, Wright IJ, Villar R (2009) Causes and consequences of variation in leaf mass per area (LMA): a meta-analysis. New Phytol 182(3):565–588
Rabia AH, Neupane J, Lin Z, Lewis K, Cao G, Guo W (2022) Principles and applications of topography in precision agriculture. Adv Agron 171:143–189. https://doi.org/10.1016/bs.agron.2021.08.005
Rani A, Taunk J, Jangra S, Yadav RC, Yadav NR, Yadav D, Yadav HP (2022) Development of advance pearl millet lines tolerant to terminal drought stress using marker-assisted selection. Vegetos 35(1):63–73. https://doi.org/10.1007/s42535-021-00284-0
Sagan V, Maimaitijiang M, Sidike P, Eblimit K, Peterson KT, Hartling S, Esposito F, Khanal K, Newcomb M, Pauli D, Ward R (2019) UAV-based high resolution thermal imaging for vegetation monitoring, and plant phenotyping using ICI 8640 P, FLIR Vue Pro R 640, and thermomapcameras. Remote Sensing 11(3):330. https://doi.org/10.3390/rs11030330
Saini DK, Chopra Y, Pal N, Chahal A, Srivastava P, Gupta PK (2021) Meta-QTLs, ortho-MQTLs and candidate genes for nitrogen use efficiency and root system architecture in bread wheat (*Triticum aestivum* L.). Physiol Mol Biol Plants 27:2245–2267
Sandhu N, Sethi M, Kumar A, Dang D, Singh J, Chhuneja P (2021) Biochemical and genetic approaches improving nitrogen use efficiency in cereal crops: a review. Front Plant Sci 12:657629. https://doi.org/10.3389/fpls.2021.657629

Shailani A, Joshi R, Singla-Pareek SL, Pareek A (2021) Stacking for future: pyramiding genes to improve drought and salinity tolerance in rice. Physiol Plant 172(2):1352–1362. https://doi.org/10.1111/ppl.13270

Simberloff D, Martin JL, Genovesi P, Maris V, Wardle DA, Aronson J, Courchamp F, Galil B, García-Berthou E, Pascal M, Pyšek P (2013) Impacts of biological invasions: what's what and the way forward. Trends Ecol Evol 28(1):58–66

Singh UM, Dixit S, Alam S, Yadav S, Prasanth VV, Singh AK, Venkateshwarlu C, Abbai R, Vipparla AK, Badri J, Ram T (2022) Marker-assisted forward breeding to develop a drought, bacterial leaf blight, and blast resistant rice cultivar. Plant Genome 15(1):e20170. https://doi.org/10.1002/tpg2.20170

Singh BK, Delgado-Baquerizo M, Egidi E, Guirado E, Leach JE, Liu H, Trivedi P (2023) Climate change impacts on plant pathogens, food security and paths forward. Nat Rev Microbiol 21:640–656. https://doi.org/10.1038/s41579-023-00900-7

Sprink T, Eriksson D, Schiemann J, Hartung F (2016) Regulatory hurdles for genome editing: process-vs. product-based approaches in different regulatory contexts. Plant Cell Rep 35:1493–1506

Stamp P (1989) Seasonal factors of correlation between climatic factors, photosynthetic and agronomic traits in maize. J Agron Crop Sci 162(4):261–267

Suarez L, Zhang P, Sun J, Wang Y, Poblete T, Hornero A, Zarco-Tejada PJ (2021) Assessing wine grape quality parameters using plant traits derived from physical model inversion of hyperspectral imagery. Agric For Meteorol 306:108445. https://doi.org/10.1016/j.agrformet.2021.108445

Sytar O, Brestic M, Zivcak M, Olsovska K, Kovar M, Shao H, He X (2017) Applying hyperspectral imaging to explore natural plant diversity towards improving salt stress tolerance. Sci Total Environ 578:90–99. https://doi.org/10.1016/j.scitotenv.2016.08.014

Takahashi F, Kuromori T, Sato H, Shinozaki K (2018) Regulatory gene networks in drought stress responses and resistance in plants. In: Survival strategies in extreme cold and desiccation: adaptation mechanisms and their applications. Springer, Singapore, pp 189–214

Theeuwen TP, Logie LL, Harbinson J, Aarts MG (2022) Genetics as a key to improving crop photosynthesis. J Exp Bot 73(10):3122–3137. https://doi.org/10.1093/jxb/erac076

Torii KU (2012) Mix-and-match: ligand–receptor pairs in stomatal development and beyond. Trends Plant Sci 17(12):711–719

Tracy SR, Nagel KA, Postma JA, Fassbender H, Wasson A, Watt M (2020) Crop improvement from phenotyping roots: highlights reveal expanding opportunities. Trends Plant Sci 25(1):105–118. https://doi.org/10.1016/j.tplants.2019.10.015

Verma A, Niranjana M, Jha SK, Mallick N, Agarwal P, Vinod (2020) QTL detection and putative candidate gene prediction for leaf rolling under moisture stress condition in wheat. Sci Rep 10(1):18696. https://doi.org/10.1038/s41598-020-75703-4

Wang W, Zhang Y, Xu C, Ren J, Liu X, Black K, Gai X, Wang Q, Ren H (2015) Cucumber ECERIFERUM1 (CsCER1), which influences the cuticle properties and drought tolerance of cucumber, plays a key role in VLC alkanes biosynthesis. Plant Mol Biol 87:219–233

Wang T, Liu Y, Wang M, Fan Q, Tian H, Qiao X, Li Y (2021) Applications of UAS in crop biomass monitoring: a review. Front Plant Sci 12:616689. https://doi.org/10.3389/fpls.2021.616689

Wani SH, Sah SK, Hossain MA, Kumar V, Balachandran SM, Al-khayri JM, Jain S, Johnson D (2016) Advances in plant breeding strategies: agronomic, abiotic and biotic stress traits. Springer, Cham

Werner T, Nehnevajova E, Köllmer I, Novák O, Strnad M, Krämer U, Schmülling T (2010) Root-specific reduction of cytokinin causes enhanced root growth, drought tolerance, and leaf mineral enrichment in Arabidopsis and tobacco. Plant Cell 22(12):3905–3920

Xiong X, Li J, Su P, Duan H, Sun L, Xu S, Sun Y, Zhao H, Chen X, Ding D, Zhang X (2023) Genetic dissection of maize (Zea mays L.) chlorophyll content using multi-locus genome-wide association studies. BMC Genomics 24(1):384. https://doi.org/10.1186/s12864-023-09504-0

Yang F, Liu Q, Cheng Y, Feng L, Wu X, Fan Y, Raza MA, Wang X, Yong T, Liu W, Liu J (2020) Low red/far-red ratio as a signal promotes carbon assimilation of soybean seedlings by increasing the photosynthetic capacity. BMC Plant Biol 20(1):1–12

Yao J, Sun D, Cen H, Xu H, Weng H, Yuan F, He Y (2018) Phenotyping of Arabidopsis drought stress response using kinetic chlorophyll fluorescence and multicolor fluorescence imaging. Front Plant Sci 9:603. https://doi.org/10.3389/fpls.2018.00603

Ye C, Ishimaru T, Lambio L, Li L, Long Y, He Z, Htun TM, Tang S, Su Z (2022) Marker-assisted pyramiding of QTLs for heat tolerance and escape upgrades heat resilience in rice (Oryza sativa L.). Theor Appl Genet 135(4):1345–1354. https://doi.org/10.1007/s00122-022-04035-w

Yu K, Miao H, Liu H, Zhou J, Sui M, Zhan Y, Xia N, Zhao X, Han Y (2022) Genome-wide association studies reveal novel QTLs, QTL-by-environment interactions and their candidate genes for tocopherol content in soybean seed. Front Plant Sci 13:1026581

Zanatta CB, Benevenuto RF, Nodari RO, Agapito-Tenfen SZ (2020) Stacked genetically modified soybean harboring herbicide resistance and insecticide rCry1Ac shows strong defense and redox homeostasis disturbance after glyphosate-based herbicide application. Environ Sci Eur 32(1):1–17. https://doi.org/10.1186/s12302-020-00379-6

Zhang YT, Zhang YQ, Yang QC, Tao LI (2019) Overhead supplemental far-red light stimulates tomato growth under intra-canopy lighting with LEDs. J Integr Agric 18(1):62–69

Zhang H, Ge Y, Xie X, Atefi A, Wijewardane NK, Thapa S (2022) High throughput analysis of leaf chlorophyll content in sorghum using RGB, hyperspectral, and fluorescence imaging and sensor fusion. Plant Methods 18(1):1–17. https://doi.org/10.1186/s13007-022-00892-0

Zhang P, Huang J, Ma Y, Wang X, Kang M, Song Y (2023) Crop/plant modeling supports plant breeding: II. Guidance of functional plant phenotyping for trait discovery. Plant Phenomics 5:0091. https://doi.org/10.34133/plantphenomics.0091

Zhou L, Ni E, Yang J, Zhou H, Liang H, Li J, Jiang D, Wang Z, Liu Z, Zhuang C (2013) Rice OsGL1-6 is involved in leaf cuticular wax accumulation and drought resistance. PLoS One 8(5):e65139

Zhu S, Li Y, Vossen JH, Visser RG, Jacobsen E (2012) Functional stacking of three resistance genes against *Phytophthorainfestans* in potato. Transgenic Res 21:89–99. https://doi.org/10.1007/s11248-011-9510-1

Zhu YH, Weiner J, Jin Y, Yu MX, Li FM (2022) Biomass allocation responses to root interactions in wheat cultivars support predictions of crop evolutionary ecology theory. Front Plant Sci 13:858636. https://doi.org/10.3389/fpls.2022.858636

Integrating Marker-Assisted (MAS) and Genomic Selection (GS) for Plant Functional Trait Improvement

11

Shulbhi Verma, A. R. S. S. H. Gupta, Santhoshi Yalla, Shreya, Parth J. Patel, Ritu Sharma, Anjana A. V, and Akhil Donga

Abstract

In plant breeding, the combination of genomic selection (GS) and marker-assisted selection (MAS) is a novel method for improving functional traits. To enable more accurate and effective selection, molecular markers linked to target qualities are identified and used in MAS. On the other side, genomic selection speeds up the breeding process by using genome-wide information to anticipate an individual's genetic merit for complex traits. By integrating the advantages of both approaches, the synergy between MAS and GS has enormous potential to further plant functional trait development. Through this integration, genetic markers for both simple and polygenic features can be exploited, resulting in a more thorough and precise selection process. The collaborative effort that results improves breeding efficiency and speeds up the development of new plant varieties with better functional features, ultimately promoting sustainable agriculture and global food security.

Keywords

Marker-assisted selection (MAS) · Genomic selection (GS) · Plant breeding · Plant functional traits · Genetic markers · Trait improvement · Sustainable agriculture

S. Verma (✉) · A. R. S. S. H. Gupta · S. Yalla · Shreya · P. J. Patel · R. Sharma · A. V. A · A. Donga
Department of Genetics and Plant Breeding, C.P. College of Agriculture, S.D. Agricultural University, S.K. Nagar, Gujarat, India
e-mail: shulbhiverma@sdau.edu.in

N. Kumar, H. Singh (eds.), *Plant Functional Traits for Improving Productivity*,
https://doi.org/10.1007/978-981-97-1510-7_11

11.1 Introduction

The global population is anticipated to reach 9.70 billion by 2050. Addressing the escalating demand for food for this growing population will necessitate a substantial increase in crop production, ranging from 25% to 70% above present levels by the year 2050 (Hunter et al. 2017). Simultaneously, the impact of climate change poses a significant hurdle to sustainable food production. Due to climate change, globally, predicted loss of 10–40% in crop productivity is estimated for 2100 (Anonymous 2022). Therefore, scientists are compelled to devise solutions especially for the crops that are threatened by climate change and environmental disturbances and conversely how the anthropogenic activities are affecting the ecosystems services like nutrient cycling, water purification, climate regulation, and cultural services. As the crop productivity is an end result of genotypic and environmental interactions, there has been an urgent need to focus on identification and assessment of the functional traits in the biodiversified arena that are affected by ecological parameter and conversely affect the ecosystem services, along with interrogation of such potentially useful climate adaptive traits into agronomically superior lines through advanced breeding and biotechnological applications like marker-assisted selection and genomic selection.

Functional traits are characteristics that show up in plants phenotype and are associated with how that plant reacts to its environment and/or influences the properties of its ecosystem. These attributes can be morphological, biochemical, physiological, structural, phonological, or behavioral (Violle et al. 2007; Kumar et al. 2020a, b, c). It is evident from this that functional traits must cross two paths. A scenario where traits play a role in shaping both natural and man-made services, thereby affecting the ecological parameters in their surroundings. These characteristics are referred to as effect traits because they focus on the ecosystem rather than the organisms' own adaptive advantages. Conversely, the other path consists of response traits that influence germplasm's ability to adapt and flourish in that ecosystem, even in the face of anthropogenic or natural environmental disturbances.

Molecular studies of genetic architecture of functional traits have become feasible largely because of the revolution in DNA marker technology. As most of the functional traits are complex and are affected by variation at multiple genetic loci and environmental factors, the identification of quantitative trait loci (QTLs) through application of different molecular markers on mapping populations has become imperative. But the work of physiologists and breeders does not end here. Efficient screening and interrogation of such potential QTLs related to functional traits into agronomically efficient cultivars is the domain need to be worked upon. Therefore, exploitation of efficient and precise application of marker-assisted selection and genomic selection comes into picture.

MAS relies on the development of DNA-based markers that are primarily in linkage disequilibrium with the underlying genes responsible for the trait of interest (Chawade et al. 2019). The success of marker-assisted selection (MAS) hinges on the degree of association between the marker and the locus of interest. A higher degree of association reduces the chance of recombination, leading to more effective selection. Once tightly linked markers to genes or quantitative trait loci (QTLs) are identified, breeders can use specific DNA markers as a diagnostic tool before field evaluation of large plant populations (Boopathi 2013). Marker-assisted selection has been extensively used in maize, sorghum, different types of wheat, and tree species like poplar.

Individual genetic components contributing to traits such as whole-plant traits (such as spinescence, branching architecture, leaf area: sapwood area ratio, root-mass fraction, and salt resistance), leaf traits (including vein density, light-saturated photosynthetic rate, leaf dark respiration, and photosynthetic pathway), and stem traits (like stem-specific density, twig dry-matter content and twig drying time, bark thickness, and xylem conductivity) can be selected independently and ultimately stabilized, given the complex inheritance patterns associated with these functional traits. Marker-assisted selection facilitates the selection of targeted alleles, a task that is considerably challenging, expensive, and time-consuming when relying solely on phenotypic scoring for characters like salt resistance, light-saturated photosynthetic rate, leaf dark respiration, photosynthetic pathway, etc. (Kumar et al. 2021a, b). However, marker-assisted selection has limitations, especially when dealing with traits controlled by a large number of genes with small effects, known as quantitative traits (Bernardo 2008). Furthermore, the use of biparental mapping populations in most QTL studies doesn't readily translate to breeding applications (Heffner et al. 2009) as the populations in biodiversity are more complex in regard to functional traits. To address these limitations, genomic selection (GS) has emerged as a method that jointly utilizes information from all molecular markers for selection purposes (Persa et al. 2021).

Genomic selection (GS) created a variety of models to deal with problems that arise when a large number of markers (p) are taken into account as control variables in comparison to the total number of observations (n) for model fitting ($p > n$). GS requires both genomic and phenotypic information for model calibration, screening candidate genotypes based solely on their marker profiles. The central process of GS involves calculating genomic estimated breeding values (GEBVs) for individuals with only genotypic data, using a model derived from individuals with both phenotypic and genotypic data. An individual's GEBV is the cumulative effect of all marker alleles present in that individual, as incorporated into the GS model applied to the population under selection (Singh and Singh 2015).

However, accuracy of genomic predictions is affected by the training and breeding population size, genetic diversity of breeding population, the heritability of the underlying trait, the impact of genotype-environment (G × E) interactions which is the principal component while dealing with functional traits, marker density which increases the cost, and the genetic relationship between the training population and the breeding population or selection candidates (Howard et al. 2014), whereas changes in gene frequencies and epistatic interactions can significantly impact genomic estimated breeding value (GEBV) estimates. Therefore, all these parameters with proper implementation provides successful genomic selection (GS).

In this chapter, we will discuss how the approaches of marker-assisted selection and genomic selection together can facilitate improvement in the functional traits for anthropogenic breeding objectives and enhancement in functionality of ecosystem services.

11.2 Improvement of Plant Functional Traits (PFTs) Using Genomic Selection

In contemporary plant breeding and agriculture, the enhancement of plant functional traits (PFTs) using genomic selection is a cutting-edge strategy. By enhancing desired features in plants through the use of genomics, this technology helps to improve agricultural systems' resilience, productivity, and sustainability. In order to find certain genomic areas linked to desired traits, a systematic investigation of an organism's complete genome is called genomic selection. This method concentrates on characteristics linked to plant physiology, structure, and ecological interactions in the setting of PFTs (Kumar et al. 2021c, d). Breeders can choose which plants to pick for additional breeding programs with greater knowledge of the genetic basis of these traits (Mir et al. 2023).

The potential of genomic selection to speed up the breeding process is a major benefit for bettering PFTs. It frequently takes several generations to obtain the required trait improvements using traditional breeding procedures. On the other side, genomic selection shortens the time required for trait enhancement by allowing the early identification of prospective individuals. Additionally, by enabling the simultaneous enhancement of several traits, this method helps breeders tackle difficult problems including disease resistance, drought resistance, and nutrient efficiency. Because of its comprehensive approach, genomic selection is a potent tool for creating plants that can flourish in a variety of environmental settings (Heffner et al. 2009; Kumar et al. 2017, Kumar et al. 2019; Krishnappa et al. 2021).

High-throughput sequencing and bioinformatics are two examples of cutting-edge technologies that must be integrated for genomic selection for PFT improvement to be successful. Large genomic datasets may be efficiently analyzed, thanks to these techniques, which also offer insightful information about the genetic changes linked to particular functional features (Bhat et al. 2016; Marsh et al. 2021).

11.3 Marker-Assisted Selection and Plant Functional Traits

Plant breeding has found a useful tool in marker-assisted selection (MAS), which makes it easier and more accurate to improve plant functional traits (PFTs). This novel method uses molecular markers connected to certain genes linked to desired qualities, enabling breeders to make well-informed choices during the selection process. Marker-assisted selection is essential for finding and improving characteristics pertaining to plant physiology, structure, and ecological interactions in PFTs. The DNA sequences linked to the presence or lack of specific traits are frequently the markers used in MAS, which helps breeders forecast a plant's expression of a trait early in development (Kadirvel et al. 2015; Wu et al. 2022).

Accelerating the breeding process is one of MAS's main benefits for bettering PFTs. Conventional breeding techniques depend on phenotypic data, which might be imprecise and require multiple growing seasons to evaluate. On the other hand, MAS makes it possible for breeders to precisely and quickly select plants that have the desired functional characteristics by identifying individuals with those traits at the molecular level. Furthermore, MAS aids in the improvement of several qualities at once. When trying to improve complex features like disease resistance, abiotic stress tolerance, and nutrient usage efficiency, this is especially helpful. Breeders can create plant varieties with a combination of functional features that enhance overall plant health and productivity by utilizing markers linked to these traits (Kumar et al. 2021; Mollier et al. 2023).

Marker-assisted selection also makes precision breeding easier by minimizing the need for lengthy field trials. This improves the predictability of trait expression in a variety of environmental circumstances while also saving time and resources. Breeders are therefore able to create cultivars that are more suited to particular agro-ecological conditions. The use of molecular biology methods in plant breeding, including genotyping and DNA sequencing, has significantly expanded the use of MAS. Breeders can use these methods to identify and validate markers linked to particular functional features, giving them important information for targeted selection (De Mori and Cipriani 2023).

11.4 Application of GS (Genomic Selection)

11.4.1 Understanding Genomic Selection

Genomic selection is the process of estimating an individual's genetic worth using their genomic profile. GS uses information from the full genome, in contrast to conventional breeding techniques, which rely on reported phenotypic traits. Breeders may now capture genetic differences linked to complex qualities like disease resistance, yield, and stress tolerance, thanks to this all-inclusive strategy (Budhlakoti et al. 2022).

11.4.2 Improving Stress Tolerance

A major obstacle in the field of agriculture is crop adaptation to environmental stresses such as heat, drought, and pests. Through the identification and selection of genomic areas linked to stress tolerance, genomic selection provides a useful remedy. Breeders can create crops that flourish in a variety of harsh conditions by concentrating on the underlying genetic elements, which will ultimately increase resilience and yield (Shikha et al. 2017; Anilkumar et al. 2022).

11.4.3 Enhancing Nutrient Use Efficiency

Plant development and yield optimization depend on effective nutrient usage. Genes linked to enhanced nutrient absorption, use, and translocation within the plant can be identified using genomic selection. Breeders can create crops with improved nutrient usage efficiency, lowering input costs and environmental impact while raising total output, by choosing individuals with favorable genetic profiles (Mallikarjuna et al. 2022; Krishna et al. 2022).

11.4.4 Optimizing Photosynthetic Efficiency

One important mechanism affecting plant productivity is photosynthesis. Identification of genes linked to carbon uptake, chlorophyll concentration, and photosynthetic efficiency is made easier by genomic selection. Breeders can generate crops with greater photosynthetic capacities, which will increase biomass and improve production potential, by choosing plants with superior genetic profiles in these areas (van Bezouw et al. 2019; Theeuwen et al. 2022).

11.4.5 Accelerating Breeding Cycles

It frequently takes several generations to notice and choose for desired qualities when using traditional breeding procedures. By enabling breeders to anticipate an individual's genetic potential early in its life cycle, genomic selection quickens this process. This shortens the time required to create new kinds with enhanced functional traits by enabling quicker and more effective breeding cycles (Voss-Fels et al. 2019a, b; Budhlakoti et al. 2022).

11.5 Application of Marker-Assisted Selection (MAS)

In plant breeding, marker-assisted selection (MAS) is a potent technique that enables the DNA-level selection of desired features. MAS can greatly speed up the breeding process by finding and using molecular markers connected to particular genes linked to crucial functional features in plants (Boopathi and Boopathi 2020; Song et al. 2023). There are various uses for enhancing plant functional characteristics to increase crop productivity. The following are the applications of MAS:

11.5.1 Disease Resistance

- Identify the molecular markers linked to the genes that provide disease resistance in crops.
- Plants resistant to specific diseases or pests can be chosen using MAS.

- In order to decrease the demand for chemical pesticides, accelerate the creation of new crop varieties with increased resistance.

11.5.2 Abiotic Stress Tolerance

- Find markers that are connected to genes that are linked to resistance to environmental stressors like salt, drought, or extremely high temperatures.
- Screen and choose plants with greater resistance to particular abiotic stressors by using MAS.
- Develop crops that can flourish in harsh environments, thereby increasing cultivation areas and assuring consistent yields.

11.5.3 Nutrient Use Efficiency

- Determine the molecular markers linked to the genes governing the efficiency of nutrition absorption and utilization.
- To choose plants with better nutrient usage efficiency and lessen the demand for fertilizers, use MAS.
- Develop crops that will flourish in soils deficient in nutrients and support sustainable agriculture.

11.5.4 Yield Enhancement

- Determine the markers linked to the genes that affect yield-related variables such fruit growth, flowering period, and seed size.
- To increase overall yield, use MAS to identify plants with the best combinations of these features.
- In order to meet the growing demand for food around the world, speed up the development of crop varieties with high yields.

11.5.5 Quality Improvement

- Determine the markers linked to the genes governing crop quality characteristics like taste, nutritional value, and shelf life.
- Utilize MAS to choose plants with enhanced quality characteristics that satisfy market demands and consumer preferences.
- Develop crops that have better postharvest qualities and more nutrient profiles.

11.6 Challenges and Future Directions of Marker-Assisted Selection for Functional Trait Improvement

Although almost all of agriculturally significant species have molecular marker maps, the foundation of any MAS effort, most species have very different map densities, but as far as concern of functional traits, there is quite under develop dark spot. One of the main obstacles to the widespread use of MAS at the current time is the high expense of marker genotyping, especially in underdeveloped nations and for the breeding of crops for functional traits. According to Ribaut et al. (2010), there are a variety of reasons why MAS adoption is so low in developing nations, such as inadequate funding, a shortage of skilled personnel, and inadequate facilities for genotyping and/or phenotyping. Improving functional traits in plants holds immense potential for enhancing agricultural productivity, resilience, and sustainability. However, this endeavor faces several significant challenges.

11.6.1 Complex Trait Architecture

Many important functional features, such as the ability to withstand drought or use of nutrients efficiently, are polygenic, which means that different genes have limited impacts on them individually. These genes are still difficult to identify and manipulate (Boyes et al. 2014; Combs and Bernardo 2013).

11.6.2 Insufficient Linkage Information

It is challenging to find closely connected markers for effective selection since many functional qualities are governed by quantitative trait loci (QTLs) with weak or complicated effects (Xu and Crouch 2008; Gupta et al. 2012).

11.6.3 Lack of Available QTLs for Particular Functional Trait

(Hassan et al. 2023). In SG sorghum, there is a need to improve the RG 03123 line for the introgressed QTL for the enhancing GLA retention and photosynthesis rate (Galyuon et al. 2019).

11.6.4 Trade-Offs and Synergies

There are frequent quests when trying to improve a particular attribute at the expense of another. It's critical to comprehend and manage these trade-offs in order to obtain the best possible attribute combinations (Mason et al. 2020; Kunstler et al. 2016).

11.6.5 Phenotyping Bottlenecks

Measurement of pertinent functional features accurately and effectively is still a key constraint, especially in field settings. Technological obstacles still exist, although high-throughput phenotyping is continually evolving (Dhondt et al. 2014; Fiorani and Schurr 2013). Accurate and high-throughput phenotyping is necessary for effective utilization, especially for complex features. It's still difficult to integrate reliable phenotyping platforms with MAS workflows (Dhondt et al. 2014; Araus et al. 2018).

11.6.6 Accuracy and Prediction

For all other treatment-by-population combinations, there was no evidence of a relationship between trait heritability and the accuracy of genomic predictions. Reduced phenotypic diversity or a significant genotype-by-environment interaction may be the cause of the poorer prediction accuracy, regardless of heritability, given the training population had the same size for all characteristics (Zhang et al. 2017; Cerrudo et al. 2018).

11.6.7 G × E Interactions

Epistasis and genotype environment interactions (GEI) restrict the use of MAS for improving complex characteristics like functional metabolic pathways which are depending on others. Although its exact contribution to the regulation of quantitative features is controversial, epistasis is known to play a part in this process (Heslot et al. 2014).

11.7 Future Directions

In the future, plant functional trait enhancement through the integration of genomic selection (GS) and marker-assisted selection (MAS) has the potential to revolutionize agricultural processes and improve crop breeding results. High-throughput genotyping technological developments will be essential for the effective and economical acquisition of massive amounts of genetic data. Breeders will be able to make better informed decisions and shorten the breeding process as a result of the identification of important genetic markers linked to desirable qualities being made easier. Our understanding of the genetic basis of complex traits will expand with the development and application of comprehensive genomic resources, such as reference genomes and pan-genomes, opening the door to more focused and successful trait enhancement approaches.

The ongoing development of data analytics and machine learning algorithms will greatly aid in the integration of MAS and GS. The utilization of sophisticated computational instruments will be crucial in handling extensive datasets, enhancing the precision of characteristic forecasts, and streamlining the process of choosing plants possessing intended functional features. Furthermore, a deeper investigation of gene function and regulation will be made possible by the synergy between MAS and GS with functional genomics and gene-editing technologies. This comprehensive approach will speed up the development of crops with increased resilience, adaptability, and overall performance in a variety of environmental conditions. It will also improve our ability to dissect polygenic traits and offer opportunities for precise and targeted modifications to plant genomes.

Successful integration of MAS and GS into mainstream breeding programs will need worldwide cooperation, public-private partnerships, and collaborative efforts across research institutions. The production of crop varieties suited to particular geographies and environmental difficulties will be made possible by the sharing of genomic data, resources, and experience, which will accelerate progress. It will also be crucial to address ethical issues and create transparent regulatory frameworks in order to ensure the ethical and responsible application of these cutting-edge genomic techniques. In order to address the increasing demands of global food security, the integration of MAS and GS is positioned to make a significant contribution to sustainable agriculture by navigating these future paths and stimulating the development of crops with better functional traits.

11.8 Conclusion

In conclusion, a revolutionary step forward in crop breeding has been made with the merging of genomic selection (GS) and marker-assisted selection (MAS) for the enhancement of functional traits in plants. Our capacity to recognize, comprehend, and control the genetic components underpinning complex functional features is about to undergo a revolutionary shift because of the convergence of high-throughput genotyping technologies, extensive genomic resources, and sophisticated data analytics. When combined with functional genomics and gene-editing technologies, the synergistic application of MAS and GS holds the potential to expedite the production of crops that exhibit increased yield, resilience to environmental stressors, and adaptability. The integration of MAS and GS stands as a beacon for sustainable and efficient crop improvement, providing a way to address global challenges like food security in the face of a changing climate and growing population demands, as cooperative efforts and ethical considerations continue to shape the landscape of agricultural innovation. A new era of precision agriculture, where crops are tailored to fit the different requirements of farmers and consumers while protecting the long-term health of our ecosystems, will only come about through the successful adoption of these cutting-edge genetic techniques.

References

Anilkumar C, Sunitha NC, Harikrishna, Devate NB, Ramesh S (2022) Advances in integrated genomic selection for rapid genetic gain in crop improvement: a review. Planta 256(5):87

Anonymous (2022) NAARM vision 2050. https://naarm.org.in. Assessed 16 Dec 2023

Araus JL, Kefauver SC, Zaman-Allah M, Cairns JE, Turner NW (2018) Translating high-throughput phenotyping into crop breeding programs. Plant Physiol 176(1):261–276

Bernardo R (2008) Molecular markers and selection for complex traits in plants: learning from the last 20 years. Crop Sci 48(5):1649–1664

Bhat JA, Ali S, Salgotra RK, Mir ZA, Dutta S, Jadon V, Tyagi A, Mushtaq M, Jain N, Singh PK, Singh GP (2016) Genomic selection in the era of next generation sequencing for complex traits in plant breeding. Front Genet 7:221

Boopathi NM (2013) Marker-assisted selection. Springer, New Delhi

Boopathi NM, Boopathi NM (2020) Marker-assisted selection (MAS). In: Genetic mapping and marker assisted selection: basics, practice and benefits. Springer, New Delhi, pp 343–388

Boyes DC, Kertesz MA, Naylor RW (2014) Navigating the complexity of the drought stress transcriptome: past successes and future challenges. Plant Sci 214:127–136

Budhlakoti N, Kushwaha AK, Rai A, Chaturvedi KK, Kumar A, Pradhan AK, Kumar U, Kumar RR, Juliana P, Mishra DC, Kumar S (2022) Genomic selection: a tool for accelerating the efficiency of molecular breeding for development of climate-resilient crops. Front Genet 13:832153

Cerrudo D, Cao S, Yuan Y, Martinez C, Suarez EA, Babu R et al (2018) Genomic selection outperforms marker assisted selection for grain yield and physiological traits in a maize doubled haploid population across water treatments. Front Plant Sci 9:366

Chawade A, van Ham J, Blomquist H, Bagge O, Alexandersson E, Ortiz R (2019) High-throughput field-phenotyping tools for plant breeding and precision agriculture. Agronomy 9(5):258

Combs E, Bernardo R (2013) Accuracy of genomewide selection for different traits with constant population size, heritability, and number of markers. Plant Genet 6:1–7. https://doi.org/10.3835/plantgenome2012.11.0030

De Mori G, Cipriani G (2023) Marker-assisted selection in breeding for fruit trait improvement: a review. Int J Mol Sci 24(10):8984

Dhondt S, Mertens MC, Van Cleemput O, Verstraete F (2014) High-throughput phenotyping: from concept to field reality. Plant Biosyst 148(4):519–530

Fiorani F, Schurr U (2013) Future trends and challenges in field-based high throughput phenotyping. Annu Rev Plant Biol 64(1):761–791

Galyuon IK, Gay A, Hash CT, Bidinger FR, Howarth C (2019) A comparative assessment of the performance of a stay-green sorghum (*Sorghum bicolor* (L) Moench) introgression line developed by marker-assisted selection and its parental lines. Afr J Biotechnol 18(26):548–563

Gupta PK, Kumar R, Tomar RS, Munshi AD, Singh NP (2012) Marker-assisted selection in crop plants: present status and future perspectives. Mol Breed 30(3):699–713

Hassan MA, Dahu N, Hongning T, Qian Z, Yueming Y, Yiru L, Shimei W (2023) Drought stress in rice: morpho-physiological and molecular responses and marker-assisted breeding. Front Plant Sci 14:1215371. https://doi.org/10.3389/fpls.2023.1215371

Heffner EL, Sorrells ME, Jannink JL (2009) Genomic selection for crop improvement. Crop Sci 49(1):1–12

Heslot N, Akdemir D, Sorrells ME, Jannink JL (2014) Integrating environmental covariates and crop modeling into the genomic selection framework to predict genotype by environment interactions. Theor Appl Genet 127:463–480

Howard R, Carriquiry AL, Beavis WD (2014) Parametric and nonparametric statistical methods for genomic selection of traits with additive and epistatic genetic architectures. G3 (Bethesda) 4(6):1027–1046

Hunter MC, Smith RG, Schipanski ME, Atwood LW, Mortensen DA (2017) Agriculture in 2050: recalibrating targets for sustainable intensification. Bioscience 67(4):386–391

Kadirvel P, Senthilvel S, Geethanjali S, Sujatha M, Varaprasad KS (2015) Genetic markers, trait mapping and marker-assisted selection in plant breeding. In: Bahadur B, Venkat Rajam M, Sahijram L, Krishnamurthy K (eds) Plant biology and biotechnology. Springer, New Delhi, pp 65–88

Krishna TPA, Maharajan T, Ignacimuthu S, Ceasar SA (2022) Improving the nutrient-use efficiency in millets by genomics approaches. In: Omics of climate resilient small millets. Springer Nature, Singapore, pp 205–220

Krishnappa G, Savadi S, Tyagi BS, Singh SK, Mamrutha HM, Kumar S, Mishra CN, Khan H, Gangadhara K, Uday G, Singh G (2021) Integrated genomic selection for rapid improvement of crops. Genomics 113(3):1070–1086

Kumar N, Suyal DC, Sharma IP, Verma A, Singh H (2017) Elucidating stress proteins in rice (Oryza sativa L.) genotype under elevated temperature: a proteomic approach to understand heat stress response. 3 Biotech 7:205

Kumar N, Jeena N, Singh H (2019) Elevated temperature modulates rice pollen structure: a study from foothill Himalayan Agro-ecosystem in India. 3Biotech 9:175

Kumar A, Dwivedi GK, Tewari S, Paul J, Anand R, Kumar N, Kumar P, Singh H, Kaushal R (2020a) Carbon Mineralization and Inorganic Nitrogen Pools under Terminalia chebula Retz.-Based Agroforestry System in Himalayan Foothills, India. For Sci 66(5):634–643

Kumar A, Kumar P, Singh H, Kumar N (2020b) Adaptation and mitigation potential of roadside trees with bioextraction of heavy metals under vehicular emissions and their impact on physiological traits during seasonal regimes. Urb For Urb Green 58:126900

Kumar A, Dwivedi GK, Tewari S, Paul J, Sah VK, Singh H, Kumar P, Kumar N, Kaushal R (2020c) Soil organic carbon pools under Terminalia chebula Retz. based agroforestry system in Himalayan foothills, India. Curr Sci 118(7):1098–1103

Kumar A, Kumar P, Singh H, Bisht S, Kumar N (2021a) Relationship of physiological plant functional traits with soil carbon stock in temperate forest of Garhwal Himalaya. Curr Sci 120(8):1368–1373

Kumar A, Kumar P, Singh H, Kumar N (2021b) Modulation of plant functional traits under essential plant nutrients during seasonal regime in natural forests of Garhwal Himalayas. Plant Soil 465:197–212

Kumar A, Tewari S, Singh H, Kumar P, Kumar N, Bisht S, Kushwaha S, Tamta N, Kaushal R (2021c) Biomass accumulation and carbon stocks in different agroforestry system prevalent in Himalayan foothills, India. Curr Sci 120(6):1083–1088

Kumar A, Kumar P, Singh H, Kumar N (2021d) Impact of plant functional traits on near saturated hydraulic conductivity of soil under different forests of Kempty watershed in Garhwal Himalaya. Ind J Soil Conser 49(1):38–44

Kunstler G, Falster D, Coomes DA, Hui F, Kooyman RM, Laughlin DC et al (2016) Plant functional traits have globally consistent effects on competition. Nature 529(7585):204–207

Mallikarjuna BP, Shettigar N, Radhika DH, Devi EL, Bhat JS, Patil BS, Lohithaswa HC, Mallikarjuna MG (2022) Genome-wide association studies and genomic selection for nutrient use efficiency in cereals. In: Next-generation plant breeding approaches for stress resilience in cereal crops. Springer Nature Singapore, Singapore, pp 161–197

Marsh JI, Hu H, Gill M, Batley J, Edwards D (2021) Crop breeding for a changing climate: integrating phenomics and genomics with bioinformatics. Theor Appl Genet 134:1677–1690

Mason CM, LaScaleia MC, De La Pascua DR, Monroe JG, Oolsby EW (2020) Learning from dynamic traits: seasonal shifts yield insights into ecophysiological trade-offs across scales from macroevolutionary to intraindividual. Int J Plant Sci 181(1):88–102

Mir ZA, Chandra T, Saharan A, Budhlakoti N, Mishra DC, Saharan MS, Mir RR, Singh AK, Sharma S, Vikas VK, Kumar S (2023) Recent advances on genome-wide association studies (GWAS) and genomic selection (GS); prospects for Fusarium head blight research in Durum wheat. Mol Biol Rep 50(4):3885–3901

Mollier M, Roychowdhury R, Tzudir L, Sharma R, Barua U, Rahman N, Pal S, Gogoi B, Kalita P, Jain D, Das R (2023) Evaluation of morpho-physiological and yield-associated traits of rice (Oryza sativa L.) landraces combined with marker-assisted selection under high-temperature stress and elevated atmospheric CO_2 levels. Plants 12(20):3655

Persa R, Ribeiro PCDO, Jarquin D (2021) The use of high-throughput phenotyping in genomic selection context. Crop Breed Appl Biotechnol 21. https://doi.org/10.1590/1984-70332021v21sa19
Ribaut J-M, de Vicente MC, Delannay X (2010) Molecular breeding in developing countries: challenges and perspectives. Curr Opin Plant Biotechnol 13:1–6
Shikha M, Kanika A, Rao AR, Mallikarjuna MG, Gupta HS, Nepolean T (2017) Genomic selection for drought tolerance using genome-wide SNPs in maize. Front Plant Sci 8:550
Singh BD, Singh AK (2015) Marker-assisted plant breeding: principles and practices. Springer, New Delhi
Song L, Wang R, Yang X, Zhang A, Liu D (2023) Molecular markers and their applications in marker-assisted selection (MAS) in bread wheat (Triticum aestivum L.). Agriculture 13(3):642
Theeuwen TP, Logie LL, Harbinson J, Aarts MG (2022) Genetics as a key to improving crop photosynthesis. J Exp Bot 73(10):3122–3137
van Bezouw RF, Keurentjes JJ, Harbinson J, Aarts MG (2019) Converging phenomics and genomics to study natural variation in plant photosynthetic efficiency. Plant J 97(1):112–133
Violle C, Navas ML, Vile D, Kazakou E, Fortunel C, Hummel I et al (2007) Let the concept of trait be functional! Oikas 116:882–892
Voss-Fels KP, Herzog E, Dreisigacker S, Sukumaran S, Watson A, Frisch M, Hayes B, Hickey LT (2019a) “SpeedGS” to accelerate genetic gain in spring wheat. In: Applications of genetic and genomic research in cereals. Woodhead Publishing, Duxford, pp 303–327
Voss-Fels KP, Cooper M, Hayes BJ (2019b) Accelerating crop genetic gains with genomic selection. Theor Appl Genet 132:669–686
Wu YP, Chang YC, Kuo HI, Lin BN, Wang SM, Tseng YC (2022) The development of two high-yield and high-quality functional rice cultivars using marker-assisted selection and conventional breeding methods. Int J Mol Sci 23(9):4678
Xu Y, Crouch JH (2008) Marker-assisted selection in plant breeding: from publications to practice. Crop Sci 48(2):391–407
Zhang A, Wang H, Beyene Y, Semagn K, Liu Y, Cao S et al (2017) Effect of trait heritability, training population size and marker density on genomic prediction accuracy estimation in 22 bi-parental tropical maize populations. Front Plant Sci 8:1916. https://doi.org/10.3389/fpls.2017.01916

Genetic Engineering and Gene Editing for Targeted Trait Modifications

12

Tannu Ruhil, Hukum Singh, and Santan Barthwal

Abstract

In the twenty-first century, the scientific community faces the challenge of ensuring food security for an increasing population while tackling climate change simultaneously. In this direction, the manipulation of functional traits has captured the interest of scientists worldwide. The arrival of genetic engineering and gene editing has revolutionized the research aiming to improve plant functional traits to enhance productivity with better precision and speed. Genetic engineering techniques like gene silencing and transgenesis alleviate the risks and raise the adaptability of the plant. Trait modifications benefit from plant architecture, flower characters, disease and stress tolerance, and herbicide resistance. This further aids in fulfilling consumer requirements and meeting global food deficiency. Newly emerged plant breeding technologies like clustered regularly interspaced short palindromic repeats, zinc finger nucleases, and transcription activator-like effector nucleases have facilitated the advancement of enhanced genetic modification in plants for trait improvement. Continuous efforts and more research in this direction can ameliorate crop losses occurring due to a wide range of factors and aid in successfully fulfilling world hunger.

Keywords

Genetic engineering · Gene editing · Trait modification · Crop

T. Ruhil (✉) · H. Singh · S. Barthwal
Plant Physiology Discipline, Genetics and Tree Improvement Division, Forest Research Institute, Dehradun, Uttarakhand, India

N. Kumar, H. Singh (eds.), *Plant Functional Traits for Improving Productivity*,
https://doi.org/10.1007/978-981-97-1510-7_12

12.1 Introduction

The population is on the surge and is projected to reach 9.9 billion by the end of 2050, thereby putting huge pressure over food supply (Das et al. 2023). Plants encounter a range of stresses that impose constraints in their agricultural productivity. However, technological advancements have led to the evolution of novel crop developments which are not only higher in yield but also have greater climate resilience (Kumar et al. 2015, 2016, 2017). For many years, crop enhancement has been achieved through conventional plant breeding methods or various physical, chemical (such as ethyl methane sulphonate, gamma radiation), and biological approaches (like transposon insertion, T-DNA), resulting in deletions, point mutations, gene duplications, and rearrangements. Although conventional breeding aids in selection and combination of plants having desirable traits for better crops, this accomplishment is exclusive to the sexual mating of plants. Conventional breeding also results in loss of genetic diversity (Louwaars 2018).

For plant's trait improvement, genetic diversity is a vital source. Therefore, for the evolution of new plant varieties, creation of variations in the gene pool is extremely important. This results in the emergence of gene editing, which accurately modifies the target DNA via the addition, removal, or modification of genetic material utilizing specifically designed site-specific nucleases (SSNs) (Arora and Narula 2017). Gene editing commenced with zinc finger nucleases (ZFNs), followed by transcription activator-like effector nucleases (TALENs), and finally the recent clustered regularly interspaced short palindromic repeats (CRISPR)/CRISPR-associated protein (Cas) technology. They paved the path for enhancing plants through a mechanism involving the excision of single nucleotides. These techniques have proved beneficial in addressing several concerns over inefficiency and unpredictability like regulation of traditionally used transgenesis and random mutagenesis and are emerging for gene editing for desirable trait modifications (Bhattacharyya et al. 2023). The sustained utilization of genetically altered crops has substantially increased global agricultural output over the past two decades (Kumar et al. 2020).

These crops were first introduced for commercial use in the year 1996 and have since achieved astonishing success. These crops have successfully reduced the need for pesticides and the environmental impact quotient. Several transgenic plants have been made till now like antibiotic-resistant tobacco and petunia, Flavr Savr' tomato, canola with modified oil composition, *Bt*maize, *Bt*Potato, *Bt*cotton, etc. Maximum transgenic events have occurred in maize. Research conducted through a global meta-analysis on the outcomes of adopting genetically engineered crops reveals an average increase of 22% in crop yields, resulting in an approximate 68% boost in grower profits. The application of gene editing is anticipated to contribute to addressing the formidable task of providing sustenance for a global population of 10 billion individuals (Kumar et al. 2020).

12.2 Genetic Engineering Technology

The progress in recombinant DNA technology has been instrumental in enhancing crop improvement, particularly through transgenesis (Tiwari et al. 2021). Presently, gene-editing technology is widely employed to induce both phenotypic and genetic variations in plants.

12.2.1 Transgenic Technology

In this approach, a segment of DNA is transported to the target cell via chemical, physical, or biological or alternative means to integrate into the target genome, facilitating expression (Lanigan et al. 2020). The initial application of transgenic technology dates to 1983, when herbicide-resistant transgenic tobacco was created using *Agrobacterium tumefaciens*. Since that time, this technology has been consistently advancing. Several technologies are currently available like virus vector-mediated technology, *A. tumefaciens* transformation technology, polyethylene glycol induction, liposome-mediated transfection, etc. Biolistic technology and *A. tumefaciens* transformation are the most used in plants. In *A. tumefaciens* transformation, T-DNA is integrated into the genome randomly of damaged plants to achieve expression. Accordingly, the target sequence can be incorporated into *A. tumefaciens* at a multi-clone site and then randomly inserted into the host cell genome. While this uncomplicated and efficient method has its merits, it is constrained by its host range (Ye et al. 2022). This limitation prompted the development of biolistic technology, employing high-pressure inert gas to propel target DNA wrapped tungsten or gold particles into recipient cells for transformation. However, the effectiveness of DNA integration is not substantial in this method, and the simultaneous entry of numerous particles can result in an elevated quantity of DNA copies, possibly resulting in a higher occurrence of genes silencing or inactivation. In a study, the maize C4-PPDK (pyruvate orthophosphate dikinase) gene was introduced into rice, a C_3 plant via bombardment method. This resulted in raised CO_2 assimilation rates in transgenic lines along with higher photosynthetic activities. There was also rise in total nitrogen in the flag leaves indicating more nitrogen absorption from soil (Singh et al. 2010a, b, 2013a, b, 2014, 2021, 2024; Singh and Verma 2013a, b). The gene also affected tillers enhancing their yield (Zhang et al. 2010; Kumar et al. 2019, 2021). Further, in a similar study done by Shen et al. (2015), transgenic lines overexpressing maize phosphor *enol* pyruvate carboxylase (PEPC) showed improved ability to preserve water and pigments during drought. While drought induces the generation of reactive oxygen species, causing harm to photosystems, particularly PSI, the increased intercellular CO_2 concentration in transgenic rice protects peptides, photosynthetic complexes, and the thylakoid membrane's ultrastructure from oxidative damage.

12.2.2 Gene Editing Technology

Genome editing tools utilize nucleases to alter particular nucleotides in the target genome where double-stranded (ds) breaks specific to the target are introduced (Fig. 12.1). There are four main categories of plant genome editing methods that rely on site-specific endonucleases. These comprise meganucleases, ZFNs, TALENs, and CRISPR/Cas9. These methodologies have significant promise for enhancing and breeding crops (Ye et al. 2022).

The procedure generally involves two main stages: firstly, nucleases cleave the target sites of ds DNA, creating double-stranded breaks (DSBs) that can be harmful in several instances. Subsequently, to guarantee the preservation of the genome's integrity, cells activate two inherent repair pathways—homologous recombination (HR) pathway and non-homologous end junction (NHEJ) (Ye et al. 2022). The latter is simpler to execute, exhibiting effective repair but constrained stability. It is utilized for the deletion or insertion of small fragments which exists in the repair site region. HR is employed to mend fractures in homologous templates. While this method surpasses the traditional homologous recombination technique in terms of effectiveness and precision for gene editing, its efficiency falls short of that achieved by NHEJ repair (Murray and Carr 2018). MegNs, ZFNs, and TALENs are less preferred in genome editing as they are time-consuming and laborious. CRISPR is relatively easier, economical, and efficient to use for genome editing (Gao 2021).

12.2.2.1 Meganucleases

These are endonucleases capable of recognizing and cleaving sequences longer than 12 base pairs that are targeted in a manner specific to the sequence. They are found in archaebacteria, algae, yeast, bacteria, fungi, and even certain plant species (Kaur et al. 2022). Although discovered in the 1980s, they were extensively characterized within the mammalian cells in the 1990s. Because of their extended recognition sites, the likelihood of their target sites occurring is low within a sizable genome. Also known as homing endonucleases, they initiate DSBs at a specific location in the genome and subsequently spread throughout the genome via host homologous repair, though their exact role remains unclear (Silva et al. 2011). Meganucleases generated breaks are repaired via NHEJ system which leads to the generation of micro-insertions or error-prone deletions at the designated target site. However, their use in genome editing is constrained as the process of modifying them to aim different DNA sequences is quite complex (Kumar et al. 2021a, b).

12.2.2.2 Zinc Finger Nucleases

These are sequence-specific and man-made where endonuclease domain (Fok I endonuclease, exhibiting unrestricted cleavage capability) is merged with zinc finger DNA-binding domains (engineered zinc finger protein, demonstrating selective binding capability) to recognize a 3-bp module. The former encompasses *Flavobacterium* okeanokoites I (FokI) and the latter have 3- to 6-zinc finger repeats, identifying a 9–18-bp-long nucleotide sequence. FokI enzyme acts in dimer form, and therefore for each target sequence, dual ZFNs are needed for cleavage (Iqbal et al. 2020). This results in DSB formation that are subsequently repaired either by HR in the presence of homologous template sequence or NHEJ in its absence. NHEJ often results in frame-shift mutations owing to nucleotides deletion or insertion (Kaur et al. 2022). ZFNs are applied in crops like corn, rice, etc. But it has its intricacies like it's hard to obtain higher efficiency via casually linking specific ZFPs; has high off-target efficiency, not suitable for multiple editing; and has high cytotoxicity. Therefore, its applications in crops are limited. Constructing large modular proteins inclusive of ZFNs pair which bring *Fok*1 in target DNA vicinity is a rather expensive and tedious process which limits its wide application. They introduce a broad range of genetic modifications in the target genome like deletions, inversions, insertions, translocations, and point mutations. ZFNs have better activity and specificity and therefore have been engineered for generation of knockouts for removing gene function and for incorporating gain of function mutations (Novak 2019)). ZFN was used for maize plants where IPK1 disruption leads to herbicide resilience along with changes in inositol phosphate within developing seeds.

12.2.2.3 TALENs

These restriction enzymes are engineered to cleave DNA at precise sites. TALENs recognize single base pairs where transcription activator-like effector domains (for DNA-binding) are linked with endonuclease domains (*Fok*1 nuclease) (Iqbal et al. 2020). The former is composed of natural TAL effector (TALE) transcription factors, being released upon infection by *Xanthomonas*. They are identified as repeat variable residues (RVDs), consisting of 33–34 amino acids that display conserved repetitive sequences, with variability occurring at the 12th and 13th positions and recognize specific DNA sequences that are TALEN substrate. The cleavage domain is composed of FokI enzyme and dimerized for action. TALENs pair will form dimer of Fok I endonuclease after subsequent target site binding for the generation of DSBs. They are less toxic and more targeted and have no such contextual effect like ZFPs. They have been used in many crops like wheat, rice, and corn. Though they are better than ZFNs, their novel array construction for specific gene is not pocket friendly for gene editing, and they are also unsuitable for multiple gene editing due to their need of a protein pair for recognizing both sense and antisense DNA strands for DSB. These are not recommended for highly methylated sites or multiple editing as their binding sites are constrained by guanine nucleotides and so limited in use. In a work done by Clasen et al. (2016), they were employed to disrupt the vacuolar invertase gene (VInv) in potatoes. This gene is responsible for protein production that catalyses the breakdown of sucrose into fructose and glucose. Tubers harvested from plants with a complete knockout exhibited undetectable levels of reducing sugars, and the chips also had less acrylamide levels and were lightly coloured. In another study, OsBADH2 (betaine aldehyde dehydrogenase) gene was disrupted. The defective badh2 allele leads to the production of 2-acetyl-1-pyrroline (2AP), a primary aromatic compound. This resulted in increase of 2AP content of rice grains from none to 0.35–0.75 mg/kg in the T1 lines exhibiting homozygous mutations (Shan et al. 2015).

12.2.2.4 CRISPR/Cas System

Archaea and bacterial species employ this system as a defence mechanism to protect themselves against invading viruses. About 40% of different bacterial species and almost all archaea sequenced genomes have this system which can degrade DNA, RNA, or both. This system is the most commonly used. Cas9 nuclease was discovered from *Streptococcus pyogenes* and paved the foundation for this system. This system contains customizable single-guide RNA (sgRNA) recognizing target DNA and binding via Watson and Crick base pairing and DNA endonuclease (monomeric Cas9). The former one is changeable based on target sequence comprising of a trans-activating crRNA (tracrRNA) and proto-spacer adjacent motif (PAM) of CRISPR RNA (crRNA) in bacteria (Rahman et al. 2022). The crRNA comprises a sequence that is 20 nucleotides long and complements the target sequence, and the tracrRNA joins pre-crRNA to form a double-stranded RNA. RNase III activates the pre-crRNA, leading to its transformation into mature crRNA (Jiang and Doudna 2017). Later it was found that dual tracrRNA and crRNA can be replaced by a sgRNA to guide Cas9 protein. Following the formation of the Cas9-sgRNA complex, it moves towards the target DNA double helix and explores for a canonical PAM adjacent to the potential gRNA complementary sequence that results in gRNA-DNA heteroduplex, further resulting in Cas9 nickase-mediated breakage of DNA. The Cas9 nuclease encompasses the following: REC I and REC II domains (recognition domains) for binding DNA and sgRNA, HNH and RuvC (nuclease domains) for complementary and non-complementary strand cleavage at target region, bridge helix domain for nuclease activity initiation, and the protospacer adjacent motif (PAM) interaction (PI) domain. Majority of Cas proteins need PAM at the 3′ site, but some require at 5′ of the target DNA (Rahman et al. 2022). The RuvC domain of the Cas9 protein cleaves the PAM sites of DNA, and the HNH domain of the Cas9 protein fractures the DNA strand that is complementary to the sgRNA.

The CRISPR/Cas9 tools are highly specific and are widely employed in plants for gene knockout, base editing, gene corrections, gene replacement, prime editing, and site-specific transgene incorporation. This system is widely used in crop enhancement. It is now a leading system for targeted mutagenesis in eukaryotes. This system is suitable for several edits and is highly efficient to generate deletions/insertions. However, it has off-target effects and targeting restriction by PAM sites which limits its efficiency. The introduction of prime editors and base editors has broadened the functionalities of this system. This system is simple, inexpensive, and adaptable, compared to other tools, and therefore has been broadly used for trait improvement. Its capacity for multiplex genome editing positions it as a top priority within the scientific community (Ye et al. 2022).

CRISPR/Cas9 used in a study to knockout *OsNAC006* gene showed a raised H_2O_2 level and superoxide radicals (O^{2-}) and simultaneously declined chlorophyll content and antioxidant enzymatic activity. This shows that this gene is engaged in heat stress tolerance via facilitating photosynthesis and limits antioxidant enzyme activity, which was triggered due to oxidative stress under elevated temperature (Singh et al. 2020; Kumar et al. 2021c, d). Another study utilized this technology to

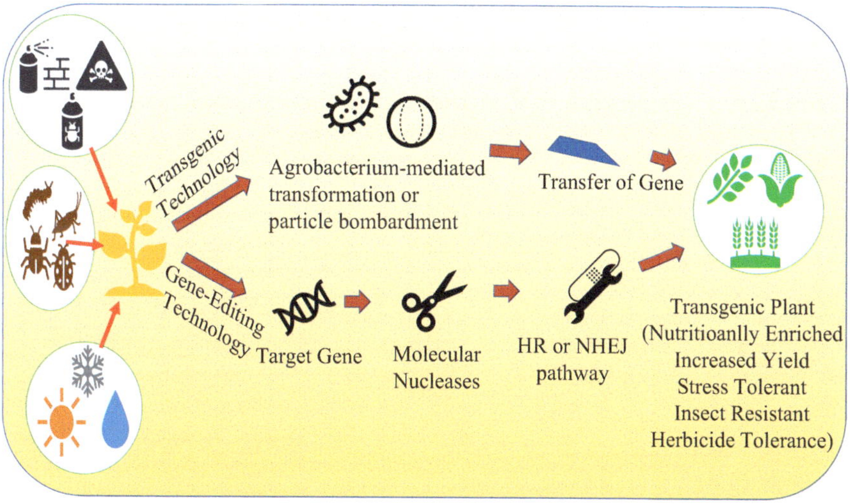

Fig. 12.1 The crop under various types of stresses can be made tolerant towards them via gene editing and transgenic technology

edit two group of pyrabactin resistance like genes in rice. Genes of group 1 (PYL1–PYL6 and PYL12) showed to have extra essential functions in seed dormancy, stomatal movement, besides growth regulation as compared to those in group 2 (PYL7–PYL11 and PYL13). pyl1/4/6 genes, among group I mutants, demonstrated optimal growth and increased grain yield (Wang et al. 2020) (Fig. 12.1).

12.3 Advantages of Manipulating Functional Traits in Plants

12.3.1 Insect Resistance

Insect pests are the crucial biotic stress agents causing significant crop loss. Farmers rely on high-priced chemical insecticides to tackle insect pests, but these are deleterious to environment as well as mankind and an economic burden on farmers (Tyagi et al. 2020). The recent expand in the horizons of gene editing strategies presents several possibilities of influencing crucial agronomic traits. Genetic engineering provides insect resistance to crop plants by incorporating specific DNA sequences or genes via either particle bombardment or *Agrobacterium*-mediated transformation. The cultivation area under insect-resistant (IR) crops ranked second in 2017 with 23.3 million hectares (ISAAA 2017). The first transgenic-resistant crop was cotton which was developed in the USA by Monsanto and Delta and Pine companies against lepidopteron insect pest (Little et al. 2017). As of now, ten transgenic crops with insect resistance have been brought into commercial use (Kumar et al. 2020). Majorly these crops have insecticidal genes like *cry* gene and *vip* gene. Few

Table 12.1 Various insect resistant crops developed via CRISPR/Cas technology

S. no.	Crop	Transgene	Target insect	Reference
1.	Cotton	*cry2AX1*	*Helicoveraarmigera*	Jadhav et al. (2020)
2.	Sweet Potato	*cry1Aa*	*Spodoptera litura*	Zhong et al. (2019)
3.	Soyabean	*cry 8 like*	*Holtrichiapanallele*	Qin et al. (2019)
4.	Sugarcane	*Vip3A*	*Chiloinfuscatellus*	Riaz et al. (2020)
5.	Cowpea	*Arcl*	Bruchids	Grazziotin et al. (2020)
6.	Rice	*cry2Aa+cry1Ca*	*Chilosuppressalis*	Qiu et al. (2019)
7.	Pigeon pea	*cry2Aa*	*H. armigera*	Singh et al. (2018)
8.	Cowpea	*Vip3Ba1*	*Marucavitrata*	Bett et al. (2017)
9.	Cotton	*cry51Aa2*	*Lygus* species	Gowda et al. (2016)
10.	Tomato	*cry1Ab*	*Tutaabsoluta*	Soliman et al. (2021)

insect resistance crops are listed in table 12.1, developed by CRISPR/Cas technology (Table 12.1).

Bacillus thuringiensis (Bt), a soil bacterium with a positive gram stain, generates crystals outside the spore that possess insecticidal proteins (ICPs). These endotoxins specially target the epithelium tissue of the midgut in larvae of coleopteran, dipteran, and lepidopteran insects and cause pore formation, resulting in ion leakage and subsequently insect mortality. Cry protein is non-toxic to mammals. In addition to ICPs, plant species have been engineered to express lectins, vegetative insecticidal proteins (VIPs), fusion proteins, and genes associated with pathogen defence (Paul and Das 2021). VIP genes derived from *Bacillus* species like *B. thuringiensis* and *B. cereus* have been isolated with *vip3Aa20* and *vip3A(a)* genes known to be heterologously expressed in maize and cotton, respectively, as per ISAAA Database (2019). Furthermore, genes encoding protease inhibitors (PIs) sourced from bacteria, plants, and fungi have also been utilized to grant plants resistance against insects. PIs are natural defence proteins produced by plants in reaction to insect assault or physical damage (Kumar et al. 2020). These inhibitors function by impeding the proteolytic enzymes found in the digestive system of insects and their larvae, thereby hindering the acquisition of essential amino acids for their growth and development. Examples include potato protease inhibitor II encoded by *pinII* gene and trypsin inhibitor encoded by *CpTI* gene that restricts insect digestive enzymes.

Transgenic crops having insect-resistant genes have shown positive impact on productivity. Chimeric toxins created by exchanging domains among toxins and utilizing combinatorial ICPs have been employed to enhance effectiveness in combating insects and to mitigate the development of resistance. RNA interference (RNAi) may be used in silencing selected genes of insect feeding and development. This leads to the trait improvement via targeted 'loss-of- function' engineering. However, the complete silencing of the target loci is absent by RNAi technology which limits its wide use. It is imperative in the current context to cultivate plant varieties with genetic resilience (Tyagi et al. 2020).

12.3.2 Herbicide Tolerance

In agricultural fields, weeds vie with crops for essential resources like water, space, and light and can also contribute to the spread of diseases and pests and therefore limit crop yield, productivity, and quality (Singh 2024). Their seeds or pollen also contain toxins and may cause poisoning in animals and human on mixing with crop seeds. Chemical herbicides, though, are an easy method for weed control, but herbicide-resistant crops are better alternative as they reduce chemicals-mediated toxicity, reduce weeding cost, and also expand the herbicidal spectrum. Conventional mutation breeding and crossbreeding can be used for introducing single-nucleotide polymorphism (SNP) for important traits, but this process is expensive, laborious, time-consuming, and sometimes even impossible.

Resistance to herbicides in crops can be attained through various mechanisms, including target site gene amplification, target site mutation, elevated detoxification of herbicides, and enhanced metabolism. Target site mutation, recognized as the principle mechanism of resistance, has proven effective in developing herbicide-resistant germplasm across numerous crops. Mainly the target genes selected are *ACCase*gene, *ALS* gene, and *EPSPS* gene for producing herbicide-resistant crops. In the past, the creation of herbicide-tolerant crops relied on conventional mutagenesis breeding methods, employing chemical mutagenesis with EMS (ethylmethanesulfonate) to induce many mutations, especially C to T transitions (Gosavi et al. 2022). Its major drawbacks were time-taking and labour-intensive which leads to the development of transgenic technology as the primary method for crop enhancement. The initial herbicide-tolerant genetically engineered crops which become commercially available were developed to resist broad-spectrum herbicides such as glufosinate and glyphosate. A list of herbicide tolerant crops, developed via numerous genome editing technologies is given in table 12.2.

The strict regulations and lengthy and expensive procedures of these crops have limited their use.

In the past, TALENs and ZFNs were extensively employed for site-directed mutagenesis. However, CRISPR/Cas9 is currently the predominant tool for gene editing. Synthetic auxins, glyphosate, PSII (Photosystem II) inhibitors (triazinones, triazines, amides, phenylcarbamate, triketones, isoxazoles, and dinitroaniline) are the frequently employed herbicides in agricultural cultivation. The recently developed cytidine base editors (CBEs) employ Cas9n-fused cytidine deaminases (hAPOBEC3A, hHID) for inducing C-to-T conversion without the generation of ds breaks and without the requirement for a donor DNA. Till now, it has been used in developing herbicide-tolerant maize, watermelon, rice, Arabidopsis, and wheat (Gosavi et al. 2022). Adenine base editors (ABEs), comprising Cas9n and adenine deaminases (specifically *Escherichia coli* TadA variants), are utilized for gene targeting as well, and they induce A-to-G conversion, conferring tolerance to herbicides. Concurrently editing four intrinsic herbicide-targeting genes (OsTubA2, OsALS1, OsACC, OsGS1) using ABEs in rice highlight the considerable potential of this approach in creating crops tolerant to a broad spectrum of herbicides. Prime editors (PEs) employ a reverse transcriptase in combination with CRISPR/Cas9

Table 12.2 Several herbicide tolerant crops developed through various genome editing technologies

S. no	Crop	Target gene	Mutations	Method	Tolerance towards	Reference
1.	Tobacco	*ALS SuRA*and*SuRB*	P191A, W568L, and S647T	ZFN	Imidazolinones and sulfonamide herbicides	Townsend et al. (2009)
2.	Rice	*OsALS*	W548L and S627I	TALEN	Bispyribac-sodium-tolerant lines	Li et al. (2016)
3.	Rice	*OsEPSPS*	DNA fragment replacement and insertion via NHEJ pathway with donors carrying TIPS	CRISPR/Cas9	Glyphosate tolerance	
4.	Rice	*TaALS genes*	Generating various nucleotide substitutions in A, B, and D genomes of wheat	rAPOBEC1-Cas9n and specific sgRNAs	Imidazolinone	Shimatani et al. (2017)
5.	Wheat	*Acetyl-coenzyme A protein*	Modification of the A1992 site	Cytidine Base Editors	Quizalofop	Zhang et al. (2019)
6.	Allotetraploid oilseed rape	*BnALSs*	P197 site	rAPOBEC1 or hAPOBEC3A fused to a Cas9 nickase	Sulphonylureas	Cheng et al. (2021)
7.	Rice	*OsACC*	C2186R substitution	Adenine base editors	Haloxyfop-R-methyl	Li et al. (2018a, b)
8.	Rice	*OsALS*	C-to-G (W548L)and C-to-T (P171S)	Prime editors	Aryloxyphenoxypropionate	Xu et al. (2020)
9.	Rice	*OsEPSPS*	G-to-C (W2125C) transversion in OsACC and TIPS change	Prime editors	Herbicide	Li et al. (2020)
10.	Maize	*ZmALS1 gene*	W542L and/or S621I	Prime editors	Multiple herbicides	Jiang et al. (2020)

nickase (Cas9n) (H840A) and a prime editing guide RNA (pegRNA) that makes the deletion, insertion, and base-to-base conversions without donor DNA need. Consequently, multiple nucleotide edits can be introduced at target sites in crops, and this technology has found extensive application in editing genes related to herbicide targeting in crops. However, the usage of Pes is constrained by its relatively low editing efficiency (Yan et al. 2021).

12.3.3 Stress Tolerance

All abiotic stress leads to severe loss in crops and therefore threatens food security globally. Drought shows around 50–70% reduced crop yield (Kumar 2020). Salinity stress is another menace, and about 20% of the irrigated farmland is impacted by elevated salt levels (Van Zelm et al. 2020). Temperature also depresses crop production (Kumar et al. 2021). Plants flourish most in an ideal temperature range, and any deviation, whether an increase or decrease, significantly impedes plants growth (Kazemi-Shahandashti and Maali-Amiri 2018). It becomes a necessity in changing climate to develop resilient crop varieties which can withstand abiotic stresses impact. Even though genetic engineering as well as conventional breeding aided remarkably in producing biotic stress-resilient crop varieties, the complex mechanisms of abiotic stresses limited their addressable. The application of genome editing has been extensively used to improve crops' resilience to abiotic stress (Table 12.3), but the rigid regulations further limit wide cultivation of such crops (Shukla et al. 2018). It can be used to study newer traits and to target trait improvement for ameliorating abiotic stress impact (Kumar et al. 2020a, b, c). The utilization of conventional breeding methods for the enhancement of crops' resilience to abiotic stress faces limitations, as it is a slow process attributed to challenges such as the incompatible transfer of traits between species, the requirement for multiple backcrosses to achieve the desired trait, the involvement of many genes, and the requirement for a suitable gene for breeding. In contrast, genetic engineering proves more

Table 12.3 The transgenic crops made tolerant from various abiotic stresses via CRISPR/Cas system

S. no.	Crop	Targeted gene	Tolerance towards	Reference
1.	Rice	*OsPQT3*	Salinity	Alfatih et al. (2020)
2.	Rice	*PYL1/4/6*	High-temperature	Miao et al. (2018)
3.	Cotton	*HB12* gene	Drought	He et al. (2020)
4.	Arabidopsis	*ACQOS*	Salinity	Kim et al. (2021)
5.	Wheat	*SAL1*	Drought	Abdallah et al. (2022)
6.	Rapeseed	*A6.RGA*	Drought	Wu et al. (2020)
7.	Lettuce	*NCED4*	Thermotolerance	Bertier et al. (2018)
8.	Tomato	*HyPRP1*	Salinity	Tran et al. (2021)
9.	Arabidopsis	*Trehalase gene*	Drought	Nuñez-Muñoz et al. (2021)
10.	Rice	*OsPIN5b*, *GS3*, and *OsMYB30*	Cold	Zeng et al. (2020)

advantageous, as it enables the modification of the desired gene within the same or even distantly related species (Singh and Singh 2014). The arrival of functional and structural genomics methodologies, coupled with insights into metabolic regulatory pathways of stress-responsive genes, and the discovery of profitable genetic traits using contemporary molecular markers, next-generation sequencing techniques, and genome-wide association studies, eases the integration of these traits via genetic engineering. This aids in the creation of plant varieties resilient to abiotic stress.

Genetic engineering has been done for osmoprotectants, aquaporin, heat shock proteins, molecular chaperones, plant transcription factors, and helicases which have roles in abiotic stress tolerance. For the regulation of osmotic adjustments under salt stress, ion transport genes are targeted by genetic engineering and genome editing tools. Plants require an optimum temperature of around 15–24 °C. Elevated heat stress leads to reduced crop yields, and it is anticipated to increase in the future. The changing climate and global warming require the target identification to enhance heat stress tolerance and development of heat tolerant varieties. PhytochromeB (PHY) has been identified as a thermosensor and therefore is an important target (Rahman et al. 2022).

12.3.4 Nutritional Improvement

The demand for food is also continuously rising with the growing population. Projections indicate about almost double rise in livestock products and in cereals. The global dietary habits are continuously changing and shifting from starch-based diets to a high-fat diet resulting an increase in lifestyle-related diseases worldwide (Nagamine and Ezura 2022). To address this, several countries have adopted improving crop nutrition among their national policy. Functional and nutritional attributes stand out as primary objectives in the enhancement of crops, focusing on the manufacturing of animal feed, food, and raw materials (Ku and Ha 2020). Time taking conventional breeding techniques have shifted the focus towards genome editing approaches like CRISPR/Cas9, TALEN, and ZFN (Table 12.4). Genome editing is widely employed in major staple crops like wheat, potato, rice, and maize. In case of oil seed crops, the most favoured genome editing target is monounsaturated fatty acid production, specifically oleic acid (18:1). Oleic acid is widely recognized as being both more healthful and stable, contributing to a longer shelf life compared to products containing polyunsaturated fatty acids such as linoleic acid (18:2). Thus, in rapeseed, camelina, and soybean, the focus is on the fatty acid desaturase 2 gene (*FAD2*), which encodes delta-12 desaturase responsible for converting oleic acid to linoleic acid. This gene is pinpointed and intentionally interrupted through genome editing to modify the fatty acid profile of these crops. Development of golden rice through genome editing was the major step in combating vitamin A deficiency-related diseases (Tiwari et al. 2021).

Table 12.4 Nutritionally enriched crops made via genetic engineering technology

S. no	Crop	Target gene	Technique	Effect	Reference
1.	Wheat	*Taα-gliadin*	CRISPR/Cas9	Less gluten content in wheat grain	Sánchez-León et al. (2018)
2.	Tomato	*ALC allele*	CRISPR/Cas9	Long- shelf-life of tomato	Yu et al. (2017)
3.	Corn	*IPK1*	ZFNs	Reduction in phytate content	Shukla et al. (2009)
4.	Rice	*BADH2*	TALEN	Production of Fragrances	Shan et al. (2015)
5.	Soybean	*FAD2-1A and FAD2-1B*	TALENs	Oil quality was improved	Clasen et al. (2016)
6.	Rice	*Waxy*	CRISPR/Cas9	Low-amylose and glutinous rice	Zhang et al. (2018)
7.	Sorghum	*K1C gene family*	CRISPR/Cas9	Increased digestibility and protein quality	Li et al. (2018a, b)
8.	Corn	*ZmSH2 &WX*	CRISPR/Cas9	Super sweet and waxy corn	Dong et al. (2019)
9.	Rice	*Cytochrome P450s and OsBADH2*	CRISPR/Cas9	High yielding fragrance rice variety	Usman et al. (2020)
10.	Banana	*Lycopene epsilon-cyclase gene*	CRISPR/Cas9	β-carotene-enriched banana	Kaur et al. (2020)

12.3.5 Enhanced Yield

Enhancing crop performance has consistently sought to attain increased yields, improved stress resistance, and nutritional quality (Ku and Ha 2020). The yield of crops is influenced by intricate signal networks encompassing intersection of genes and transcription factors, along with plant hormones. In hybrid japonica rice, inadequate grain filling is considered a significant bottleneck, restricting the widespread cultivation of hybrid rice, and is believed to be one of the primary factors affecting yield. This was improved via CRISPR/Cas9 for the target allele for grain-filling rate (GFR1) to improve the grain-filling rate. CRISPR/Cas9 tool is employed to identify casual genes within vital quantitative trait loci (Tiwari et al. 2021). In a genome-wide association study carried out on a natural population of rice, a newfound gene related to grain size, OsSNB, was discovered using CRISPR/Cas9 technology, and it is suggested to be a pivotal target for enhancing rice yield. An additional investigation pinpointed 129 potential genes linked to high yield across 30 rice cultivars, including IR8 and its descendants and ancestors as landmarks. Among these, 57 genes were subjected to mutation using the CRISPR/Cas9 to elucidate their respective functions. The analysis indicated that the locus Os10g0555100, encoding glucosyltransferase, regulates growth. Consequently, gene editing has significantly contributed to the identification of new alleles or genes responsible for increased

Table 12.5 The yield of crops raised through gene editing technologies

S. no	Crop	Targeted gene	Technology	Effect	Reference
1.	Rice	*OsGFR1*	CRISPR/Cas	High grain filling in japonica hybrid	Liu et al. (2019)
2.	Rice	*OsAH2*	CRISPR/Cas	Improved hull development and grain size	Ren et al. (2019)
3.	Rice	OsAAP5	CRISPR/Cas system	High yield and tiller number	Wang et al. (2019)
4.	Tomato	*SlENO*	CRISPR/Cas system	Enhanced fruit number and size	Yuste-Lisbona et al. (2020)
5.	Wheat	*TmNAS3*	Transgenic technology	Increase in the iron content and size of grains, thus increased yield	Wang et al. (2022)
6.	Rice	*OsAAP3*	CRISPR/Cas	Increase tiller number for higher yield	Lu et al. (2018)
7.	Rice	*PYL1-4-6*	CRISPR/Cas	Yield was increased by 25–31%	Miao et al. (2018)
8.	Wheat	*TaGW7-B1*	CRISPR/Cas	Increase grain weight for higher yield	Wang et al. (2021b)
9.	Wheat	*TaSTT3b-2B*	Transgenic technology	Increase grain weight for higher yield	Zhu et al. (2022)
10.	Rice	microRNA168	RNA silencing technology	Increase the number of ears for higher yield	Wang et al. (2021a)

productivity, as exemplified by the discovery that amino acid permease (OsAAP5) controls grain quality, tiller number, and yield (Wang et al. 2019). In conclusion, genetic engineering technology, particularly gene editing, holds immense potential for enhancing both crop yield and quality (Ye et al. 2022). The yield of various crops have been enhanced through gene editing technologies (Table 12.5).

12.4 Intricacies Associated with the Technology and Public Acceptance

The genetically modified crops have not yet been fully accepted owing to their health and environment concerns. There is a need for long-term feeding trails for thorough evaluation of transgenic crops safety. The *Cry9c* expressing 'Starlink' maize's protein is highly stable and so can be allergic to humans. In 2000, there was a global recall of this particular variety because of the identification of remnants of the Cry protein in food items. Another concern involves the potential horizontal transfer of antibiotic-resistant marker genes from genetically modified food to the gut microorganisms of humans and animals, raising concern about the development of antibiotic resistance in gut microflora. While such occurrences are infrequent, efforts have been undertaken to create transgenic crops that are free of markers.

Environmental concerns include the transfer of transgenes through pollen from genetically modified crops to traditional varieties and wild counterparts of cropslike creeping bent grass, barley, rice, rapeseed, and beans (Yan et al. 2015). Gene transfer from transgenic crop to their wild relative results in biodiversity loss and flow to weeds, or weedy relatives possess the risks of arrival of herbicide-resistant weeds also known as 'superweeds', and advent of new weed species like glyphosate-resistant weeds has been documented in 16 out of 24 transgenic cropping systems; among them, *Conyza canadensis* is the most prevalent, and *Amaranthus tuberculatus* and *A. palmeri* are recognized as the most financially detrimental glyphosate-resistant weeds on a global scale (Heap and Duke 2018). The high selection pressure forced by the widespread planting of transgenic IR and HT crops may lead to rise in resistance in the targeted weeds and insect population. This might result in evolution of new insect biotypes, thus leading to superweed emergence which has resistance towards transgenic technology. The main obstacle to the long-haul viability of Bt crops is the development of resistance among pests to Bt toxins. A significant instance of control failure in Bt cotton, attributed to the development of pink bollworm resistance to Cry1Ac, has been documented in Gujarat, India. Similarly, in the United States, the extensive adoption of herbicide-resistant (HR) crops resulted in the hastened evolution of weed resistance to glyphosate. Pyramiding of several insect-resistant genes has been successfully used as a strategy for delaying the resistance mechanism (Kumar et al. 2020). The elimination of primary pests may create conditions for secondary pests to thrive as the main one as reported in China where the introduction of Bt cotton has led to a substantial increase in the population of the mirid bug, which was previously a minor pest but has become a major pest over the last decade. Sometimes, off-target effects are triggered during genome editing which limits the benefits of CRISPR. The unwanted mutations of genes like deletions, insertions, inversion, and translocation comes under off-target effects. The main strategy to minimize these risks is via improving sgRNA targets to reduce mismatches and limiting Cas9 levels in cells for minimizing unintended binding (Das et al. 2023).

The costly safety assessment and the intricate, time-consuming regulatory approval process needed for commercialization serve as constraints factors that impede the advancement and implementation of transgenic crop products. The estimated time is generally 13 years for the whole process starting from transgenic crop project development till its launch. Longer and expensive regulatory approval process limits entry of smaller biotechnology firms into transgenic crop sector. Patentable nature of transgenic crop genes limits access of important genetic elements to others. It is necessary to conduct a risk assessment of GM crops before obtaining approval for commercialization to ensure the safety of the environment and food/feed (Kumar et al. 2020).

12.5 Future Perspectives

The public concern and unacceptance over transgenic crops have introduced several alternative technologies for introducing enhanced crops in recent times. The methods like intragenesis and cisgenesis make use of genetic modifications, and the modified crop genome does not possess any foreign genetic material. Consequently, the GM crop plants are comparable to those created through conventional breeding. Moreover, these methods assist in producing improved crop plants that may be challenging to achieve using traditional breeding approaches (Kumar et al. 2020). Cisgenesis entails genetically modifying organisms by introducing a complete, unaltered gene sequence, including introns and its native regulatory elements (such as the native promoter and terminator), in the same orientation. The cisgenesis improved several crop traits as high phytase activity in barley and scab resistance in apple. Intragenesis involves the insertion of preferred genes from the identical crop species or sexually compatible plant species, either in the sense or antisense orientation (Kumar et al. 2020). The regulatory components, including the promoter and terminator, can be obtained from genes other than the target gene itself, creating a novel genetic arrangement that modifies the gene's expression patterns. This method has been successfully utilized to attain targeted suppression of the asparagine synthase-1 (StAst1) gene in potatoes, specifically targeting tuber tissues. The aim is to decrease acrylamide levels (a potential carcinogen) following processing, with reported field trials demonstrating a 70% reduction in the acrylamide-forming potential of potatoes. Importantly, this reduction did not have any adverse effects on tuber yield or shape (Chawla et al. 2012). There is a need for the update of regulatory process for precise gene-editing methods that may not need the identical review as conventional GE. There is a need for GMO-like regulation for genome editing that is based on approaches used for plant production. The next wave of genetically modified products can boost public attitude towards these crops, and the advancement of technologies and development of unintended safety evaluation methods can ameliorate the potential risks associated with these crops (Chen and Lin 2013). This opens the door for the future sustainable advancement of genetically modified crops. The HT and IR traits of transgenic crops dominated the GM industry for a long time, but now many upcoming generations of GM crops having more health benefits are coming to the market. Genome editing has its own hurdles, and its future depends upon overcoming those challenges and issues. Various ethical questions need to be answered by scientists as well as society.

Even after the advancement of genome editing technology, there is still limitation in generating all the wanted changes within the genome. Along with that, for better trait improvement in crops, precise genome editing is extremely needed. The accuracy of CRISPR-Cas can be improved by temporarily expressing the editing components, as shown in wheat and maize. This is achieved through the use of carefully designed guide RNAs or by utilizing engineered, precise versions of Cas12a, Cas9, and deaminases (Jin et al. 2020). Nevertheless, additional research is required for addressing the potential for unintended editing, chiefly in the development of more responsive approaches for detecting genome-wide unintended mutations in plants and in the identification of upgraded or novel editors with greater specificity. Concerns about unintended effects of genome editing in plants will be addressed through technological advancements. More efforts for ensuring regulatory transparency and open dialog are still needed (Gao 2021). A science-based public communication should also be ensured.

12.6 Conclusion

The advancements in genome editing platforms have opened new horizons of plant research. Designing plants having multiple attributes along with the climatic resilience is no longer just in theory. The continuous efforts and more research in this direction can ameliorate crop losses occurring due to wide range of factors and aids in successfully fulfilling the world hunger. However, the intricacies specially by the regulatory authorities need to be tackled by the respective governments across countries for transgenic crops wide cultivation. The public concern over their toxicity towards environment and humans also need to be looked upon by the scientific community for these crops' wide acceptance. The rising human population, arable land shrinkage, and climate change have raised the need for the creation of crop varieties with high yields. These varieties should not only be nutritionally enhanced but also possess resilience against stresses. These crops also reduce the usage of harmful chemicals over plants, thereby reducing environmental footprint. The advancements in genome editing have opened novel opportunities for plant breeding. This will revolutionize agriculture future.

References

Abdallah NA, Elsharawy H, Abulela HA, Thilmony R, Abdelhadi AA, Elarabi NI (2022) Multiplex CRISPR/Cas9-mediated genome editing to address drought tolerance in wheat. GM Crops Food. https://doi.org/10.1080/21645698.2022.2120313

Alfatih A, Wu J, Jan SU, Zhang ZS, Xia JQ, Xiang CB (2020) Loss of rice PARAQUAT TOLERANCE 3 confers enhanced resistance to abiotic stresses and increases grain yield in field. Plant Cell Environ 43(11):2743–2754

Arora L, Narula A (2017) Gene editing and crop improvement using CRISPR-Cas9 system. Front Plant Sci 8:1932

Bertier LD, Ron M, Huo H, Bradford KJ, Britt AB, Michelmore RW (2018) High-resolution analysis of the efficiency, heritability, and editing outcomes of CRISPR/Cas9-induced modifications of NCED4 in lettuce (*Lactuca sativa*). G3 (Bethesda) 8(5):1513–1521

Bett B, Gollasch S, Moore A, James W, Armstrong J, Walsh T et al (2017) Transgenic cowpeas (*Vigna unguiculata* L. Walp) expressing Bacillus thuringiensis Vip 3Ba protein are protected against the Maruca pod borer (Marucavitrata). Plant Cell Tissue Organ Culture 131:335–345

Bhattacharyya N, Anand U, Kumar R, Ghorai M, Aftab T, Jha NK et al (2023) Phytoremediation and sequestration of soil metals using the CRISPR/Cas9 technology to modify plants: a review. Environ Chem Lett 21(1):429–445

Chawla R, Shakya R, Rommens CM (2012) Tuber-specific silencing of asparagine synthetase-1 reduces the acrylamide-forming potential of potatoes grown in the field without affecting tuber shape and yield. Plant Biotechnol J 10(8):913–924

Chen H, Lin Y (2013) Promise and issues of genetically modified crops. Curr Opin Plant Biol 16(2):255–260

Cheng H, Hao M, Ding B, Mei D, Wang W, Wang H et al (2021) Base editing with high efficiency in allotetraploid oilseed rape by A3A-PBE system. Plant Biotechnol J 19(1):87–97

Clasen BM, Stoddard TJ, Luo S, Demorest ZL, Li J, Cedrone F et al (2016) Improving cold storage and processing traits in potato through targeted gene knockout. Plant Biotechnol J 14(1):169–176

Das T, Anand U, Pal T, Mandal S, Kumar M, Radha et al (2023) Exploring the potential of CRISPR/Cas genome editing for vegetable crop improvement: an overview of challenges and approaches. Biotechnol Bioeng 120(5):1215–1228

Dong Z, Hu Z, Qin Q, Dong F, Huang L, Long J, Chen P, Lu C, Pan M (2019) CRISPR/Cas9-mediated disruption of the immediate early-0 and 2 as a therapeutic approach to *Bombyx mori* nucleopolyhedrovirus in transgenic silkworm. Insect Mol Bio 28(1):112–122

Gao C (2021) Genome engineering for crop improvement and future agriculture. Cell 184(6):1621–1635

Gosavi G, Ren B, Li X, Zhou X, Spetz C, Zhou H (2022) A new era in herbicide-tolerant crops development by targeted genome editing. ACS Agric Sci Technol 2(2):184–191

Gowda A, Rydel TJ, Wollacott AM, Brown RS, Akbar W, Clark TL, Flasinski S, Nageotte JR, Read AC, Shi X, Werner BJ (2016) A transgenic approach for controlling Lygus in cotton. Nat Comm 7(1):12213

Grazziotin MA, Cabral GB, Ibrahim AB, Machado RB, Aragão FJ (2020) Expression of the Arcelin 1 gene from Phaseolus vulgaris L. in cowpea seeds (*Vigna unguiculata* L.) confers bruchid resistance. Ann App Bio 176(3):268–274

He X, Luo X, Wang T, Liu S, Zhang X, Zhu L (2020) GhHB12 negatively regulates abiotic stress tolerance in Arabidopsis and cotton. Environ Exp Bot 176:104087

Heap I, Duke SO (2018) Overview of glyphosate-resistant weeds worldwide. Pest Manag Sci 74(5):1040–1049

Iqbal Z, Iqbal MS, Ahmad A, Memon AG, Ansari MI (2020) New prospects on the horizon: genome editing to engineer plants for desirable traits. Curr Plant Biol 24:100171

ISAAA (2017) Global status of commercialized biotech/GM crops in 2017: biotech crop adoption surges as economic benefits accumulate in 22 years. ISAAA brief no. 53. ISAAA, Ithaca

ISAAA Database (2019) GM approval database. ISAAA, Ithaca

Jadhav MS, Rathnasamy SA, Natarajan B, Duraialagaraja S, Varatharajalu U (2020) Study of expression of indigenous Bt cry2AX1 gene in T 3 progeny of cotton and its efficacy against *Helicoverpa armigera* (Hubner). Braz Arch Biol Technol 63:e20180428

Jiang F, Doudna JA (2017) CRISPR–Cas9 structures and mechanisms. Annu Rev Biophys 46:505–529

Jiang YY, Chai YP, Lu MH, Han XL, Lin Q, Zhang Y et al (2020) Prime editing efficiently generates W542L and S621I double mutations in two ALS genes in maize. Genome Biol 21:1–10

Jin S, Fei H, Zhu Z, Luo Y, Liu J, Gao S et al (2020) Rationally designed APOBEC3B cytosine base editors with improved specificity. Mol Cell 79(5):728–740

Kaur N, Alok A, Shivani, Kumar P, Kaur N, Awasthi P, Chaturvedi S, Pandey P, Pandey A, Pandey AK, Tiwari S (2020) CRISPR/Cas9 directed editing of lycopene epsilon-cyclase modulates metabolic flux for beta-carotene biosynthesis in banana fruit. Metab Eng 59:76–86. https://doi.org/10.1016/j.ymben.2020.01.008

Kaur N, Sharma S, Hasanuzzaman M, Pati PK (2022) Genome editing: a promising approach for achieving abiotic stress tolerance in plants. Int J Genomics 2022:5547231

Kazemi-Shahandashti SS, Maali-Amiri R (2018) Global insights of protein responses to cold stress in plants: signaling, defence, and degradation. J Plant Physiol 226:123–135

Kim ST, Choi M, Bae SJ, Kim JS (2021) The functional association of ACQOS/VICTR with salt stress resistance in Arabidopsis thaliana was confirmed by CRISPR-mediated mutagenesis. Int J Mol Sci 22(21):11389

Ku HK, Ha SH (2020) Improving nutritional and functional quality by genome editing of crops: status and perspectives. Front Plant Sci 11:577313

Kumar S (2020) Abiotic stresses and their effects on plant growth, yield and nutritional quality of agricultural produce. Int J Food Sci Agric 4:367–378

Kumar N, Kumar N, Shukla A, Shankhdhar SC, Shankhdhar D (2015) Impact of terminal heat stress on pollen viability and yield attributes of rice (*Oryza sativa* L.). Cereal Res Commun 43(4):616–626

Kumar N, Shankhdhar SC, Shankhdhar D (2016) Impact of elevated temperature on antioxidant activity and membrane stability in different genotypes of rice (*Oryza sativa* L.). Indian J Plant Physiol 21(1):37–43

Kumar N, Suyal DC, Sharma IP, Verma A, Singh H (2017) Elucidating stress proteins in rice (*Oryza sativa* L.) genotype under elevated temperature: a proteomic approach to understand heat stress response. 3 Biotech 7:205

Kumar N, Jeena N, Singh H (2019) Elevated temperature modulates rice pollen structure: a study from foothill Himalayan agro-ecosystem in India. 3 Biotech 9:175

Kumar K, Gambhir G, Dass A, Tripathi AK, Singh A, Jha AK, Yadava P, Choudhary M, Rakshit S (2020) Genetically modified crops: current status and future prospects. Planta 251:1–27

Kumar A, Dwivedi GK, Tewari S, Paul J, Anand R, Kumar N, Kumar P, Singh H, Kaushal R (2020a) Carbon mineralization and inorganic nitrogen pools under *Terminalia chebula* Retz.-based agroforestry system in Himalayan foothills, India. For Sci 66(5):634–643

Kumar A, Kumar P, Singh H, Kumar N (2020b) Adaptation and mitigation potential of roadside trees with bio-extraction of heavy metals under vehicular emissions and their impact on physiological traits during seasonal regimes. Urban For Urban Green 58:126900

Kumar A, Dwivedi GK, Tewari S, Paul J, Sah VK, Singh H, Kumar P, Kumar N, Kaushal R (2020c) Soil organic carbon pools under Terminalia chebula Retz. based agroforestry system in Himalayan foothills, India. Curr Sci 118(7):1098–1103

Kumar N, Jeena N, Kumar A, Khairakpam R, Singh H (2021) Comparative response of rice cultivars to elevated air temperature in Bhabar region of Indian Himalaya: status on yield attributes. Heliyon 7:e07474

Kumar A, Kumar P, Singh H, Bisht S, Kumar N (2021a) Relationship of physiological plant functional traits with soil carbon stock in temperate forest of Garhwal Himalaya. Curr Sci 120(8):1368–1373

Kumar A, Kumar P, Singh H, Kumar N (2021b) Modulation of plant functional traits under essential plant nutrients during seasonal regime in natural forests of Garhwal Himalayas. Plant Soil 465:197–212

Kumar A, Tewari S, Singh H, Kumar P, Kumar N, Bisht S, Kushwaha S, Tamta N, Kaushal R (2021c) Biomass accumulation and carbon stocks in different agroforestry system prevalent in Himalayan foothills, India. Curr Sci 120(6):1083–1088

Kumar A, Kumar P, Singh H, Kumar N (2021d) Impact of plant functional traits on near saturated hydraulic conductivity of soil under different forests of Kempty watershed in Garhwal Himalaya. Indian J Soil Conserv 49(1):38–44

Lanigan TM, Kopera HC, Saunders TL (2020) Principles of genetic engineering. Genes 11(3):291

Li T, Liu B, Chen CY, Yang B (2016) TALEN-mediated homologous recombination produces site-directed DNA base change and herbicide-resistant rice. J Genet Genomics 43(5):297–305

Li A, Jia S, Yobi A, Ge Z, Sato SJ, Zhang C, Angelovici R, Clemente TE, Holding DR (2018a) Editing of an alpha-kafirin gene family increases, digestibility and protein quality in sorghum. Plant Physiol 177(4):1425–1438

Li C, Zong Y, Wang Y, Jin S, Zhang D, Song Q et al (2018b) Expanded base editing in rice and wheat using a Cas9-adenosine deaminase fusion. Genome Biol 19:1–9

Li H, Li J, Chen J, Yan L, Xia L (2020) Precise modifications of both exogenous and endogenous genes in rice by prime editing. Mol Plant 13(5):671–674

Little NS, Catchot AL, Allen KC, Gore J, Musser FR, Cook DR, Luttrell RG (2017) Supplemental control with diamides for Heliothines1 in Bt cotton. Southwestern Entomol 42(1):15–26

Liu E, Zeng S, Zhu S, Liu Y, Wu G, Zhao K, Liu X, Liu Q, Dong Z, Dang X, Xie H, Li D, Hu X, Hong D (2019) Favorable alleles of grain-filling rate1 increase the grain-filling rate and yield of rice. Plant Physiol 181:1207–1222. https://doi.org/10.1104/pp.19.00413

Louwaars NP (2018) Plant breeding and diversity: a troubled relationship? Euphytica 214(7):114

Lu K, Wu B, Wang J, Zhu W, Nie H, Qian J, Huang W, Fang Z (2018) Blocking amino acid transporter Os AAP 3 improves grain yield by promoting outgrowth buds and increasing tiller number in rice. Plant Biotechnol J 16(10):1710–1722

Miao C, Xiao L, Hua K, Zou C, Zhao Y, Bressan R A, Zhu J K (2018) Mutations in a subfamily of abscisic acid receptor genes promote rice growth and productivity. PNAS 115(23):6058–6063

Murray JM, Carr AM (2018) Integrating DNA damage repair with the cell cycle. Curr Opin Cell Biol 52:120–125

Nagamine A, Ezura H (2022) Genome editing for improving crop nutrition. Front Genome Ed 4:850104

Novak S (2019) Plant biotechnology applications of zinc finger technology. Methods Mol Biol 1864:295–310

Nuñez-Muñoz L, Vargas-Hernández B, Hinojosa-Moya J, Ruiz-Medrano R, Xoconostle-Cázares B (2021) Plant drought tolerance provided through genome editing of the trehalase gene. Plant Signal Behav 16(4):1877005

Paul S, Das S (2021) Natural insecticidal proteins, the promising bio-control compounds for future crop protection. Nucleus 64:7–20

Qin D, Liu XY, Miceli C, Zhang Q, Wang PW (2019) Soybean plants expressing the *Bacillus thuringiensis* cry8-like gene show resistance to *Holotrichia parallela*. BMC Biotechnol 19:1–12

Qiu L, Sun Y, Jiang Z, Yang P, Liu H, Zhou H, Wang X, Zhang W, Lin Y, Ma W (2019) The midgut V-ATPase subunit A gene is associated with toxicity to crystal 2Aa and crystal 1Ca-expressing transgenic rice in *Chilo suppressalis*. Insect Mol Biol 28(4):520–527

Rahman MU, Zulfiqar S, Raza MA, Ahmad N, Zhang B (2022) Engineering abiotic stress tolerance in crop plants through CRISPR genome editing. Cells 11(22):3590

Ren D, Cui Y, Hu H, Xu Q, Rao Y, Yu X, Zhang Y, Wang Y, Peng Y, Zeng D, Hu J (2019) AH 2 encodes a MYB domain protein that determines hull fate and affects grain yield and quality in rice. Plant J 100(4):813–824

Riaz S, Nasir IA, Bhatti MU, Adeyinka OS, Toufiq N, Yousaf I, Tabassum B (2020) Resistance to *Chilo infuscatellus* (Lepidoptera: Pyraloidea) in transgenic lines of sugarcane expressing *Bacillus thuringiensis* derived Vip3A protein. Mol Biol Rep 47:2649–2658

Sánchez-León S, Gil-Humanes J, Ozuna CV, Giménez MJ, Sousa C, Voytas DF, Barro F (2018) Low-gluten, nontransgenic wheat engineered with CRISPR/Cas9. Plant Biotechnol J 16(4):902–910

Shan Q, Zhang Y, Chen K, Zhang K, Gao C (2015) Creation of fragrant rice by targeted knockout of the Os BADH 2 gene using TALEN technology. Plant Biotechnol J 13(6):791–800

Shen WJ, Chen GX, Xu JG, Jiang Y, Liu L, Gao ZP, Ma J, Chen X, Chen TH, Lv CF (2015) Overexpression of maize phospho enol pyruvate carboxylase improves drought tolerance in rice by stabilization the function and structure of thylakoid membrane. Photosynthetica 53(3):436–446

Shimatani Z, Kashojiya S, Takayama M, Terada R, Arazoe T, Ishii H et al (2017) Targeted base editing in rice and tomato using a CRISPR-Cas9 cytidine deaminase fusion. Nat Biotechnol 35(5):441–443

Shukla VK, Doyon Y, Miller JC, DeKelver RC, Moehle EA, Worden SE, Mitchell JC, Arnold NL, Gopalan S, Meng X, Choi VM (2009) Precise genome modification in the crop species Zea mays using zinc-finger nucleases. Nature 459(7245):437–441

Shukla M, Al-Busaidi KT, Trivedi M, Tiwari RK (2018) Status of research, regulations and challenges for genetically modified crops in India. GM Crops Food 9(4):173–188

Silva G, Poirot L, Galetto R, Smith J, Montoya G, Duchateau P, Pâques F (2011) Meganucleases and other tools for targeted genome engineering: perspectives and challenges for gene therapy. Curr Gene Ther 11(1):11–27

Singh HP, Singh BP (2014) Genetic engineering of field, industrial and pharmaceutical crops. Am J Plant Sci 5(26):3974

Singh H, Verma A (2013a) Physiological responses of rice cultivars to various nitrogen levels. Int J Agric Environ Biotechnol 6(3):383–388

Singh H, Verma A (2013b) Characterization and screening of high nitrogen efficient rice genotype to curtail environmental pollution. Int J Agric Environ Biotechnol 6(1):23–30

Singh H, Verma A, Shukla A (2010a) Comparative study of yield and yield components of hybrid and inbred genotypes of rice (*Oryza Sativa* L.). Int J Agric Environ Biotechnol 3:355–360

Singh H, Verma A, Krishnamoorthy Mand Shukla A (2010b) Consequence of diverse nitrogen levels on leaf pigments in five rice genotypes under field emergent circumstance. Int J Bioresour Stress Manage 1:189–193

Singh H, Verma A, Rai SK (2013a) Biochemical evaluation of different rice genotypes grown at various nitrogen levels using SDS-PAGE. Curr Adv Agric Sci 5(1):144–146

Singh H, Verma A, Shukla A (2013b) Guttation fluid as a physiological marker for selection of nitrogen efficient rice (*Oryza sativa* L.) genotypes. Afr J Biotechnol 12(44):6276–6281

Singh H, Verma A, Ansari MW, Shukla A (2014) Physiological response of rice (*Oryza sativa* L.) genotypes to elevated nitrogen applied under field conditions. Plant Signal Behav 9:e29015. https://doi.org/10.4161/psb.29015

Singh S, Kumar NR, Maniraj R, Lakshmikanth R, Rao KYS, Muralimohan N, Arulprakash T, Karthik K, Shashibhushan NB, Vinutha T, Pattanayak D (2018) Expression of Cry2Aa, a *Bacillus thuringiensis* insecticidal protein in transgenic pigeon pea confers resistance to gram pod borer, *Helicoverpa armigera*. Sci Rep 8(1):8820

Singh H, Yadav M, Kumar N, Kumar A, Kumar M (2020) Assessing adaptation and mitigation potential of roadside trees under the influence of vehicular emissions: a case study of *Grevillea robusta* and *Mangifera indica* planted in an urban city of India. PLoS One 15(1):e0227380

Singh S, Singh H, Sharma SK, Nautiyal R (2021) Seasonal variation in biochemical responses of bamboo clones in the sub-tropical climate of Indian Himalayan foothills. Heliyon 7(4):e06859. https://doi.org/10.1016/j.heliyon.2021.e06859

Singh H (2024) Exploring adaptive modulation in plant functional traits and its impact on the productivity of *Acacia auriculiformis* under CO_2-enriched environment. Ind Crop Prod 210. https://doi.org/10.1016/j.indcrop.2024.118186

Singh M, Singh H, Kumar A, Kumar M, Barthwal S, Thakur A (2024) Soil nitrogen availability determines the CO_2 fertilization effect on tree species (Neolamarckia cadamba): growth and physiological evidence. Environ Sustain. https://doi.org/10.1007/s42398-023-00300-w

Soliman HI, Abo-El-Hasan FM, El-Seedy AS, Mabrouk YM (2021) Agrobacterium-mediated transformation of tomato (*Lycopersicon esculentum* mill.) using a synthetic *cry1ab* gene for enhanced resistance against Tutaabsoluta (Meyrick). J Microbiol Biotechnol Food Sci 2021:67–74

Tiwari M, Kumar Trivedi P, Pandey A (2021) Emerging tools and paradigm shift of gene editing in cereals, fruits, and horticultural crops for enhancing nutritional value and food security. Food Energy Secur 10(1):e258

Townsend JA, Wright DA, Winfrey RJ, Fu F, Maeder ML, Joung JK, Voytas DF (2009) High-frequency modification of plant genes using engineered zinc-finger nucleases. Nature 459(7245):442–445

Tran MT, Doan DTH, Kim J, Song YJ, Sung YW, Das S et al (2021) CRISPR/Cas9-based precise excision of SlHyPRP1 domain (s) to obtain salt stress-tolerant tomato. Plant Cell Rep 40:999–1011

Tyagi S, Kesiraju K, Saakre M, Rathinam M, Raman V, Pattanayak D, Sreevathsa R (2020) Genome editing for resistance to insect pests: an emerging tool for crop improvement. ACS Omega 5(33):20674–20683

Usman B, Nawaz G, Zhao N, Liu Y, Li R (2020) Generation of high yielding and fragrant rice (*Oryza sativa* L.) lines by CRISPR/Cas9 targeted mutagenesis of three homoeologs of cytochrome P450 gene family and *OsBADH2* and transcriptome and proteome profiling of revealed changes triggered by mutations. Plants 9(6):788. https://doi.org/10.3390/plants9060788

Van Zelm E, Zhang Y, Testerink C (2020) Salt tolerance mechanisms of plants. Annu Rev Plant Biol 71:403–433

Wang J, Wu B, Lu K, Wei Q, Qian J, Chen Y, Fang Z (2019) The amino acid permease 5 (*OsAAP5*) regulates tiller number and grain yield in rice. Plant Physiol 180(2):1031–1045

Wang B, Zhong Z, Wang X, Han X, Yu D, Wang C, Song W, Zheng X, Chen C, Zhang Y (2020) Knockout of the OsNAC006 transcription factor causes drought and heat sensitivity in rice. Int J Mol Sci 21(7):2288

Wang H, Li Y, Chern M, Zhu Y, Zhang LL, Lu JH et al (2021a) Suppression of rice miR168 improves yield, flowering time and immunity. Nat Plants 7(2):129–136

Wang W, Tian B, Pan Q, Chen Y, He F, Bai G et al (2021b) Expanding the range of editable targets in the wheat genome using the variants of the Cas12a and Cas9 nucleases. Plant Biotechnol J 19(12):2428–2441

Wang H, Liao S, Li M, Wei J, Zhu B, Gu L, Li L, Du X (2022) TmNAS3 from *Triticum monococum* directly regulated by *TmbHLH47* increases Fe content of wheat grain. Gene 811:146096

Wu J, Yan G, Duan Z, Wang Z, Kang C, Guo L et al (2020) Roles of the *Brassica napus* DELLA protein *BnaA6*. RGA, in modulating drought tolerance by interacting with the ABA signaling component BnaA10. ABF2. Front Plant Sci 11:577

Xu R, Li J, Liu X, Shan T, Qin R, Wei P (2020) Development of plant prime-editing systems for precise genome editing. Plant Commun 1(3):100043

Yan S, Zhu J, Zhu W, Li Z, Shelton AM, Luo J, Cui J, Zhang Q, Liu X (2015) Pollen-mediated gene flow from transgenic cotton under greenhouse conditions is dependent on different pollinators. Sci Rep 5(1):15917

Yan D, Ren B, Liu L, Yan F, Li S, Wang G, Sun W, Zhou X, Zhou H (2021) High-efficiency and multiplex adenine base editing in plants using new TadA variants. Mol Plant 14(5):722–731

Ye R, Yang X, Rao Y (2022) Genetic engineering technologies for improving crop yield and quality. Agronomy 12(4):759

Yu QH, Wang B, Li N, Tang Y, Yang S, Yang T, Xu J, Guo C, Yan P, Wang Q, Asmutola P (2017) CRISPR/Cas9- induced targeted mutagenesis and gene replacement to generate long-shelf life tomato lines. Sci Rep 7:11874. https://doi.org/10.1038/s41598-017-12262-1

Yuste-Lisbona FJ, Fernandez-Lozano A, Pineda B, Bretones S, Ortiz-Atienza A, Garcia-Sogo B, Muller NA, Angosto T, Capel J, Moreno V, Jimenez-Gomez JM, Lozano R (2020) ENO regulates tomato fruit size through the floral meristem development network. Proc Natl Acad Sci U S A 117:8187–8195. https://doi.org/10.1073/pnas.1913688117

Zeng Y, Wen J, Zhao W, Wang Q, Huang W (2020) Rational improvement of rice yield and cold tolerance by editing the three genes *OsPIN5b, GS3*, and *OsMYB30* with the CRISPR–Cas9 system. Front Plant Sci 10:1663

Zhang J, Bandyopadhyay A, Sellappan K, Wang G, Xie HA, Datta K, Datta SK (2010) Characterization of a C_4 maize pyruvate orthophosphate dikinase expressed in C_3 transgenic rice plants. Afr J Biotechnol 9(2):234–242

Zhang J, Zhang H, Botella JR, Zhu JK (2018) Generation of new glutinous rice by CRISPR/Cas9-targeted mutagenesis of the Waxy gene in elite rice varieties. J Integr Plant Biol 60(5):369–375

Zhang X, Mogel KJHV, Lor VS, Hirsch CN, De Vries B, Kaeppler HF et al (2019) Maize sugary enhancer1 (*se1*) is a gene affecting endosperm starch metabolism. Proc Natl Acad Sci 116(41):20776–20785

Zhong Y, Ahmed S, Deng G, Fan W, Zhang P, Wang H (2019) Improved insect resistance against *Spodoptera litura* in transgenic sweetpotato by overexpressing Cry1Aa toxin. Plant Cell Rep 38:1439–1448

Zhu X, Rong W, Wang K, Guo W, Zhou M, Wu J et al (2022) Overexpression of *TaSTT3b-2B* improves resistance to sharp eyespot and increases grain weight in wheat. Plant Biotechnol J 20(4):777–793

Plant Functional Traits Assisted Crop Adaptation to Abiotic and Biotic Stress

13

Manish Singh, Jyoti, Narendra Kumar, and Hukum Singh

Abstract

In the era of climate change, it is crucial to prioritize the resistance and adaptation of crops to different stresses to ensure food security. This chapter examines the role of plant functional traits in facilitating crop adaptation under both abiotic and biotic stresses. Plant functional traits, encompassing morphological, physiological, and biochemical characteristics, are crucial in determining a plant's response to its environment. It describes the role of distinct functional characteristics in promoting increased stress resilience in crops, facilitating their successful adaptation to adverse environmental circumstances. Examining abiotic stress factors, such as drought, salinity, and extreme temperatures, highlights the importance of traits like root architecture, water use efficiency, and heat tolerance in crop resilience. Additionally, the study delves into the interplay between plant traits and biotic stress, including pests and diseases, elucidating mechanisms by which traits such as resistance genes, secondary metabolites, and leaf morphology impact crop defense. Furthermore, the study explores the potential for trait-based breeding strategies to develop stress-tolerant crop varieties. Identifying key functional traits associated with stress adaptation provides valu-

M. Singh (✉)
Forest Ecology and Climate Change Division, Forest Research Institute, Dehradun, Uttarakhand, India

Jyoti
Forest Biotechnology Division, Genetics and Tree Improvement Division, Forest Research Institute, Dehradun, Uttarakhand, India

N. Kumar
College of Agriculture, Central Agriculture University (I), Krydemkulai, Meghalaya, India

H. Singh
Plant Physiology Discipline, Genetics and Tree Improvement Division, Forest Research Institute, Dehradun, Uttarakhand, India

N. Kumar, H. Singh (eds.), *Plant Functional Traits for Improving Productivity*, https://doi.org/10.1007/978-981-97-1510-7_13

able insights for crop improvement programs aimed at mitigating the impact of environmental challenges on global agriculture. Understanding the intricate relationship between plant functional traits and stress adaptation is pivotal for advancing sustainable agricultural practices and ensuring food security in a changing climate.

Keywords

Adaptation · Food security · Abiotic · Biotic · Stress resilience · Sustainable agriculture

13.1 Introduction

Agricultural crops face increased number of abiotic and biotic stress combinations as a result of global warming and possible climatic anomalies linked to it, which negatively impacts their growth, development, and finally yield (Hartfield and Prueger 2015; Kumar et al. 2015, 2016). Stress is the most prevalent environmental factor limiting crop productivity, and frequency of stress condition is rising as a result of global climate change. Plants in their natural habitats adapt to stress in the environment through a variety of mechanisms, ranging from transient responses to low soil nutrient to major survival mechanisms of escape by early flowering in the absence of seasonal rainfall (Singh et al. 2010a, b; Basu et al. 2016).

Crop adaptation under stress is a critical aspect of ensuring food security and agricultural sustainability in the face of changing environmental conditions (Singh et al. 2014, 2021, 2024). Continued research and implementation of adaptive strategies are crucial for meeting the demands of a growing global population and addressing the uncertainties associated with climate change (Maheswari et al. 2012). The interplay between external stresses and the plant functional traits determines trajectory of development and survival of plants. It may be directly or indirectly accountable for the disturbance of the regular physiological processes and activities in the plant (Hartfield and Prueger 2015; Mantri et al. 2012; Meena et al. 2016). Understanding and enhancing crop adaptation to these stresses are essential for maintaining and improving yield stability (Raza et al. 2019; Kumar et al. 2017, 2019).

Abiotic and biotic stress factors play a crucial role in the complex interaction between plants and their environment (Devi et al. 2022). Abiotic stresses, such as drought, salinity, and extreme temperatures, as well as biotic stresses like pests and diseases, pose significant challenges to global crop production (Umar et al. 2021). Jenks and Hasegawa (2005) stated that abiotic stress is potentially relevant in agricultural crops since it imposes significant constraints on crop production in global agriculture. Abiotic stress encompasses nonliving factors such as drought, salinity, extreme temperatures, and nutrient deficiencies, which can adversely affect plant growth, development, and overall productivity. These stressors disrupt the physical and chemical balance essential for optimal plant functioning (Umar et al. 2021; Kumar et al. 2020a, b, c). Dehydration stress, caused by drought, salt, and extreme temperatures, is the most common abiotic stress that restricts plant growth and output (Jaleel et al. 2009; Thakur et al. 2010). From the relentless challenges of drought and the demanding constraints of high temperatures to the complex dynamics of nutrient availability, abiotic stressors pose formidable hurdles for plant adaptation (Oshunsanya et al. 2019). Salinity, soil pollution, and extreme climatic conditions further contribute to the diverse spectrum of challenges that plants navigate to maintain their metabolic equilibrium (Tuteja 2007).

Conversely, biotic stress factors involve living organisms, including pests, pathogens, and competing flora, creating a dynamic interplay within ecosystems. The relentless competition for resources and the intricate dance between plants and their biotic adversaries shape the complex tapestry of biotic stress (Gimenez et al. 2018). From microscopic pathogens causing diseases to visible herbivores consuming plant tissues, the adaptive responses of plants are finely tuned to recognize, signal, and deploy defense mechanisms against biotic stressors (Jones and Dangl 2006).

The ability to navigate these biotic interactions not only ensures individual plant survival but also influences the broader ecological balance within plant communities. Biotic stress involves living organisms, including pests, diseases, and competing plants, which impose detrimental effects on crops (Atkinson and Urwin 2012). Pathogens and herbivores can compromise plant health by causing infections, reducing nutrient availability, and inducing physical damage (Bezemer et al. 2006). Plants adapt their responses to several stress conditions and have various distinct responses, as well as certain shared responses (Ramegowda and Senthil-Kumar 2015). Hence, to comprehend the nature of interactions between combined abiotic and biotic pressures on plants, it is crucial to grasp their influence. Mittler and his colleagues devised a "stress matrix" to consolidate the relationships between different abiotic and biotic stressors on plant development and productivity. The impact of simultaneous stressors on crops is not consistently cumulative, as the result is primarily determined by the nature of the interactions between the stressors.

As we delve into the fascinating realm of abiotic and biotic stress factors, the exploration of plant adaptations unfolds, revealing the intricate strategies that have evolved over millennia to enable plants to confront and overcome the challenges posed by their dynamic and ever-changing environments (Shahzad et al. 2021). This chapter aims to enhance and rectify the existing knowledge on stress combinations in plants, based on the already accessible research. This chapter presents a comprehensive analysis of various stress combinations and their repercussions on agriculture. Specifically, it delves into the specific consequences of the simultaneous occurrence of drought and pathogen infection on significant crops. The significance of conducting simulation studies to evaluate the effects of coupled pressures on plants is also emphasized. Based on significant research on individual stresses, some possible characteristics are identified that may be used to enhance crop quality when faced with both abiotic and pathogen infection (Fig. 13.1).

The responses of plants to abiotic and biotic stresses often overlap, as the physiological and molecular mechanisms activated by one type of stress can influence the plant's ability to cope with the other. Understanding the intricate interplay between abiotic and biotic stress factors is crucial for developing resilient and adaptive crop

Fig. 13.1 Abiotic and biotic stresses represent two distinct yet interconnected challenges that plants face in their natural environments

varieties that can thrive in challenging environmental conditions. Efforts in plant breeding and biotechnology aim to enhance the tolerance and resistance of crops to both abiotic and biotic stressors, ensuring sustainable and secure food production in the face of a changing climate and evolving pest pressures. Based on significant research on individual stress, several potential traits that can be employed for enhancing crop performance in the face of combined abiotic and biotic infestation. There are various abiotic and biotic plant functional traits, which are direct impacts on the crop yield (Table 13.1).

13.2 Plant Functional Traits and Abiotic Stress

13.2.1 Overview of Key Plant Functional Traits Relevant to Abiotic Stress Adaptation

Plant adaptation has a range of important functional characteristics that are essential for dealing with abiotic stresses. Characteristics such as the structure of the root system, the control of stomata, and the production of osmoprotectants are crucial

Table 13.1 Quantitative assessment of selected trait's impact on crop performance

Trait	Measurement method	Impact on yield	Research reference(s)
Abiotic stress traits			
Root depth	Soil coring, rhizotron systems	Increased water absorption, improved drought tolerance	Qian et al. (2018), Postma and Lynch (2012)
Stomatal regulation	Gas exchange measurements, thermal imaging	Enhanced water use efficiency, reduced transpiration	Flexas et al. (2016), Franks and Beerling (2009)
Ion exclusion	Ion-selective electrodes, tissue analysis	Improved salt tolerance through efficient ion regulation	Munns and Tester (2008), Zhang et al. (2019)
Temperature resilience	Thermocouples, gene expression analysis	Enhanced heat stress tolerance, improved growth	Wahid et al. (2007), Mittler (2006)
Water use efficiency	Gas exchange measurements, stable isotope analysis	Optimized water consumption for photosynthesis	Condon et al. (2004), Flexas et al. (2008)
Biotic stress traits			
Insect resistance	Insect feeding assays, gene expression analysis	Reduced damage, increased resistance to insect pests	Howe and Jander (2008), Gatehouse (2002)
Disease resistance	Disease resistance assays, molecular markers	Enhanced defense against specific pathogens	Flor (1971), Dangl and Jones (2001)
Allelopathic chemicals	Chemical analysis, bioassays	Suppression of weed growth, improved crop performance	Macías et al. (2003), Duke (2005)
Pollination efficiency	Pollination observations, fruit set analysis	Increased seed and fruit production	Klein et al. (2007)

for the ability of plants to withstand drought (McDowell et al. 2008). Additionally, leaf morphology, including cuticle thickness and pubescence, influences a plant's ability to mitigate water loss and withstand high temperatures (Bale et al. 2002; Kumar et al. 2021a, b, c, d, e). The allocation of resources to antioxidant defense mechanisms, as explored by Zhu (2016), becomes paramount in combating oxidative stress induced by factors like salinity or pollution. These adaptive traits collectively enable plants to thrive in diverse environmental conditions, underscoring the multifaceted strategies employed by flora in the face of abiotic challenges. Plant functional traits play a crucial role in the adaptation of crops to abiotic stress. These traits are the characteristics of plants that influence their growth, reproduction, and survival in specific environments. Here's an overview of key plant functional traits relevant to abiotic stress adaptation:

13.2.1.1 Root Architecture

The structure of plant roots plays a crucial role on how plants adapt to nonliving factors that might affect their growth, such as extreme weather conditions. It determines how well plants can obtain water and nutrients in difficult surroundings. Deep root systems allow plants to reach deeper soil layers with higher water content, enhancing drought tolerance (Lynch 2013). Lateral root proliferation is a fundamental aspect of root architecture that plays a crucial role in a plant's ability to explore and exploit soil resources, especially nutrients and water. Increased lateral root density improves nutrient acquisition by expanding the root's exploration area in nutrient-deficient soils (Postma and Lynch 2011). The development of root hairs further facilitates water and nutrient uptake (Bates and Lynch 1996). Traits such as aerenchyma formation assist in oxygen transport in waterlogged soils, preventing root damage (Colmer 2003). Understanding and manipulating these aspects of root architecture offer valuable strategies for breeding crops with improved resilience to abiotic stresses, crucial for sustaining agricultural productivity in the face of environmental challenges.

13.2.1.2 Leaf Morphology

Leaf morphology plays a crucial role in the adaptation of plants to abiotic stress, serving as a key interface between the plant and its environment. Various aspects of leaf structure contribute to the plant's ability to cope with challenges such as drought, high temperatures, and nutrient deficiencies. Small leaf size, increased leaf thickness, and reduced leaf surface area are common morphological adaptations that minimize water loss through transpiration, thus enhancing drought tolerance (Chaves et al. 2009). Additionally, alterations in stomatal density, size, and behavior regulate gas exchange and water loss, optimizing water use efficiency under stress conditions (Dow et al. 2014). The presence of trichomes, or small hairlike structures on the leaf surface, can provide a protective barrier against excessive sunlight and reduce water loss (Werker 2000). Understanding and manipulating these leaf morphological traits are integral to developing crops with improved resilience to abiotic stresses, contributing to sustainable agriculture in the face of changing environmental conditions.

13.2.1.3 Stomatal Traits

Stomatal traits are pivotal components of a plant's adaptive strategy to abiotic stress, influencing its water use efficiency and gas exchange capabilities. Under conditions of water scarcity, plants often exhibit a dynamic response by adjusting stomatal density, size, and conductance. Decreased stomatal density and smaller stomata reduce water loss through transpiration, enhancing drought tolerance (Engineer et al. 2014). Concurrently, controlled stomatal closure helps conserve water while maintaining essential gas exchange functions. The ability to modulate stomatal conductance contributes to the optimization of water use efficiency, particularly crucial during episodes of high temperature stress (Pillitteri and Torii 2012). Stomatal traits collectively represent a finely tuned system that enables plants to balance their water requirements and physiological functions, demonstrating their significance in the plant's ability to thrive in challenging environmental conditions (Kumar et al. 2021a, b).

13.2.1.4 Photosynthetic Pathway

The photosynthetic pathway is a crucial aspect of plant physiology, and its modulation is central to a plant's adaptation to various abiotic stresses. In response to water scarcity and high temperatures, certain plants employ the C_4 photosynthetic pathway, characterized by the spatial separation of carbon fixation and the Calvin cycle in different cell types, enhancing water use efficiency and photosynthetic performance in arid conditions (Sage and Zhu 2011; Singh et al. 2020). Additionally, plants may undergo adjustments in the C_3 photosynthetic pathway, optimizing stomatal conductance and transpiration rates to balance water availability and carbon assimilation (Flexas and Medrano 2002; Kumar et al. 2021c, d). The dynamic regulation of these photosynthetic pathways contributes significantly to a plant's ability to cope with environmental challenges, ensuring efficient energy capture and utilization even under stress conditions.

13.2.2 Potential Mechanisms of Trait-Mediated Adaptation

Trait-mediated adaptation in plants involves intricate mechanisms that allow them to thrive in challenging environments. One crucial mechanism lies in the modulation of gene expression to favor traits conducive to survival under specific abiotic stresses. For instance, the upregulation of genes associated with drought resistance or heat tolerance enhances a plant's adaptive capacity (McDowell et al. 2008). Epigenetic modifications, including DNA methylation and histone acetylation, also play a pivotal role in shaping adaptive traits by influencing gene expression patterns in response to environmental cues (Herman and Sultan 2011). Moreover, the dynamic interplay between plant-microbe interactions, as explored by Singh et al. (2017), contributes to adaptation by facilitating nutrient acquisition and enhancing stress tolerance. These mechanisms collectively highlight the sophisticated strategies employed by plants to adapt and thrive in the face of challenging environmental conditions.

13.3 Plant Functional Traits and Biotic Stress

13.3.1 Plant Functional Traits Associated with Resistance to Pests and Diseases

Plants have evolved an array of functional traits that confer resistance to pests and diseases, providing a fascinating glimpse into the intricate world of plant defense mechanisms. Resistance to herbivores often involves chemical defenses, with secondary metabolites such as alkaloids, phenolics, and terpenoids deterring feeding (Berenbaum 1995). Trichomes, or hairlike structures on leaves, act as physical barriers and can secrete toxic substances, contributing to resistance against herbivorous insects (Karban and Myers 1989). Meanwhile, disease resistance is often linked to traits like pathogen recognition receptors, which enable plants to detect and respond to microbial invaders, and the synthesis of antimicrobial compounds, including phytoalexins and pathogenesis-related proteins (Jones and Dangl 2006). A deeper exploration into these multifaceted traits reveals the dynamic strategies plants employ to ward off pests and diseases. While specific traits can vary among plant species, some common functional traits associated with resistance to pests and diseases include the following:

13.3.1.1 Chemical Defenses

- *Secondary metabolites:* Plants produce a variety of secondary metabolites such as alkaloids, terpenoids, and phenolic compounds that can deter or inhibit herbivores and pathogens (Agrawal et al. 2004).
- *Phytochemicals:* Compounds like flavonoids and tannins can act as natural pesticides and inhibit the growth of pathogens (Agrawal et al. 2004).

13.3.1.2 Morphological Traits

- *Trichomes:* Hairs on the plant surface can act as physical barriers, reducing herbivore feeding and limiting the spread of pathogens.
- *Leaf toughness:* Tougher leaves can deter herbivores and make it more challenging for pathogens to penetrate plant tissues.

13.3.1.3 Physiological Traits

- *Induced resistance:* Plants can activate defense mechanisms in response to pest or pathogen attack, leading to the production of proteins and other compounds that hinder further damage (Durrant and Dong 2004).
- *Systemic acquired resistance (SAR):* SAR is a plant's ability to induce resistance throughout its tissues following exposure to a pathogen, providing broad-spectrum defense (Durrant and Dong 2004).

13.3.1.4 Genetic Traits

- *Resistant varieties:* Some plant species or cultivars possess genetic resistance to specific pests or diseases, often due to the presence of resistance genes (Mundt 2014).
- *Rapid response genes:* These genes enable a quick and effective response to pathogen attack, limiting the extent of damage (Mundt 2014).

13.3.1.5 Ecological Traits

- *Associations with beneficial organisms:* Plants that foster relationships with beneficial insects or microorganisms, such as mycorrhizal fungi, may experience reduced pest and pathogen pressure (Bezemer et al. 2006).
- *Biodiversity effects:* Higher plant diversity in ecosystems can contribute to reduced pest and disease incidence through various ecological mechanisms (Bezemer et al. 2006).

13.3.2 Interactions Between Abiotic and Biotic Stresses and Their Impact on Plant Traits

The intricate interplay between abiotic and biotic stresses profoundly influences plant traits, shaping the very essence of their survival strategies. Abiotic stressors like drought or extreme temperatures can compromise a plant's resilience, rendering it more susceptible to biotic challenges such as pest infestations or diseases (Devi et al. 2022). In turn, biotic stressors can exacerbate the impact of abiotic stress by exploiting weakened defenses. This dynamic interaction is exemplified in studies such as McDowell et al. (2008), which illuminates the mechanisms behind plant survival and mortality during drought, underscoring the delicate balance between environmental pressures and biotic adversaries in shaping plant traits and responses. Here's an elaboration on how abiotic and biotic stress factors interact and impact plant traits:

13.3.2.1 Shared Signaling Pathways

Plants employ common signaling pathways to respond to both abiotic and biotic stresses. The abscisic acid (ABA) pathway, traditionally associated with abiotic stress responses, also plays a role in plant defense against pathogens (Ramegowda and Senthil-Kumar 2015).

13.3.2.2 Cross-Tolerance Mechanisms

Cross-tolerance mechanisms can enhance a plant's resistance to both abiotic and biotic stresses. Priming, induced by prior exposure to stress, leads to a more robust response in subsequent stress events (Balmer et al. 2015).

13.3.2.3 Nutrient Availability and Disease Resistance

Nutrient imbalances can influence a plant's susceptibility to diseases. For example, nitrogen deficiency may compromise a plant's ability to produce secondary metabolites, impacting its defense against pathogens (Singh et al. 2013a, b; Singh and Verma 2013a, b; Meng et al. 2019).

13.3.2.4 Impact on Plant Physiology and Morphology

Abiotic stressors can affect plant physiology and morphology, indirectly influencing plant-pathogen interactions. Drought-induced reductions in stomatal conductance may compromise physical barriers against pathogens (Kissoudis et al. 2014).

13.3.2.5 Indirect Effects on Pest Dynamics

Abiotic stress can alter pest dynamics by influencing plant traits. Water-stressed plants may produce altered chemical profiles that attract or repel certain herbivores, impacting the plant's interaction with pests (Birkett et al. 2000).

13.3.2.6 Genetic Regulation and Trait Expression

Both abiotic and biotic stresses induce changes in gene expression patterns, leading to the activation or suppression of specific plant traits. This genetic regulation can

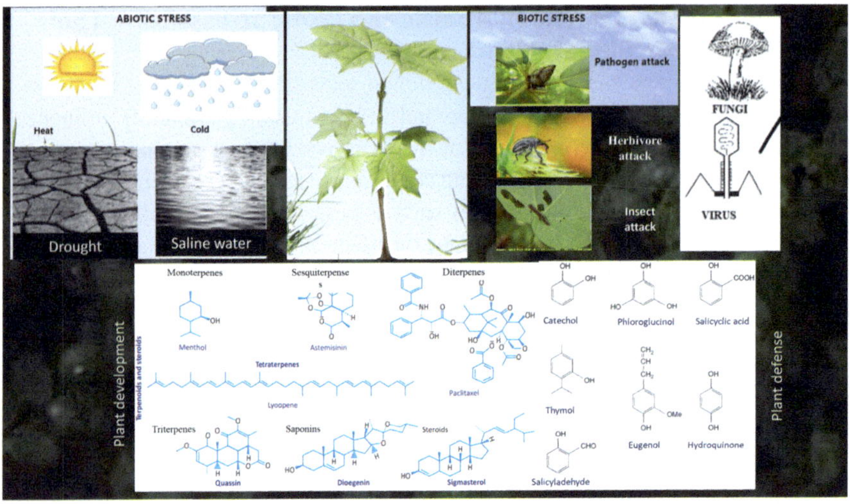

Fig. 13.2 Schematic representation of abiotic and biotic stresses and their impact on plant traits

result in altered root architecture, production of secondary metabolites, and the induction of pathogenesis-related proteins (Atkinson and Urwin 2012).

13.3.3 The Significance of Understanding Both Biotic and Abiotic Stresses in Combination

This integrated approach is essential because the responses and adaptations triggered by plants to combat one type of stress may impact their susceptibility or resilience to the other. For instance, a plant undergoing drought stress might become more susceptible to certain pathogens. Therefore, a holistic understanding of the interplay between biotic and abiotic stress factors is essential for developing sustainable and effective strategies in agriculture and ecological management. This perspective aligns with studies emphasizing the need for integrated pest and stress management practices (Bita and Gerats 2013; Atkinson and Urwin 2012) (Fig. 13.2).

13.4 Integrating Plant Functional Traits for Crop Improvement

13.4.1 Breeding Approaches to Enhance Desirable Traits

Some breeding approaches for crop improvement are the following:

13.4.1.1 Conventional Breeding

Traditional methods involve selecting and crossing plants with desirable traits over several generations. This approach has been successful in developing many high-yielding crop varieties (Acquaah 2009).

13.4.1.2 Marker-Assisted Selection (MAS)

This method involves using molecular markers linked to specific genes associated with desirable traits. It allows for more precise and efficient selection of plants in early stages (Collard and Mackill 2008).

13.4.1.3 Genomic Selection

This approach utilizes genomic information to predict the breeding value of plants. It can accelerate the breeding process by selecting individuals based on their entire genetic makeup (Heffner et al. 2009.

13.4.1.4 Transgenic Breeding

Genetic engineering involves introducing specific genes into plants to confer desirable traits, such as resistance to pests or tolerance to environmental stresses (James 2011).

13.4.1.5 CRISPR-Cas9 Technology

This revolutionary gene-editing tool enables precise modification of specific genes to enhance desired traits without introducing foreign DNA (Hsu et al. 2014).

13.4.1.6 Gene Pyramiding

Gene pyramiding is a strategic approach in plant breeding that involves the simultaneous incorporation of multiple genes or alleles into a single crop variety to enhance its resistance or tolerance to various stresses. This method aims to stack desirable traits in a single genetic background, resulting in a more robust and resilient crop. Gene pyramiding is particularly valuable in addressing complex and dynamic challenges such as biotic and abiotic stresses (Fig. 13.3).

13.4.2 Specific Examples of Crops Successfully Adapting to Stress Through Functional Traits

One notable example of a crop successfully adapting to stress through functional traits is rice (*Oryza sativa*). In response to flooding, which can severely impact rice yields, certain varieties have evolved the Sub1A gene that enhances tolerance to submergence. Sub1A-positive rice varieties exhibit reduced elongation of stems and leaves during flooding, allowing for better survival and subsequent recovery once the water recedes. This functional trait has been integrated into various rice cultivars through molecular breeding techniques, contributing significantly to the resilience of rice crops in flood-prone regions (Xu et al. 2006; Septiningsih et al. 2009).

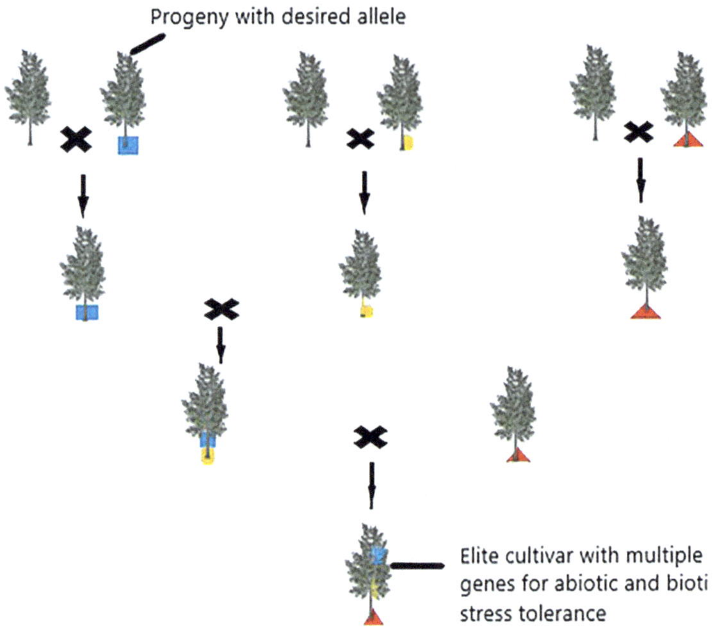

Fig. 13.3 Representation of gene pyramiding against biotic and abiotic stresses

13.5 Conclusion

Many different types of stress have a disastrous impact on crop yield and global food productivity worldwide. It is vital to protect crop plants against biotic and abiotic stresses, which result in yield and meet the demands of the world's expanding population. The integration of plant functional traits in crop breeding strategies has emerged as a promising and holistic approach to enhance adaptation under both abiotic and biotic stressors. This multidimensional framework, encompassing morphological, physiological, and biochemical traits, provides a comprehensive understanding of plant responses to environmental challenges. The identification and selection of key functional traits associated with stress tolerance offer a targeted and effective means of developing crops resilient to diverse and evolving threats.

Research in this field has demonstrated the potential of functional trait-based approaches to improve crop performance under adverse conditions. By unraveling the intricate relationships between traits and stress responses, breeders can make informed decisions to prioritize and manipulate specific characteristics for enhanced adaptability. Moreover, the incorporation of advanced technologies, such as genomics and precision breeding, holds promise for accelerating the development of stress-resistant crop varieties. A complete understanding of the physiological and biochemical mechanisms, particularly the signaling cascades, which occur in tolerant plants in response to stress would enable the manipulation of sensitive crop plants and ultimately enhance agricultural output in the foreseeable future.

References

Acquaah G (2009) Principles of plant genetics and breeding. Wiley, Hoboken

Agrawal AA, Conner JK, Stinchcombe JR (2004) Evolution of plant resistance and tolerance to frost damage. Ecol Lett 7(12):1199–1208

Atkinson NJ, Urwin PE (2012) The interaction of plant biotic and abiotic stresses: from genes to the field. J Exp Bot 63(10):3523–3543

Bale JS, Masters GJ, Hodkinson ID, Awmack C, Bezemer TM, Brown VK, Butterfield J, Buse A, Coulson JC, Farrar J, Good JEG, Harrington R, Hartley S, Jones TH, Lindroth RL, Press MC, Symrnioudis I, Watt AD, Whittaker JB (2002) Herbivory in global climate change research: direct effects of rising temperature on insect herbivores. Glob Chang Biol 8(1):1–16

Balmer A, Pastor V, Gamir J, Flors V, Mauch-Mani B (2015) The 'prime-ome': towards a holistic approach to priming. Trends Plant Sci 20(7):443–452

Basu S, Ramegowda V, Kumar A, Pereira A (2016) Plant adaptation to drought stress. F1000Res 5:F1000

Bates TR, Lynch JP (1996) Stimulation of root hair elongation in Arabidopsis thaliana by low phosphorus availability. Plant Cell Environ 19(5):529–538

Berenbaum M (1995) Phototoxicity of plant secondary metabolites: insect and mammalian perspectives. Arch Insect Biochem Physiol 29(2):119–134

Bezemer TM, Harvey JA, Kowalchuk GA, Korpershoek H, van der Putten WH (2006) Interplay between Senecio jacobaea and plant, soil, and aboveground insect community composition. Ecology 87(8):2002–2013

Birkett MA, Campbell CA, Chamberlain K, Guerrieri E, Hick AJ, Martin JL, Woodcock CM (2000) New roles for cis-jasmone as an insect semiochemical and in plant defense. Proc Natl Acad Sci 97(16):9329–9334

Bita CE, Gerats T (2013) Plant tolerance to high temperature in a changing environment: scientific fundamentals and production of heat stress-tolerant crops. Front Plant Sci 4:273

Chaves MM, Flexas J, Pinheiro C (2009) Photosynthesis under drought and salt stress: regulation mechanisms from whole plant to cell. Ann Bot 103(4):551–560

Collard BC, Mackill DJ (2008) Marker-assisted selection: an approach for precision plant breeding in the twenty-first century. Philos Trans R Soc B Biol Sci 363(1491):557–572

Colmer TD (2003) Long-distance transport of gases in plants: a perspective on internal aeration and radial oxygen loss from roots. Plant Cell Environ 26(1):17–36

Condon JT, Boyce P, Corkindale CJ (2004) The first-time fathers study: a prospective study of the mental health and wellbeing of men during the transition to parenthood. Aust N Z J Psychiatry 38(1–2):56–64

Dangl JL, Jones JD (2001) Plant pathogens and integrated defence responses to infection. Nature 411(6839):826–833

Devi B, Kumar MN, Chutia M, Bhattacharyya N (2022) Abiotic and biotic stress challenges of cassava in changing climate and strategies to overcome: a review. Sci Hortic 305:111432

Dow GJ, Bergmann DC, Berry JA (2014) An integrated model of stomatal development and leaf physiology. New Phytol 201(4):1218–1226

Duke SO (2005) Taking stock of herbicide-resistant crops ten years after introduction. Pest Manage Sci 61(3):211–218

Durrant WE, Dong X (2004) Systemic acquired resistance. Annu Rev Phytopathol 42:185–209

Engineer CB, Ghassemian M, Anderson JC, Peck SC, Hu H, Schroeder JI (2014) Carbonic anhydrases, EPF2 and a novel protease mediate CO_2 control of stomatal development. Nature 513(7517):246–250

Flexas J, Medrano H (2002) Drought-inhibition of photosynthesis in C_3 plants: stomatal and non-stomatal limitations revisited. Ann Bot 89(2):183–189

Flexas J, Ribas-Carbo M, Diaz-Espejo A, Galmés J, Medrano H (2008) Mesophyll conductance to CO_2: current knowledge and future prospects. Plant Cell Environ 31(5):602–621

Flexas J, Díaz-Espejo A, Conesa MA, Coopman RE, Douthe C, Gago J, Niinemets Ü (2016) Mesophyll conductance to CO_2 and Rubisco as targets for improving intrinsic water use efficiency in C_3 plants. Plant Cell Environ 39(5):965–982

Flor HH (1971) Current status of the gene-for-gene concept. Annu Rev Phytopathol 9(1):275–296

Franks PJ, Beerling DJ (2009) Maximum leaf conductance driven by CO_2 effects on stomatal size and density over geologic time. Proc Natl Acad Sci 106(25):10343–10347

Gatehouse JA (2002) Plant resistance towards insect herbivores: a dynamic interaction. New Phytol 156(2):145–169

Gimenez E, Salinas M, Manzano-Agugliaro F (2018) Worldwide research on plant defense against biotic stresses as improvement for sustainable agriculture. Sustainability 10(2):391

Hartfield JL, Prueger JH (2015) Temperature extremes: effect on plant growth and development. Weather Clim Extrem 4:4–10

Heffner EL, Sorrells ME, Jannink JL (2009) Genomic selection for crop improvement. Crop Sci 49(1):1–12

Herman JJ, Sultan SE (2011) Adaptive transgenerational plasticity in plants: case studies, mechanisms, and implications for natural populations. Front Plant Sci 2:102

Howe GA, Jander G (2008) Plant immunity to insect herbivores. Annu Rev Plant Biol 59:41–66

Hsu PD, Lander ES, Zhang F (2014) Development and applications of CRISPR-Cas9 for genome engineering. Cell 157(6):1262–1278

Jaleel CA, Manivannan P, Wahid A, Farooq M, Al-Juburi HJ, Somasundaram R, Panneerselvam R (2009) Drought stress in plants: a review on morphological characteristics and pigments composition. Int J Agric Biol 11(1):100–105

James C (2011) Global status of commercialized biotech/GM crops, 2011, vol 44. ISAAA, Ithaca

Jenks M, Hasegawa PM (2005) Plant abiotic stress, biological science series. Blackwell Publishing, Ames

Jones JD, Dangl JL (2006) The plant immune system. Nature 444(7117):323–329

Karban R, Myers JH (1989) Induced plant responses to herbivory. Annu Rev Ecol Syst 20(1):331–348

Kissoudis C, van de Wiel C, Visser RG, van der Linden G (2014) Enhancing crop resilience to combined abiotic and biotic stress through the dissection of physiological and molecular crosstalk. Front Plant Sci 5:207

Klein AM, Vaissière BE, Cane JH, Steffan-Dewenter I, Cunningham SA, Kremen C, Tscharntke T (2007) Importance of pollinators in changing landscapes for world crops. Proc R Soc B Biol Sci 274(1608):303–313

Kumar N, Kumar N, Shukla A, Shankhdhar SC, Shankhdhar D (2015) Impact of terminal heat stress on pollen viability and yield attributes of rice (Oryza sativa L.). Cereal Res Commun 43(4):616–626

Kumar N, Shankhdhar SC, Shankhdhar D (2016) Impact of elevated temperature on antioxidant activity and membrane stability in different genotypes of rice (Oryza sativa L.). Indian J Plant Physiol 21(1):37–43

Kumar N, Suyal DC, Sharma IP, Verma A, Singh H (2017) Elucidating stress proteins in rice (Oryza sativa L.) genotype under elevated temperature: a proteomic approach to understand heat stress response. 3 Biotech 7:205

Kumar N, Jeena N, Singh H (2019) Elevated temperature modulates rice pollen structure: a study from foothill Himalayan agro-ecosystem in India. 3 Biotech 9:175

Kumar A, Dwivedi GK, Tewari S, Paul J, Anand R, Kumar N, Kumar P, Singh H, Kaushal R (2020a) Carbon mineralization and inorganic nitrogen pools under Terminalia chebula Retz.-based agroforestry system in Himalayan foothills, India. For Sci 66(5):634–643

Kumar A, Dwivedi GK, Tewari S, Paul J, Sah VK, Singh H, Kumar P, Kumar N, Kaushal R (2020b) Soil organic carbon pools under Terminalia chebula Retz. based agroforestry system in Himalayan foothills, India. Curr Sci 118(7):1098–1103

Kumar A, Kumar P, Singh H, Kumar N (2020c) Adaptation and mitigation potential of roadside trees with bio-extraction of heavy metals under vehicular emissions and their impact on physiological traits during seasonal regimes. Urban For Urban Green 58:126900

Kumar A, Kumar P, Singh H, Bisht S, Kumar N (2021a) Relationship of physiological plant functional traits with soil carbon stock in temperate forest of Garhwal Himalaya. Curr Sci 120(8):1368–1373

Kumar A, Kumar P, Singh H, Kumar N (2021b) Modulation of plant functional traits under essential plant nutrients during seasonal regime in natural forests of Garhwal Himalayas. Plant Soil 465:197–212

Kumar A, Kumar P, Singh H, Kumar N (2021c) Impact of plant functional traits on near saturated hydraulic conductivity of soil under different forests of Kempty watershed in Garhwal Himalaya. Indian J Soil Conserv 49(1):38–44

Kumar A, Tewari S, Singh H, Kumar P, Kumar N, Bisht S, Kushwaha S, Tamta N, Kaushal R (2021d) Biomass accumulation and carbon stocks in different agroforestry system prevalent in Himalayan foothills, India. Curr Sci 120(6):1083–1088

Kumar N, Jeena N, Kumar A, Khairakpam R, Singh H (2021e) Comparative response of rice cultivars to elevated air temperature in Bhabar region of Indian Himalaya: status on yield attributes. Heliyon 7:e07474

Lynch JP (2013) Steep, cheap and deep: an ideotype to optimize water and N acquisition by maize root systems. Ann Bot 112(2):347–357

Macías FA, Marín D, Oliveros-Bastidas A, Varela RM, Simonet AM, Carrera C, Molinillo JM (2003) Allelopathy as a new strategy for sustainable ecosystems development. Biol Sci Space 17(1):18–23

Maheswari M, Yadav SK, Shanker AK, Kumar MA, Venkateswarlu B (2012) Overview of plant stresses: mechanisms, adaptations and research pursuit. In: Crop stress and its management: perspectives and strategies. Springer, Dordrecht, pp 1–18

Mantri N, Patade V, Penna S, Ford R, Pang E (2012) Abiotic stress in plants: present and future. In: Ahmad P, Prasad MNV (eds) Abiotic stress responses in plants: metabolism, productivity and sustainability. Springer, New York, pp 1–19

McDowell N, Pockman WT, Allen CD, Breshears DD, Cobb N, Kolb T, Yepez EA (2008) Mechanisms of plant survival and mortality during drought: why do some plants survive while others succumb to drought? New Phytol 178(4):719–739

Meena H, Meena RS, Singh B, Kumar S (2016) Response of bio-regulators to morphology and yield of clusterbean [Cyamopsis tetragonoloba (L.) Taub.] under different sowing environments. J Appl Nat Sci 8(2):715–718

Meng C, Tian D, Zeng H, Li Z, Yi C, Niu S (2019) Global soil acidification impacts on belowground processes. Environ Res Lett 14(7):074003

Mittler R (2006) Abiotic stress, the field environment and stress combination. Trends Plant Sci 11(1):15–19

Mundt CC (2014) Durable resistance: a key to sustainable management of pathogens and pests. Infect Genet Evol 27:446–455

Munns R, Tester M (2008) Mechanisms of salinity tolerance. Annu Rev Plant Biol 59:651–681

Oshunsanya SO, Nwosu NJ, Li Y (2019) Abiotic stress in agricultural crops under climatic conditions. In: Sustainable agriculture, forest and environmental management. Springer, Singapore, pp 71–100

Pillitteri LJ, Torii KU (2012) Mechanisms of stomatal development. Annu Rev Plant Biol 63:591–614

Postma JA, Lynch JP (2011) Theoretical evidence for the functional benefit of root cortical aerenchyma in soils with low phosphorus availability. Ann Bot 107(5):829–841

Postma JA, Lynch JP (2012) Complementarity in root architecture for nutrient uptake in ancient maize/bean and maize/bean/squash polycultures. Ann Bot 110(2):521–534

Qian F, Zhou W, Kaluvan S, Zhang H, Zuo L (2018) Theoretical modeling and experimental validation of a torsional piezoelectric vibration energy harvesting system. Smart Mater Struct 27(4):045018

Ramegowda V, Senthil-Kumar M (2015) The interactive effects of simultaneous biotic and abiotic stresses on plants: mechanistic understanding from drought and pathogen combination. J Plant Physiol 176:47–54

Raza A, Razzaq A, Mehmood SS, Zou X, Zhang X, Lv Y, Xu J (2019) Impact of climate change on crops adaptation and strategies to tackle its outcome: a review. Plan Theory 8(2):34

Sage RF, Zhu XG (2011) Exploiting the engine of C4 photosynthesis. J Exp Bot 62(9):2989–3000

Septiningsih EM, Pamplona AM, Sanchez DL, Neeraja CN, Vergara GV, Heuer S et al (2009) Development of submergence-tolerant rice cultivars: the Sub1 locus and beyond. Ann Bot 103(2):151–160

Shahzad A, Ullah S, Dar AA, Sardar MF, Mehmood T, Tufail MA, Haris M (2021) Nexus on climate change: agriculture and possible solution to cope future climate change stresses. Environ Sci Pollut Res 28:14211–14232

Singh H, Verma A (2013a) Physiological responses of rice cultivars to various nitrogen levels. Int J Agric Environ Biotechnol 6(3):383–388

Singh H, Verma A (2013b) Characterization and screening of high nitrogen efficient rice genotype to curtail environmental pollution. Int J Agric Environ Biotechnol 6(1):23–30

Singh H, Verma A, Shukla A (2010a) Comparative study of yield and yield components of hybrid and inbred genotypes of rice (*Oryza Sativa L.*). Int J Agric Environ Biotechnol 3:355–360

Singh H, Verma A, Krishnamoorthy Mand Shukla A (2010b) Consequence of diverse nitrogen levels on leaf pigments in five rice genotypes under field emergent circumstance. Int J Bioresour Stress Manage 1:189–193

Singh H, Verma A, Rai SK (2013a) Biochemical evaluation of different rice genotypes grown at various nitrogen levels using SDS-PAGE. Curr Adv Agric Sci 5(1):144–146

Singh H, Verma A, Shukla A (2013b) Guttation fluid as a physiological marker for selection of nitrogen efficient rice (*Oryza sativa* L.) genotypes. Afr J Biotechnol 12(44):6276–6281

Singh H, Verma A, Ansari MW, Shukla A (2014) Physiological response of rice (*Oryza sativa* L.) genotypes to elevated nitrogen applied under field conditions. Plant Signal Behav 9:e29015. https://doi.org/10.4161/psb.29015

Singh RK, Singh P, Li HB, Yang LT, Li YR (2017) Soil–plant–microbe interactions: use of nitrogenfixing bacteria for plant growth and development in sugarcane. In: Plant-microbe interactions in agro-ecological perspectives, Fundamental mechanisms, methods and functions, vol 1. Springer, pp 35–59

Singh H, Yadav M, Kumar N, Kumar A, Kumar M (2020) Assessing adaptation and mitigation potential of roadside trees under the influence of vehicular emissions: a case study of Grevillea robusta and Mangifera indica planted in an urban city of India. PLoS One 15(1):e0227380

Singh S, Singh H, Sharma SK, Nautiyal R (2021) Seasonal variation in biochemical responses of bamboo clones in the sub-tropical climate of Indian Himalayan foothills. Heliyon 7(4):e06859. https://doi.org/10.1016/j.heliyon.2021.e06859

Singh M, Singh H, Kumar A, Kumar M, Barthwal S, Thakur A (2024) Soil nitrogen availability determines the CO_2 fertilization effect on tree species (*Neolamarckia cadamba*): growth and physiological evidence. Environ Sustain. https://doi.org/10.1007/s42398-023-00300-w

Thakur P, Kumar S, Malik JA, Berger JD, Nayyar H (2010) Cold stress effects on reproductive development in grain crops: an overview. Environ Exp Bot 67(3):429–443

Tuteja N (2007) Mechanisms of high salinity tolerance in plants. Methods Enzymol 428:419–438

Umar OB, Ranti LA, Abdulbaki AS, Bola AL, Abdulhamid AK, Biola MR, Victor KO (2021) Stresses in plants: biotic and abiotic. In: Current trends in wheat research. IntechOpen, London

Wahid A, Gelani S, Ashraf M, Foolad MR (2007) Heat tolerance in plants: an overview. Environ Exp Bot 61(3):199–223

Werker E (2000) Trichome diversity and development. Adv Bot Res 31:1–35

Xu K, Xu X, Fukao T, Canlas P, Maghirang-Rodriguez R, Heuer S et al (2006) Sub1A is an ethylene-response-factor-like gene that confers submergence tolerance to rice. Nature 442(7103):705–708

Zhang M, Liang X, Wang L, Cao Y, Song W, Shi J, Jiang C (2019) A HAK family Na^+ transporter confers natural variation of salt tolerance in maize. Nat Plants 5(12):1297–1308

Zhu J (2016) Abiotic signaling and response in plants. Plant Cell 167:313–324

Managing Crop Adaptation to Changing Environment

14

Amit Bijlwan, Rajeev Ranjan, Mukesh Kumar, Siddhant Gupta, and Ankita Jha

Abstract

The pressing trends in climate change that are already noticeable, the likelihood of further climate shifts, and the growing magnitude of potential climate repercussions underscore the urgent need for a more cohesive approach to address agricultural adaptation. Climate change affects agriculture in diverse ways, including altered annual rainfall patterns, shifting temperatures, increased heat waves, changes in weed and pest behaviour, and fluctuations in atmospheric CO_2 and ozone levels. Adaptation in agriculture is vital to reduce vulnerability and mitigate climate-related consequences. Particularly, susceptible small-holder farmers are already battling changing rainfall patterns and rising temperatures in accordance with the current climate trends. Due to the heterogeneity of small-holder agricultural systems, local-specific adaptation techniques are crucial in areas that primarily depend on small-scale agriculture. Soil conservation, water management, and crop diversification are examples of farmer-led initiatives that fall under these categories. Supplemental irrigation, crop rotation, intercropping, fertilizer application, and mixed farming are important adaptation strategies. In many developing countries, agriculture is a primary source of livelihood, especially for resource-constrained rural communities. Adapting to climate change will enhance food security, reduce vulnerability, and positively impact land productivity, ensuring sustained production. Technological advancements, such as

A. Bijlwan · R. Ranjan (✉) · S. Gupta
Department of Agrometeorology, G.B. Pant University of Agriculture and Technology, Pantnagar, Uttarakhand, India

M. Kumar
Department of Vegetable Science, G.B. Pant University of Agriculture and Technology, Pantnagar, Uttarakhand, India

A. Jha
ICAR-Indian Institute of Water Management, Bhubaneswar, Odisha, India

N. Kumar, H. Singh (eds.), *Plant Functional Traits for Improving Productivity*, https://doi.org/10.1007/978-981-97-1510-7_14

remote sensing, provide critical tools for adaptive management. With its ability to deliver predictive analytics and real-time data, remote sensing technology is an essential tool for this endeavour. It empowers farmers to adapt to changing climatic conditions, adopt sustainable practices, and mitigate potential risks to crop yields and food production. Adapting to climate change will secure food supplies, reduce vulnerability, and support sustainable agricultural production. Leveraging technology, like remote sensing, and evaluations of crop germplasm to integrate beneficial traits into existing crops further enhance adaptive capabilities for farmers.

Keywords

Climate change · Crop adaptation · Remote sensing · Advance technology and crop germplasm

14.1 Introduction

The agricultural process and produces are foremost susceptible to the variations of climate variability or change, marked by varied weather events, shifting precipitation patterns, and altered temperatures, all of which significantly influence crop development, growth, and yields. Adaptation to climate change is essential to mitigate vulnerabilities in both social and biological systems. The principal objective of adaptation strategies is to enhance resilience in societies facing climate change (Smit and Wandel 2006). Crop adaptation, crucial for food production, pertains to the intricate relationship between environmental factors and the growth response of crop plants. At the end of current epoch/millennial era, crop productivity is about to decline globally, as the maximum temperature rises from 2.6 to 4 °C. A decline in the production of major agricultural crops is the primary concern to food security, especially governing the intense growth of population, globally (Rogelj et al. 2016).

The Earth is anticipated to experience a norm temperature increase by the end of the twenty-first century, with this period considered the most warming phase (Pachauri 2014). Extreme precipitation events, causing floods, and prolonged drought conditions due to irregular rainfall patterns are anticipated consequences of global warming, driven predominantly by industrialization (Kanojia and Dijkwel 2018). These environmental alterations exert pervasive effects on diverse biotic entities, encompassing flora, fauna, aquatic organisms, and the human populace. Within the Indian Himalayas, the confluence of economic deceleration, natural resource depletion, and climate change has adversely impacted both agrarian output and employment prospects. A case study conducted in the Upper Kosi drainage basin within the Kumaon division Himalayas, India, elucidates a noteworthy decline in annual agricultural productivity, approximating 25% over the preceding three decades. This decrement has manifested as a yearly food deficit of 65% and a substantial reduction in food production/capita. Furthermore, discernible diminishing trends have been observed in local employment opportunities other than at agricultural farms within diverse traditional rural sectors (Tiwari and Joshi 2012).

The phenomenon of climate change has induced a discernible warming trend on a global scale, measured from the preindustrial era (1850–1900) to the contemporary period (1998–2018). As documented by the Intergovernmental Panel on Climate Change (IPCC), the average global temperature over land has escalated by 1.41 °C (IPCC 2019). This escalation has manifested in an augmented frequency of elevated temperatures during both diurnal and nocturnal periods, concomitant with a reduction in the occurrences of cold days/length of winter season and nights (Orlowsky and Seneviratne 2012; Mishra et al. 2015). Additionally, there have been discernible alterations in the extreme weather phenomena either in terms of prevalence and intensity, including droughts and precipitation events, at a regionally specific scale (Spinoni et al. 2019). The confluence of industrialization and burgeoning global population has perpetuated processes like deforestation and escalated land requisites, contributing to a sustained elevation in greenhouse gas (GHG) emissions.

With each incremental rise of global temperature (~1 °C), a consequential reduction in wheat yields by an estimated 4–6% is anticipated (Asseng et al. 2015; Liu et al. 2016). Analogously, projections indicate a comparable adverse impact on major productivity regions of maize and by the close of the century with regions responsible for the world's maize production (~56%) expected to undergo a decline in yield (Bassu et al. 2014; Pugh et al. 2016; Zhao et al. 2017). Wheat cultivation is notably susceptible to temperature extremes induced by climate change across diverse nations, potentially resulting in about 6% reduction in crop yield for each unit increment in temperature (Thornton et al. 2014). Drought conditions and elevated temperatures emerge as pivotal stressors significantly impacting cereal yields (Asseng et al. 2015; Kumar et al. 2015, 2016, 2017). The central enzyme of photosynthesis, Rubisco, faces disruption if temperatures surpass 35 °C, leading to a cessation of the photosynthetic process. As sessile organisms, plants have evolved adaptive mechanisms encompassing escape, avoidance, and tolerance strategies to contend with temporary water limitations, ensuring their survival and reproductive success. Climate change-induced water deficits and temperature extremes notably influenced the propagative and reproductive phase of plant growth. Water stress during cereals flower initiation and inflorescence phases has been reported to adversely affect these developmental processes. Similarly, sterility in cereals is noticed during propagative development and may be induced by elevated temperatures reaching approximately 30 °C (Barnabás et al. 2008; Kumar et al. 2019, Kumar et al. 2021a, b, c, d, e).

The fluctuations in climate already have a significant effect on biological systems and the smallholders, communities, and nations that rely on them. Many stressors (biotic or abiotic) on plants are increased due to unpredictability in cropping season climatic/weather events and environmental extremes (Pugh et al. 2016; Singh et al. 2020). As Earth's temperature continues to rise, the climate undergoes significant changes, leading to abiotic stressors. These environmental alterations pose numerous threats to prevalent crop growth dynamics. In the field conditions, scarcity of soil moisture and available water and heat emerges as the most prevalent stresses, significantly impacting plant growth. Optimal temperatures are essential for normal plant growth and flowering, making temperature fluctuations a crucial factor in plant physiology. Grain production in cereals and other crops are adversely affected primarily by heat stress, while stress imposed by cold temperature leads to sterility, and plant morpho-physiology is negatively affected by drought stress. These climatic challenges profoundly disturb plant development, resulting in extensive responses at molecular, biochemical, physiological, and morphological levels. Cereal crops like maize, rice, and wheat, vital for meeting global food demands, face considerable challenges. Wheat, occupying an approximately 39% of total cultivated land globally and boasting a high protein concentration (8–20% per gram), is a leading staple crop. Despite its extensive cultivation, wheat productivity lags behind maize and rice. An increase of 2 °C in temperature across various projection scenarios is expected to result in a notable decline in wheat crop productivity. This is anticipated to lead to a 6% reduction in yield due to environmental fluctuations. (Sharma et al. 2022).

Under changing average weather conditions over long period of time (Challinor et al. 2007), the decrease in the length of grain-filling phase is a major contributor to reduced crop productivity. Sustaining crop yield becomes a critical task in contemporary farming practices, emphasizing the need for the development of new stress-tolerant cultivar and advanced farming operations. In pursuit of sustainable agriculture and ensuring food security for the burgeoning global population, the cultivation of stress-tolerant plants and a comprehensive understanding of their responses to diverse stress conditions are imperative. Plants exhibit variations in gene expression, physiology, and metabolism in response to various climatic stresses (Kumar et al. 2021a, b). In the face of distinct stresses, plant organs and tissues undergo damage, eliciting context-specific transcriptional responses in specific cells or tissues, particularly in the root system. Salinity, drought, and chemical effluence considered as primary stress induce the production of proteins which are stress-responsive in nature (aquaporin, cold shock proteins, cyclophilins, etc.). These primary stresses, in turn, give rise to subsidiary stresses, including oxidative and osmotic. Under drought conditions, an elevated level of carbon dioxide (CO_2) in leaves instigates the generation of reactive oxygen species (ROS), triggering multifaceted stresses in crops. The closure of stomata during drought impedes the manoeuvre of CO_2 within the leaf, leading to ROS production due to heightened oxygen levels (Kumar et al. 2021c, d). Fundamental physiological processes such as respiration, plant development, and photosynthesis are disrupted by membrane breakdown induced by ROS production. Crucial cellular components such as carbohydrates, lipids, proteins, and biopolymers of life (genetic material) are compromised under drought-induced ROS stress (Ahmad et al. 2018; Carvalho and Amâncio 2019).

14.2 Changing Environmental Conditions

The IPCC designates the forthcoming 25 years as the era of committed climate change, emphasizing the anticipation of approximately 1 °C warming across all projected scenarios (Intergovernmental Panel on Climate Change 2013). This temporal framework underscores a lag in climate action, indicating a delay between mitigation intentions and the tangible realization of emissions reductions. The inevitability of climate change is attributed to several factors: (1) the momentum in emissions resulting from concurrent income and population growth; (2) the persistence of long-lived, costly capital facilities influencing emissions over an extended period; (3) delayed physical responses, exemplified by the time required for the ocean to achieve equilibrium; and (4) the prolonged atmospheric lifespan of certain greenhouse gases. The repercussions of this inevitable climate change hold significant implications for agriculture. The 1 °C of committed climate change essentially necessitates agricultural adaptation within the next quarter-century. This adaptation involves the development of practices and management systems to sustain productive agriculture, mitigating potential food security issues. Unlike historical agricultural adaptations with enduring effects, contemporary adaptation must be an ongoing process as climate change progresses. Additionally, the capacity to adapt is constrained in certain regions by financial, human, and physical capital limitations, as well as cultural and lifestyle barriers. Moreover, traditional knowledge passed down through generations may prove insufficient in the face of climate-induced shifts, necessitating alternative forms of learning. Because of excessive utilization of fossil fuels and the continuous deforestation, the concentration of CO_2 has escalated up to 400 μmol^{-1} in the atmosphere. At the end of this century, the CO_2 concentration will elevate up to 800 μmol^{-1} (Vaughan et al. 2018; Kumar et al. 2020a, b, c).

For instance, areas traditionally cultivated may find optimal use in livestock farming, lands predominantly dedicated to wheat may yield greater productivity with corn cultivation, and areas cultivating corn may benefit from transitioning to sorghum. Such substantial shifts in agricultural enterprises may introduce unfamiliar circumstances, requiring enhanced extension services and other information dissemination methods to facilitate a smooth transition.

Impacts of climate change are divided into the following classes.

14.2.1 Biophysical Impacts

(a) **Physiological and Anatomical Effects**
  - Impacting agricultural produce, rangelands, woodland, and livestock at farms in terms of both quantity and quality

(b) **Changes in Natural Resources**
  - Alterations in land, soil, and hydrological resources, affecting both their quantity and quality

(c) **Biotic Challenges**
  - Increased challenges posed by weeds and pests, influencing agricultural ecosystems

(d) **Spatial and Temporal Distribution Shifts**
  - Changes in the spatial and temporal distribution of impacts, potentially disrupting ecological and agricultural systems

(e) **Oceanic Changes**
Goal 14 of the United Nations aims to conserve and sustainably use oceans, seas, and marine resources for sustainable development. In pursuit of this goal, attention is directed towards addressing challenges such as rising sea levels and alterations in ocean salinity. The increasing sea temperatures are inducing noteworthy shifts in the habitats of various fish species, necessitating a comprehensive approach to mitigate the impacts on marine ecosystems.

14.2.2 Social and Economic Impacts

(a) **Agricultural Productivity**
There has been a notable decrease in both yields and production, which has had a considerable impact on overall agricultural productivity, thereby hindering progress toward Sustainable Development Goals (SDG) 2, specifically Target 2.2, aimed at doubling the incomes of small-scale agricultural producers and enhancing productivity further.

(b) **Economic Consequences**
Reduced marginal GDP derived from agriculture, affecting economic stability.

(c) **Market Dynamics**
The fluctuations in world market prices, posing challenges for economic planning, are particularly relevant in the context of SDG2 and Target 2.7. The inherent volatility in global market prices presents a formidable obstacle to effective economic planning, emphasizing the need for strategic measures to navigate and mitigate the impacts on sustainable development initiatives.

(d) **Trade Distribution Changes**
Alterations in the geographical distribution of trade regimes, affecting global trade dynamics.

(e) **Food Security and Hunger**
Addressing the SDG specifically Target 2.2, which emphasizes the end of all forms of malnutrition, is crucial in mitigating the risk of hunger and food supply crisis in underdeveloped society. This challenge poses significant social challenges, highlighting the urgency and importance of comprehensive efforts to eradicate malnutrition and ensure access to adequate, safe, and wholesome food for all.

(f) **Migration and Civil Unrest**
Target 16a, of SDG which focuses on strengthening relevant national institutions for capacity building, including through international cooperation, holds implications for migration patterns and potential civil unrest. As communities adapt to changing conditions, the need to prevent violence, combat terrorism, and address crime becomes intertwined with broader socio-political challenges, emphasizing the interconnected nature of security, stability, and migration dynamics.

14.2.3 Climate Change Adaptation Approaches

Two primary categories of adaptation strategies are recognized: autonomous adaptation and planned adaptation.

14.2.3.1 Autonomous/Spontaneous Adaptation

Autonomous adaptation entails the spontaneous response of individuals, such as farmers, to shifting climate conditions (Datta et al. 2022). This response may involve alterations in agricultural practices, such as changing cultivated species or simulating planting and harvesting dates in response to varied patterns of precipitation.

14.2.3.2 Planned Adaptation

These measures for adaptation are deliberate, conscious policy options designed to enhance the flexibility of agricultural systems or facilitate specific adaptations. These measures are often comprehensive and span multiple sectors, with examples including intentional selection of new crop cultivars and multi-environmental growth pattern tailored to diverse agro-climatic zones. Resource substitution induced by scarcity is also considered a planned adaptation, encompassing actions like replacing old crops with new ones.

Research at the farm level has demonstrated substantial reductions in adverse impacts from climate change through the full implementation and applications of adaptation measures (Mendelsohn and Dinar 1999). For advancement and implementation of short-term adjustments, considered as autonomous adaptations, do not require any involvement from other sectors such as policy or research. In contrast, long-term adaptations involve significant changes in farming practices structurally aimed at overcoming adversity. Examples include alterations in cultivations practices and operations to maximize yields under future climatic conditions, the application of innovative farm technologies and biotechnologies, adoption of new land management techniques, and the utilization of water more effectively such as micro- or sprinkler irrigation. These long-term adaptations are essential for addressing the enduring challenges posed by climate change. Reilly and Schimmelpfennig (1999) delineate distinct primary classes of adaptation, encompassing a range of strategies to address the impacts of climate change. This classification provides a systematic framework for understanding diverse avenues of adaptation, reflecting the multifaceted nature of approaches required to address the complex challenges imposed by climate change scenario in agriculture and forestry.

(a) **Seasonal Changes and Sowing Dates**

Adaptation involves adjustments to the timing of agricultural activities, particularly modifications in seasonal patterns and sowing dates.

(b) **Different Variety or Species**

Adoption of alternative crop varieties or species represents a key adaptive measure to navigate changing environmental conditions.

(c) **Water Supply and Irrigation System**

Adaptation strategies include the optimization and modification of water supply systems and irrigation practices to contend with altered climatic conditions.

(d) **Other Inputs**

Implementing adaptive agricultural practices necessitates an exhaustive approach that contemplates a range of inputs. Components such as fertilizer application, methods of tillage, grain drying procedures, and various field operations collectively contribute to embellishing the adaptability and sustainability of agricultural ecosystems. A holistic consideration of these elements ensures informed decision-making for optimized agricultural outcomes.

(e) **New Crop Varieties**

The introduction and cultivation of novel crop varieties, specifically bred or engineered to withstand changing climate conditions, serve as a proactive adaptation strategy.

(f) **Forest Fire Management and Agroforestry Promotion**

Adaptive measures extend beyond traditional agriculture to encompass forest fire management strategies, the promotion of agroforestry practices, and the implementation of adaptive management techniques utilizing suitable tree species and silvicultural practices (FAO 2005).

14.3 Crop Selection and Breeding for Adaptations

Genetic heterogeneity is a fundamental prerequisite in plant genetic programmes, denoting the aggregate of all genes within a crop species, collectively referred to as germplasm. Resilience against emerging challenges such as diseases and pests and adaptation to changing environmental conditions the genetic diversity, or the variance in genes within a specific population, plays a pivotal role in species survival (Snowdon et al. 2021). Plant breeders leverage genetic diversity as the primary material to initiate and enhance their breeding inheritance, allowing for the creation of new and improved strains with desired attributes, including those favoured by farmers and breeders alike, such as resistance against pest and disease.

Significance of crop genetic diversity for agricultural resilience:

(a) **Resilience and Adaptation**: A broader genetic base within a crop species equips it with a diverse array of traits, enabling adaptation to changing environmental conditions and providing a defence against diseases, pests, and pathogens.
(b) **Crop Improvement**: For a crop improvement initiative, genetic diversity serves as a constructive stock. Plant breeders can selectively choose desirable traits from diverse varieties to enhance crop performance in terms of harvest products, its quality, and resistance against both biotic and abiotic stresses.
(c) **Sustainability**: Diverse crops contribute to the sustainability of agriculture by minimizing vulnerability to pests, diseases, and environmental fluctuations, thereby ensuring long-term food security.

Factors influence genetic diversity in crops:

(a) **Natural Variation**: Genetic diversity arises naturally through processes such as sexual reproduction, mutations, and gene flow among populations.
(b) **Human Selection**: Historical and contemporary agricultural practices, including selective breeding, have impacted the genetic diversity of crops. Intensive cultivation of specific varieties may reduce overall diversity.
(c) **Environmental Factors**: Environmental pressures and geographical isolation also influence genetic diversity by favouring certain traits or limiting gene flow.

The maintenance and utilization of genetic diversity involve strategic measures:

(a) **Gene Banks**: Repositories such as seed banks and gene banks play a crucial role in preserving diverse germplasm for future use.
(b) **Breeding Programmes**: Crop breeding efforts focus on incorporating diverse traits from wild relatives or landraces into cultivated varieties, thereby expanding their genetic base.
(c) **Conservation Efforts**: Initiatives aimed at conserving and protecting traditional landraces and wild relatives contribute significantly to the preservation of genetic diversity, ensuring its continued availability for future generations.

14.3.1 Breeding for Resilience

14.3.1.1 Drought

Drought, a meteorological phenomenon, is commonly specified as an interval characterized by a considerable lack of rainfall, resulting in water stress for plants. Drought is characterized as any period during which deficiencies in plant and/or soil water adversely impact the growth and development of crops. These deficiencies can arise from either insufficient moisture supply or excessive moisture demand. The duration and timing of the drought period play a crucial role in determining the extent of damage inflicted on the plant's growth and development. The India Meteorological Department (IMD) classifies agricultural drought as a situation where the weekly rainfall is half of the normal or less, with a normal weekly rainfall of 5.0 mm or more. If 4 consecutive weeks with such conditions occur from mid-May to October, it is deemed agricultural drought (Murthy and Sesha Sai 2010).

Understanding Plant Drought Tolerance: Insights into Physiology, Biochemistry, and Molecular Mechanisms

(a) Physiological Responses:

Physiological responses in plants exposed to drought stress encompass a sequence of events vital for survival. These include the recognition of root signals, leading to adaptive adjustments such as the loss of turgor, osmotic adjustment, and a subsequent reduction in leaf water potential. Additionally, drought stress induces a decrease in stomatal conductance to CO_2, resulting in reduced concentration of internal CO_2. These physiological adjustments contribute to a decline in net photosynthesis and ultimately lead to reduced growth rates.

(b) Biochemical Responses:

In response to drought stress, plants exhibit intricate biochemical adaptations aimed at maintaining cellular functions. These adaptations involve a transient decrease in photochemical efficiency and a reduction in the efficiency of the Rubisco enzyme, impacting carbon fixation (Kumar et al. 2022). Furthermore, plants accumulate stress metabolites such as α-tocopherol, polyamines, glutathione, and MDHA. Concurrently, there is an upregulation of antioxidative enzymes, including SOD, CAT, APX, POD, GR, and MDHAR. This orchestrated response serves to mitigate the accumulation of ROS, preventing oxidative damage that could lead to cell death.

(c) Molecular Responses:

Plants deploy sophisticated mechanisms to enhance drought tolerance at the molecular level. Stress-responsive gene expression is activated, leading to an increased accumulation of congenital traits responsible for abscisic acid (ABA) biosynthesis. The expression of ABA-responsive genes further contributes to the plant's adaptive response. Synthesis of distinct proteins, such as late embryogenesis abundant (LEA), desmoplakin (DSP), RAB, and dehydrins, is triggered to conserve cellular structures and macromolecules from desiccation-induced damage. Notably, LEA proteins play a pivotal role in safeguarding other proteins from aggregation during drought stress, ensuring cellular integrity.

Understanding and manipulating these responses at the physiological, biochemistry, and polynucleotide ladder levels are pivotal for developing crops with enhanced drought stress resilience, thereby contributing to sustainable agriculture towards the ever-changing environmental conditions.

Drought Resistance in Crop Plants

Resistance against drought in crop plants denotes their ability to thrive, undergo normal growth and development, and reproduce effectively in the face of insufficient soil moisture conditions.

(a) **Mechanisms of Drought Resistance:**

(i) **Drought Escape:**

Plants regulate their life cycle to avoid stress, allowing them to complete their stages of the development during periods of competent soil moisture. Desert ephemeral plants, characterized by their brief lifespan, rapidly germinate, develop, and flower in response to intermittent rains, completing their life cycle before the advent of the dry season.

(ii) **Drought Avoidance:**

Under soil moisture stress conditions, plants possess the ability to maintain a propitious internal water balance. Due to this mechanism, some plant enables a longer development period in crops through reduced water use or increased water uptake.

(iii) **Drought Tolerance:**

Drought tolerance encompasses a plant's acclimation to arid or drought-prone environments. Within arid regions, plants confront unpredictable drought occurrences, precluding the possibility of evading unfavourable circumstances. Consequently, these plants demonstrate the capacity to endure water stress through biochemical and morphological adaptations, coupled with strategies to mitigate cellular damage. Drought tolerance mechanisms involve the maintenance of turgor pressure via osmotic adjustment (Azhar and Rehman 2018), augmentation of cell elasticity, and reduction of cell size through protoplasmic resistance.

Sources of Drought Resistance in Plant Breeding

(i) **Cultivated Varieties for Drought Resistance:**

A straightforward approach in plant breeding is the transfer of drought resistance traits from cultivated genotype and germplasm to cultivated species. This method is advantageous due to its ease of implementation without encountering issues of cross incompatibility. Cultivated varieties serve as valuable sources of traits, allowing for the efficient integration of drought resistance into breeding programmes. This approach streamlines the breeding process, facilitating the development of crop cultivars better equipped to withstand moisture deficit conditions.

(ii) **Wild Relatives and Wild Species for Drought Resistance:**

Exploring wild relatives and wild species presents challenges in transferring drought resistance due to issues like cross incompatibility, hybrid inviability, hybrid sterility, and undesired gene linkage. Despite these challenges, wild sources of drought resistance have been identified in several important crops, including wheat, sugarcane, and tomato. Overcoming these obstacles is crucial for harnessing the valuable genetic traits found in wild species, contributing to the development of more resilient and drought-resistant crop varieties.

(iii) **Breeding Methods for Drought Resistance:**

Breeding for drought resistance employs common methods applied in the development of yield and economic traits. The standard approaches include introduction, selection, hybridization, and mutations. These methods provide avenues for developing crop cultivars with enhanced tolerance to drought conditions, contributing to sustainable agriculture in water-limited environments.

Genetic Resources and Their Role in Crop Improvement

For the development of novel and improved crop cultivars, pool of genetic resources plays a pivotal role in advancing crop improvement and serving as the fundamental basis for development of new verities for major and minor crops as well. These resources encompass the genetic material, comprising genes and traits, inherent in diverse plant species and their wild counterparts. The rich genetic diversity within these resources offers a vast spectrum of traits that can be harnessed to augment crop productivity, adaptability, and resilience amid changing environmental conditions.

(a) **Wild Relatives in Crop Improvement:**

Wild relatives, closely linked to cultivated crops, represent a valuable subset of genetic resources. Termed crop wild relatives (CWR), these wild plant taxa share genetic traits with domesticated crops but have not undergone extensive selective breeding. Thriving in natural habitats, they exhibit traits such as disease resistance, tolerance to environmental stresses (e.g. drought or extreme temperatures), and heightened nutritional content. CWRs are indispensable for plant breeding and genetic enhancement initiatives aimed at cultivating more robust and productive crops. Their exploration and preservation are vital for ensuring food security and availability, particularly in the context of changing climatic scenario and evolving agricultural challenges.

Key Features of Wild Relatives

- **Genetic Diversity**: Encompassing a broad range of genetic variations, CWRs contribute to breeding programmes by offering traits such as pest and disease resistance, environmental stress tolerance, and adaptability to diverse growing conditions.
- **Adaptability**: Having evolved in various, often challenging, environments, wild relatives may possess traits facilitating adaptation to different climatic conditions, making them valuable sources of resilience for cultivated crops.
- **Undeveloped Traits**: Wild relatives harbour characteristics yet to be explored or developed in cultivated crops, encompassing traits related to taste, nutritional content, or other desirable agronomic properties.
- **Natural Resistance**: With inherent resistance to pests, diseases, or environmental stressors, wild relatives offer a reservoir of natural defences transferrable to domesticated varieties, enhancing their resilience.
- **Genetic Compatibility:** Often genetically compatible with cultivated crops, wild relatives allow successful hybridization and the transfer of desired traits through crossbreeding.
- **Conservation of Genetic Resources:** Preserving wild relatives is critical for maintaining the genetic diversity necessary for future breeding efforts, especially given the challenges posed by changing climates and evolving agricultural landscapes.

(b) **Gene Banks for Germplasm Conservation:**

Gene banks, also known as germplasm banks, serve as repositories for conserving genetic material in a living state. They come in two main types:

- **Seed Gene Banks**: These banks conserve germplasm in the form of seeds, a convenient and space-efficient storage method suitable for orthodox species whose seeds can be dried without losing variability.
- **Field Gene Banks**: Also known as plant gene banks, these areas assemble germplasm collections of growing plants. Field gene banks, representing ex situ conservation, maintain recalcitrant species or those not readily producing seeds. This form of conservation is vital for species such as coconut,

rubber, mango, cassava, yam, and cocoa. Field gene banks, established in numerous countries, provide a living, permanent collection for experimental and research purposes.
- The integration of gene banks and the preservation of wild relatives collectively contribute to sustaining genetic diversity, essential for global food security and the continued advancement of agriculture.

14.3.1.2 Breeding Strategies for Climate-Resilient Crops

Climate-resilient crop breeding stands as a crucial imperative in ensuring global food security amid escalating climate threats. This section delves into various breeding strategies, encompassing traditional techniques, molecular breeding methods, and genomic and transgenic approaches, all of which exhibit promise in enhancing crop tolerance to both abiotic and biotic challenges.

Conventional Breeding Techniques: Traditional Approaches

Traditional breeding methods have been integral in the development of climate-resilient crop varieties. These time-tested techniques involve the careful selection and crossing of plants with desirable traits to yield offspring endowed with improved attributes. Notable traditional breeding approaches for enhancing crop stress tolerance include selection, hybridization, and mutation breeding (Breseghello and Coelho 2013).

(a) **Selection**

Selection is a fundamental process in plant breeding, involving the isolation of plants with superior characteristics, such as disease resistance or drought tolerance, from a diverse population. This meticulous procedure aims to identify and cultivate individuals with desirable traits, allowing for the development of new crop varieties with enhanced stress tolerance. The application of selection spans multiple generations, as breeders consistently choose plants exhibiting the desired features, gradually refining and improving the genetic makeup of the population. Through this iterative process, plant breeders can create resilient and adapted crop varieties that thrive in specific environmental conditions, contributing to the sustainable improvement of agricultural outcomes.

(b) **Hybridization**

Hybridization is a pivotal breeding method characterized by the intentional crossbreeding of genetically dissimilar parents. This strategic combination results in the generation of hybrids that often surpass their progenitors in performance, including heightened stress resistance. By bringing together diverse genetic traits, hybridization plays a crucial role in the development of crop varieties with superior resilience to various environmental stresses. This approach leverages the complementary strengths of different parental lines, producing offspring with enhanced adaptive traits and overall stress tolerance. As a key component of plant breeding, hybridization continues to be instrumental in the creation of crops that can thrive in challenging conditions, contributing to sustainable and resilient agricultural practices.

(c) **Mutation Breeding**

Mutation breeding is a significant approach in plant breeding that incorporates the deliberate induction of genetic mutations using mutagens such as irradiation or chemical/synthesized agents. This technique has proven instrumental in the advancement of climate-adaptive crop varieties. By subjecting plants to controlled mutagenic treatments, researchers aim to introduce genetic variability, leading to the emergence of novel traits, particularly stress tolerance. The mutation breeding allows for the creation of unique genetic variations that can be selectively incorporated into breeding programmes. This method harnesses the power of induced mutations to enhance the adaptability and resilience of crops, contributing to the ongoing efforts in developing agriculturally robust and climate-resilient plant varieties.

These conventional breeding techniques have successfully contributed to the creation of climate-resilient crop varieties, by alleviating the impact of abiotic stresses, such as drought and heat. The continued integration of traditional breeding methods, alongside emerging molecular and genomic approaches, reinforces the quest for sustainable and resilient agriculture in the face of climate uncertainties.

Advancements Techniques in Molecular Breeding for Crop Improvement Programme

Significant progress in crop development has been driven by the adoption of molecular breeding techniques, empowering breeders to precisely target desired traits at the molecular level. Leveraging insights into plant genomes and the identification of genes associated with stress resistance, these techniques mark a transformative era in agriculture. A significant advancement in molecular breeding involves utilizing molecular markers located in specific genomic regions that are easily identifiable and associated with particular desirable traits.

The integration of molecular markers, such as those associated with stress resistance, into breeding programmes has revolutionized the precision of trait selection. Through marker-assisted selection (MAS), breeders can efficiently identify and choose plants with desired stress tolerance features. This targeted approach facilitates the identification of individuals with the most promising traits for further breeding, allowing breeders to accurately predict the performance of potential progeny. Molecular breeding extends its reach to harnessing genetic resources, including landraces and wild crop relatives. These genetic resources serve as repositories of valuable genes for stress tolerance (Varshney et al. 2018; Razzaq et al. 2021). By combining conventional breeding with molecular markers, breeders can strategically transfer stress-tolerant genes from these resources into elite cultivars. This innovative approach broadens the genetic diversity accessible to breeding programmes, significantly enhancing the potential for developing climate-resilient crops.

The marriage of traditional breeding practices with cutting-edge molecular techniques not only expedites the identification of desirable traits but also augments the genetic diversity at the breeder's disposal. This amalgamation holds the promise of cultivating crops better equipped to withstand the challenges posed by a dynamic and unpredictable climate, ensuring sustainable and resilient agriculture for the future.

(a) **Genomic and Transgenic Innovations for Climate-Resilient Crop Breeding:**
The advent of genomic technologies has ushered in a new era in the quest for climate-resilient crops. Leveraging quantitative trait locus (QTL) mapping and genome-wide association studies (GWAS), researchers can pinpoint genomic regions linked to stress tolerance features (Voss-Fels et al. 2019). By analysing diverse genetic markers across various plant populations, these methods provide valuable insights into the genetic variants associated with stress tolerance (Tong and Nikoloski 2021).

(b) **Genomic Selection for Accelerated Breeding:**
A paradigm shift in breeding strategies comes with the integration of genomic selection, where a plant's breeding value is predicted based on its genomic information. This approach enables breeders to select plants based on their genetic profile rather than relying solely on phenotypic assessment. By reducing the time needed for phenotypic evaluation and selection, genomic selection accelerates the breeding process, enhancing efficiency.

(c) **Transgenic Approaches and Genetic Engineering:**
In the realm of genetic engineering, transgenic approaches open new horizons for developing climate-resilient crops. This involves the introduction of specific genes into plants to confer desired traits. For instance, the incorporation of genes encoding insecticidal proteins like *Bacillus thuringiensis* (Bt) toxins has proven effective in imparting insect resistance. Similarly, the introduction of genes associated with drought or salinity tolerance holds promise for bolstering crop resilience against abiotic stresses.

(d) **Revolutionizing Crop Improvement with Genome Editing:**
Genome editing technologies, notably CRISPR-Cas9, stand as transformative tools in crop improvement. CRISPR-Cas9 allows precise modifications of specific genes, empowering breeders to make targeted changes in the plant genome (Ortigosa et al. 2019). This revolutionary capability offers unprecedented opportunities to enhance stress tolerance by precisely manipulating key genes involved in stress response pathways.

The synergy of genomic insights, transgenic approaches, and genome editing technologies not only expands our understanding of stress tolerance but also equips breeders with powerful tools to craft climate-resilient crops. These innovations mark a crucial stride towards sustainable and adaptive agriculture in the face of evolving climate challenges.

(e) **Rapid Generation Advance (RGA) in Crop Breeding:**
Rapid generation advance (RGA) constitutes a comprehensive suite of breeding methodologies strategically designed to expedite the derivation of stable lines from segregating generations in plant breeding programmes. The manipulation of crucial growth conditions, including temperature, relative humidity, photoperiod, and daylength, serves as the cornerstone of RGA. This manipulation facilitates the attainment of multiple generations within a single

year by accelerating flowering and seed set, often achieved through the cultivation of crops with shortened growth cycles and the harvesting of immature seeds (Gudi et al. 2022). The principal objective of RGA is the substantial reduction of the temporal investment traditionally associated with line development in breeding initiatives. Notably surpassing alternative approaches in terms of speed, technical simplicity, cost-effectiveness, and logistical demands, RGA stands as an instrumental tool. This accelerated process not only expedites variety development through rigorous selection cycles but also augments genetic gain within the breeding paradigm. An array of methods is employed for RGA, with notable approaches being:

(i) **Single-Seed Descent (SSD) Method**

The single-seed descent method, an adaptation of the bulk method, was initially proposed by Goulden in 1939 and later refined by Brim in 1966. In this approach, instead of combining the entire seed lot of selected plants, a single seed is randomly chosen from each selected plant to form the bulk. This method streamlines record-keeping (as opposed to the pedigree method) and is particularly effective when the primary emphasis is on enhancing quantitative traits like yield and earliness, as opposed to qualitative traits such as flesh colour and disease resistance. Key features of this scheme include the absence of specific efforts for artificial selections, or even the possibility of natural selection, until the F5 or F6 stages when the population achieves a reasonable level of homozygosity. Subsequent generations, from F3 onward, are developed from a bulk containing one seed from each F2 plant, ensuring equal representation of each F2 plant in the final population. The single-seed descent method is gaining popularity among breeders due to its efficiency and cost-effectiveness (Parmar et al. 2021).

(ii) **Speed Breeding**

Speed breeding is a novel approach applicable to day-neutral crops, involving the extension of photoperiod and controlled conditions in glasshouses to shorten breeding cycles. This technique, inspired by NASA's initiative to grow crops quickly in space, enables the growth of three to nine generations in a year. By manipulating factors such as temperature, photoperiod regime, soil moisture, planting density, nutrient composition, and CO_2, speed breeding facilitates the rapid development of stable, homozygous genotypes. It is highly amenable to phenotypic selection for various traits, though it requires a departure from established selection methods like pedigree, bulk, and incessant selection. The technique of speed breeding involves subjecting plants to continuous light conditions lasting 20–22 h, promoting extended periods of photosynthesis and consequently accelerating plant growth. This method enables the cultivation of four to six generations of wheat plants annually, a notable increase compared to the typical two generations under standard growth conditions (Wanga et al. 2021).

RGA methods, including SSD and speed breeding, represent innovative strides in crop breeding, promising faster and more efficient development of climate-resilient and improved crop varieties. These advancements are pivotal in addressing the challenges posed by evolving climates and ensuring global food security.

(iii) **Shuttle Breeding**

Shuttle breeding, an off-season experiment approach, involves the growth of genetic material in contrasting environments, enabling the turnover of two plant generations annually (Alahmad et al. 2022). This method employs segregating populations for screening, selection, and simultaneous generation advancement. To achieve two generations each year, materials are cultivated at two distinct locations varying in precipitation, altitude, and latitude. Crop plants experience inevitable exposure to diverse stress conditions, such as salinity, drought, nutrient imbalance, and biotic factors, throughout their developmental stages. Evaluating breeding lines for stress tolerance becomes crucial, prompting the exposure of early-generation materials to authentic stress environments.

When stress conditions are not readily available in the breeding station, the strategic use of shuttle breeding becomes invaluable for the rapid assessment of lines. This method entails transporting materials to different locations, facilitating the swift development of cultivars tailored for specific environments. The shuttle breeding process involves crossing selected improved varieties or robust parental lines with locally adapted parents, resulting in the generation of F1 progenies. These progenies undergo cultivation at experimental stations, followed by the development of the F2 generation in the targeted environment, subjecting materials to desired stress conditions. Subsequent selection of better-performing adapted lines yields the F3 generation. Across successive generations, a rigorous evaluation of quality for both grain and yield enables the selection of segregants exhibiting notable stress tolerance. The effectiveness of shuttle breeding is exemplified by its successful application in rice breeding for lowland rice ecosystems in eastern India (Borlaug 2007; Tadesse et al. 2019).

(iv) **Doubled Haploidy**

The creation of a doubled haploid (DH) genotype involves the doubling of chromosomes in haploid cells (n). This technology offers a notable advantage by significantly shortening the breeding cycle, attributed to the efficiency in developing inbred lines in just 1–2 years through colchicine treatment to double the chromosome sets. Tissue culture techniques like anther and pollen culture or ovule culture sometimes used to obtain haploid plants. However, the success of this method is contingent on the availability of tissue culture and regeneration protocols for specific crops. The artificial duplication of chromosomes is achieved by treating haploid seedlings with chemicals like colchicine, inhibiting mitotic activity and preventing chromosome separation during cell division. Colchicine-treated seedlings, lacking microtubules in meristematic cells, showcase the success of chromosome doubling, contributing to the production of homozygous lines (Hooghvorst and Nogués 2021).

(v) **Genomic Selection**

Genomic selection (GS), a marker-assisted selection (MAS) approach, has emerged as a powerful tool for accelerating genetic improvement in livestock breeding programmes. The primary advantage of GS lies in its ability to significantly reduce the duration of the breeding cycle compared to traditional methods, thereby minimizing the costs associated with extensive phenotyping. GS operates by estimating the effects associated with all marker loci across the entire genome simultaneously. Initially developed for livestock breeding, GS selects individuals based on genomic estimated breeding values (GEBVs), regardless of the significance of marker effects (Bhat et al. 2016). The process involves creating a 'training population' of genotyped individuals, with phenotyping contributing to building a model for estimating marker effects used to predict GEBVs. The accuracy of GS prediction methods is influenced by various factors, challenging the conventional role of phenotyping.

Machine learning methods play a crucial role in GS, estimating GEBVs and capturing higher-order interactions. GS's potential is evident in its ability to rapidly select high breeding value individuals from early-generation populations without extensive phenotyping. In cereals, GS has proven effective in improving tolerance against biotic stresses. In rice, GS has identified blast-tolerant lines (Huang et al. 2019), while in maize, it successfully selected lines for tolerance to *Stenocarpella maydis* causing ear rot (dos Santos et al. 2016). In barley, GS utilized markers and prediction models to address *Fusarium* head blight severity (Sallam et al. 2015). The sequential and scientific application of GS holds promise for revolutionizing breeding strategies and enhancing food and nutritional security (Ghosh et al. 2018).

(vi) **Marker-Assisted Breeding**

Conventional breeding heavily relies on phenotypic evaluation for desired traits, which can be inefficient and time-consuming. Marker-assisted selection provides an indirect selection method where markers tightly linked to desirable alleles or quantitative trait loci (QTLs) are utilized for selection (Collard and Mackill 2008; Das et al. 2021). This approach streamlines the selection process, significantly reducing the time required compared to conventional methods. Marker-assisted backcrossing is an efficient process that incorporates traits into established varieties using molecular markers for the target trait while simultaneously recovering the recurrent parental genome. This method has been extensively used in integrating various traits, such as grain yield and submergence tolerance, fertility restorer genes, and disease resistance in crops like rice. Marker-assisted backcrossing accelerates the selection of desired plants, expediting the varietal improvement process and enabling gene pyramiding for abiotic and biotic stress tolerance or resistance.

(vii) **Genome Editing in Plant Breeding**

Genome editing involves precisely altering DNA sequences within a living cell's genome, a challenge that has been addressed with the advent of advanced technologies. Utilizing RNA-guided CRISPR-associated (Cas) nucleases, researchers have achieved precision genome editing across various major crops. This process primarily induces double-strand breaks, triggering cellular repair responses that result in edited DNA sequences. The edited lines can then be employed to study gene functions across species. Genome editing, with its ability to make precise modifications in specific genes, has become a versatile tool for accelerating crop improvement worldwide. Techniques such as CRISPR-Cas9 have enabled the targeted alteration of single bases, contributing to advancements in cereals, pulses, oilseeds, and horticultural crops (Paschon et al. 2019; Li et al. 2020).

(f) **Adaptive Agronomic Practices**

(i) **Crop Rotation and Diversification**

Crop rotation, defined as the repeated cultivation of crops or alternating between crops and fallow land in a specific sequence, plays a crucial role in maintaining soil fertility, preventing erosion, and conserving soil moisture. This practice significantly contributes to the climate resilience of agricultural production systems. Crop rotation enhances soil water content, reduces water evaporation, and minimizes surface water runoff, ensuring high crop water use efficiency. By retaining water in crops or soil, it diminishes groundwater levels, thereby reducing the occurrence and severity of flood disasters and enhancing the overall climate resilience of the farming system.

Crop diversification involves introducing new crops or cropping systems to a farm, considering diverse returns from value-added crops. It serves as a vital instrument for economic growth, minimizing climate and market risks while providing diverse and nutritious food at the household level. Governments are urged to explore the potential of crop diversification fully, aligning enhanced productivity, profitability, and sustainability. Advantages of crop diversification include improvement in overall farming systems and behaviour, enhanced and balanced dietary variation, adoption of new technologies and practices, introduction of new plant varieties and animal breeds, income generation, and increased labour productivity (Vernooy 2015).

(ii) **Sustainable Soil Management**

Sustainable soil management is essential for effective soil functioning, contributing significantly to climate change adaptation, mitigation, desertification combat, and biodiversity promotion. This practice involves implementing operations, practices, and treatments to protect soil and enhance its performance. Key components include controlling traffic; minimizing tillage; efficient pest and nutrient management; selecting appropriate crops and rotations; maintaining soil cover; increasing diversity in crops, soil, and landscape; efficient irrigation management; and adding organic matter.

14.4 Emerging Technologies and Research for Crop Adaptation

The burgeoning global population, expected to reach approximately 9.5 billion by the year 2050, necessitates a substantial upsurge in agricultural production, estimated at up to a 70% increase from current levels (Ehrlich and Harte 2015). Concurrently, over 35% of the global population presently contends with malnourishment and enduring hunger. The compounding challenges of climate change, characterized by increased greenhouse gas emissions, water and soil pollution, and natural resource depletion, exacerbate the stress conditions faced by crop plants, resulting in diminished annual yields. The prevailing climate shifts, manifested in irregular rainfall patterns, prolonged droughts, and extreme heatwaves, contribute to a discernible reduction, up to 7%, in global cereal and pulse yields per unit increase in the average global temperature (Zhao et al. 2017). The Food and Agriculture Organization (FAO) warns of a potential 45% decline in the global production of major cereal crops by the next century if current climate change trends persist.

Modern agriculture faces formidable constraints in adapting to these challenges, necessitating innovative technologies to meet growing food demands. Biotechnological tools, including recombinant DNA technology, molecular breeding, hybrid seed production, micropropagation, and genome engineering, play a pivotal role in enhancing crop resilience and productivity. These technologies leverage plant genetic resources, facilitating hereditary-mediated gene transfer to integrate economically significant traits into modern crop varieties. Region-specific adaptations along with cultivar to climate change involve the introgression of desirable agronomic traits from diverse gene pools, coupled with genetic manipulation employing novel heterologous genes. The commercialization of transgenic crop events, addressing stress-related concerns such as weed and insect management, has significantly increased crop yields. However, public apprehensions regarding human health, environmental impact, and economic ramifications of genetically engineered (GE) plants persist (Pal et al. 2020; Ye et al. 2021). Biotechnological tools are able to address these issues through genetic manipulation, increasing photosynthesis in plants to increase CO_2 fixation, and combining biotic and abiotic stress tolerance (Bhatt et al. 2021).

14.4.1 Harnessing Wild Genetic Diversity for Crop Adaptation to Changing Climates

Crop plants have undergone a transformative journey from their wild progenitor forms through generations of human selection and classification during the domestication process. This prolonged period of domestication focused on specific traits of interest, leading to the morphological transformation of wild progenitors into modern elite cultivars. Notable alterations in phenotypes occurred under domestication.

Examples include (i) reduced grain shattering/crumbling in cereals, (ii) foliar expansion and axillary bud growth in maize (*Zea mays L.*), (iii) improved quality of fibre in cotton (*Gossypium hirsutum L.*), and (iv) modulated fruit shape, size, and colour in tomato (*Solanum lycopersicum L.*) (Fuller and Stevens 2019). Domestication also influenced traits like indehiscent pods with dormancy reduction in peas (*Pisum sativum L.*) and increased leaf number around the primary meristem and terminal bud in cabbage (*Brassica oleracea L.*) (Dempewolf et al. 2014).

The process of domestication, while integral for crop improvement, is not without limitations. The selective evolution of specific traits during domestication may inadvertently compromise existing traits associated with resistance and tolerance. This is a primary reason for reduction in the genetic diversity found in wild relatives, as the domesticated crops become more specialized. Despite its success in numerous crop species, the incorporation of wild species as genetic reservoirs for desired traits into modern cultivars encounters challenges, particularly when the evolutionary gap between wild and cultivated forms is extensive. This can hinder the production of productive crossbreed due to the non-pairing of alike genomes during reproductive cell formation. Nevertheless, the systematic and strategic use of wild specimens for trait transfer, termed 'introgression', presents a viable avenue, particularly in the context of addressing challenges posed by climate change. This approach enhances current breeding programmes, addressing challenges related to yield in the face of evolving environmental conditions (Guarino and Lobell 2011). Through introgression, the genetic diversity inherent in wild relatives becomes a valuable asset, contributing resilience and adaptability to modern cultivars, thereby aiding in the sustainable development of crops for future agricultural needs.

The adapting agriculture towards the changing climate is pivotal in safeguarding and facilitating access to the often-overlooked diversity within crop wild relatives (CWR). This aims to provide plant breeders with valuable genetic resources for enhancing the resilience of cultivated crops. Notably, recent efforts have demonstrated the practical application of CWR diversity in crop improvement. For instance, the integration of a wild rice species (*Oryza officinalis*) has successfully altered the flowering time of the rice cultivar Koshihikari (*O. sativa*), strategically avoiding the hottest periods of the day (Hajjar and Hodgkin 2007). Given the anticipated shifts in the ranges of plant pathogens due to climate change, the utilization of CWR becomes crucial in mitigating unforeseen disease pressures in various regions. Intriguingly, CWR have proven instrumental in augmenting both yields and quality traits in cultivated crops, despite the wild species themselves exhibiting comparatively modest values for these traits.

With an estimated 50,000 to 60,000 CWR species, approximately 10,000 are deemed of high potential value for agricultural policies and global food security, with 1000 closely related to key food crops. Alarming statistics suggest that up to 75% of these species are currently under threat in their natural habitats, and climate change is anticipated to exacerbate these pressures. Recognizing the urgency, the conservation of CWR is increasingly prioritized, emphasizing the importance of preserving these species in their native environments for ongoing evolutionary processes. However, the efficacy of in situ conservation efforts may be impacted by the dynamic nature of climate change. The utilization of CWR in crop improvement is exemplified through various instances, such as drought-tolerant varieties of rice (*O. longistaminata*), salinity-tolerant barley (*Hordeum coarctata*), and disease-resistant forms of sunflower (*Helianthus paradoxus*). These examples underscore the historical and potential future contributions of CWR to crop enhancement in the face of climate change, emphasizing their indispensable role in sustainable agriculture (Miller and Seiler 2003; Pickering et al. 2006; Canci and Toker 2009).

In the case of millets, resistance to rust and *Pyricularia grisea* has been successfully integrated through the process of introgression from wild relatives. While the resistance to rust was swiftly overcome, the resistance against *Pyriculariam grisea* remains effective. Furthermore, efforts to identify and transfer resistance against *Striga*, a parasitic plant, are underway within millet's primary gene pool, although the progress is currently in its early stages. Similarly, in sorghum, recent advancements in hybridization involving *Sorghum macrospermum* and *S. bicolor* show promise for introducing multiple traits related to pest and disease resistance to cultivated varieties (Wilson et al. 2000; Price et al. 2005). This breakthrough holds significant potential for enhancing the resilience of sorghum cultivars against various pests and diseases, marking a positive stride in the ongoing efforts to improve crop sustainability.

14.4.2 Biotechnology for Crop Adaptation

The pursuit of improved crop performance under environmental stresses through breeding involves the strategic accumulation of favourable alleles, representing unique and specific forms of genes that contribute to different biotic and abiotic stress tolerance. Bioengineering interventions in crop adaptation to climate change extend beyond the conventional notion of introducing genes from non-breeding organisms, as seen in transgenic crops. Rather, biotechnological tools are centred on the direct detection and transfer of genes of interest from various plant lines or organisms into the target crop. This approach eliminates the need to rely solely on the plant's phenotype, or its observable characteristics and stress responses, as a proxy for the presence of the desired gene.

While phenotyping, the measurement of a plant line's response in a given environment, remains integral to the selection process, biotechnological advancements have revolutionized the speed and efficiency of gene transfer. Identification of a genetic region associated with a conferred adaptive advantage enables the swift and efficient transfer of this genetic material, even across species barriers. This accelerated gene transfer process represents a significant enhancement over traditional methods, addressing the urgent need for rapid and precise adaptation strategies in the face of evolving environmental challenges associated with climate change.

The procurement of genes conferring abiotic stress tolerance involves a strategic approach, drawing from diverse sources such as collections of germplasm, wild crop relatives, and life forms known for their resilience to environmental challenges like induced drought stress, excess mineralizing, or extreme temperatures. The widespread distribution of these materials has left a lasting impact on breeding programmes in Asia and Africa, influencing commercial seed companies engaged in the production of superior hybrid lines.

Noteworthy breakthroughs in this field include the release of a drought and temperature-tolerant chickpea cultivar, 'BG1103', developed by the IARI, India. The drought and temperature tolerance traits in 'BG1103' are attributed to the introgression of wild genes from *Cicer reticulatum*, positioning it as a leading cultivar in the Himalayan states of India. Similarly, in 2004, the International Center for Agricultural Research in the Dry Areas released six barley cultivars with drought tolerance derived from *Hordeum spontaneum* in Syria. Exploiting the genes of *Oryza rufipogon Griff.* has enhanced tolerance to soils with high acidic-sulphate content in Vietnam, while genes from *O. longistaminata* A.Chev.&Roehrich have been employed to confer drought tolerance to rice cultivars in the Philippines, thereby expanding rice production (Brar 2005) to previously unusable lands. In the realm of tomatoes, genes sourced from *Lycopersicon chilense* and *Lycopersicon pennellii* have been instrumental in elevating drought and salinity tolerance (Nguyen et al. 2003). These scientific advancements underscore the potential of harnessing genetic diversity from diverse sources to fortify crops against environmental stresses, paving the way for sustainable and resilient agriculture in diverse ecosystems.

The integration of biotechnological solutions capable of mitigating the adverse impacts of climate change holds immense potential for sustaining food and fibre production in deteriorating environments while also enabling the expansion of cultivable areas into currently marginal lands. It is essential to note that while environmental remediation efforts remain crucial, biotechnological interventions offer a valuable buffer in addressing the urgency of these challenges. Notably, molecular breeding has demonstrated remarkable progress, particularly in the past 15 to 20 years, offering an accelerated approach for identifying, capturing, and incorporating multiple traits into plants. This expeditious process enhances the speed at which plant lines can be tested for stability and efficacy.

Genetic engineering technologies further broaden the scope of capabilities beyond the natural range of our crop plants. The precision and accuracy achieved in targeted gene insertions, coupled with advanced verification methods, instil higher confidence in the safety of newly developed plant lines. This technological advancement ensures that the insertion process does not disrupt other functional plant genes, marking a significant improvement over previous methodology. A notable example of the benefits of genetic modification is evidenced by genetically modified cotton in India, as outlined in Box 4.30 (Clements et al. 2011). The advantages derived from these advancements extend beyond cotton to encompass a diverse array of DNA rearrangement products, including those utilizing genetic engineering, promising substantial benefits in the near future.

14.4.3 Multiscale Crop Modelling

Crop modelling emerges as an invaluable instrument for evaluating the efficacy of agricultural adaptation strategies to climate change, especially considering the intricate interplay of genetics/heredity (G), environment (E), and management/cultivation practices (M). The intricacy of G × E × M interactions presents difficulties in devising empirical experiments that thoroughly encompass this multifaceted domain. Process-based crop simulation models (CMs) provide a distinctive solution by incorporating biophysical principles of crop growth to mechanistically replicate the utilization and distribution of resources, facilitating the evaluation of diverse adaptation strategies.

The utilization of crop modelling emerges as a highly effective means to evaluate the benefits associated with genetic solutions for crop enhancement. The incorporation of principles from genetics and genomics into CMs represents a pivotal step forward in the realms of crop breeding and genetic engineering. Within this framework, two distinctive methodologies, known as the 'top-down' and 'bottom-up' approaches, play a key role. These approaches leverage an in-depth understanding of genetic controls on anatomical processes to augment the prediction of phenotypic outcomes. The 'top-down' approach involves modelling the entire plant systems without explicitly simulating individual hormones, providing a comprehensive view of system dynamics and phenotypic consequences at a coarse granularity while ensuring robust prediction accuracy (Marrou et al. 2021; Messina et al. 2022). However, for establishing effective connections between complex traits like yield and genetic regulation, it becomes imperative to integrate detailed physiological processes into CMs. This integration allows for a more nuanced and accurate representation of the intricate relationships between genetics, environment, and management practices. The prediction of cultivar-specific model parameters is a complex task, involving a range of mathematical approaches. These approaches span from analytical or statistical models based on multiple loci to more sophisticated methods such as prediction of genome-wide association assessment which utilizes Bayesian and advanced neural networks and deep learning techniques. Such advancements in crop modelling open up new avenues for optimizing agricultural practices, ensuring sustainable crop production.

The bottom-up approach in crop modelling takes advantage of systems biology, offering a holistic integration of plant biological processes across various temporal and spatial scales. Models such as the e-photosynthesis model (Zhu et al. 2022) transcend traditional boundaries by intricately simulating individual metabolic reactions and their regulatory mechanisms. This sophisticated modelling approach facilitates the nuanced connection between high-throughput genomic data and observable phenotypes. When applied at the whole-plant scale, these bottom-up models unlock the potential to assess the intricate impact of genetic manipulation on specific proteins, with a particular focus on critical components like photosynthetic enzymes. This comprehensive method plays a pivotal role in identifying precise targets for enhancing crop genetics, providing a deepened apprehension of the intricate biochemical processes governing plant repercussions to dynamic environmental changes (Hernandez-Ochoa et al. 2019; Peng et al. 2020; Xiao et al. 2020).

14.5 Conclusion

Adapting agriculture to climate change demands a multi-pronged approach. Preserving genetic diversity through gene banks and wild relatives forms the bedrock, providing the raw material for breeding resilient crops. Autonomous adaptation by farmers and planned interventions like breeding for drought and heat tolerance play crucial roles. Emerging technologies like crop modelling bridge the gap between genetics, environment, and management, guiding the development of optimal adaptation strategies. By amalgamating these diverse strategies, this chapter advocates for a resilient and sustainable agricultural future, ensuring global food security amidst the challenges posed by climate change.

References

Ahmad Z, Anjum S, Waraich EA et al (2018) Growth, physiology, and biochemical activities of plant responses with foliar potassium application under drought stress—a review. J Plant Nutr 41:1734–1743. https://doi.org/10.1080/01904167.2018.1459688

Alahmad S, Rambla C, Voss-Fels KP, Hickey LT (2022) Accelerating breeding cycles. In: Wheat improvement. Springer International Publishing, Cham, pp 557–571

Asseng S, Ewert F, Martre P et al (2015) Rising temperatures reduce global wheat production. Nat Clim Chang 5:143–147. https://doi.org/10.1038/nclimate2470

Azhar MT, Rehman A (2018) Overview on effects of water stress on cotton plants and productivity. In: Biochemical, physiological and molecular avenues for combating abiotic stress tolerance in plants. Elsevier, London, pp 297–316

Barnabás B, Jäger K, Fehér A (2008) The effect of drought and heat stress on reproductive processes in cereals. Plant Cell Environ 31:11–38. https://doi.org/10.1111/J.1365-3040.2007.01727.X

Bassu S, Brisson N, Durand J-L et al (2014) How do various maize crop models vary in their responses to climate change factors? Glob Chang Biol 20:2301–2320. https://doi.org/10.1111/gcb.12520

Bhat JA, Ali S, Salgotra RK et al (2016) Genomic selection in the era of next generation sequencing for complex traits in plant breeding. Front Genet 7:221. https://doi.org/10.3389/fgene.2016.00221

Bhatt P, Gangola S, Bhandari G et al (2021) New insights into the degradation of synthetic pollutants in contaminated environments. Chemosphere 268:128827. https://doi.org/10.1016/j.chemosphere.2020.128827

Borlaug NE (2007) Sixty-two years of fighting hunger: personal recollections. Euphytica 157:287–297. https://doi.org/10.1007/s10681-007-9480-9

Brar D (2005) Broadening the genepool and exploiting heterosis in cultivated rice. In: Rice is life: scientific perspectives for the 21st century. Proceedings of the World Rice Research Conference, Tokyo and Tsukuba, Japan

Breseghello F, Coelho ASG (2013) Traditional and modern plant breeding methods with examples in rice (*Oryza sativa* L.). J Agric Food Chem 61:8277–8286. https://doi.org/10.1021/jf305531j

Canci H, Toker C (2009) Evaluation of annual wild Cicer species for drought and heat resistance under field conditions. Genet Resour Crop Evol 56:1–6. https://doi.org/10.1007/s10722-008-9335-9

Carvalho LC, Amâncio S (2019) Cutting the Gordian Knot of abiotic stress in grapevine: from the test tube to climate change adaptation. Physiol Plant 165:330–342. https://doi.org/10.1111/ppl.12857

Challinor AJ, Wheeler TR, Craufurd PQ et al (2007) Adaptation of crops to climate change through genotypic responses to mean and extreme temperatures. Agric Ecosyst Environ 119:190–204. https://doi.org/10.1016/j.agee.2006.07.009

Clements R, Haggar J, Quezada A, Torres J (2011) Technologies for climate change adaptation—agriculture sector. UNEP Risø Centre, Roskilde

Collard BCY, Mackill DJ (2008) Marker-assisted selection: an approach for precision plant breeding in the twenty-first century. Philos Trans R Soc B Biol Sci 363:557–572. https://doi.org/10.1098/rstb.2007.2170

Das DN, Karuthadurai T, Gnanasekaran S (2021) Genomic selection. In: Advances in animal genomics. Elsevier, London, pp 141–163

Datta P, Behera B, Rahut DB (2022) Climate change and Indian agriculture: a systematic review of farmers' perception, adaptation, and transformation. Environ Chall 8:100543. https://doi.org/10.1016/j.envc.2022.100543

Dempewolf H, Eastwood RJ, Guarino L et al (2014) Adapting agriculture to climate change: a global initiative to collect, conserve, and use crop wild relatives. Agroecol Sustain Food Syst 38:369–377. https://doi.org/10.1080/21683565.2013.870629

dos Santos JPR, Pires LPM, de Castro Vasconcellos RC et al (2016) Genomic selection to resistance to Stenocarpella maydis in maize lines using DArTseq markers. BMC Genet 17:86. https://doi.org/10.1186/s12863-016-0392-3

Ehrlich PR, Harte J (2015) To feed the world in 2050 will require a global revolution. Proc Natl Acad Sci 112:14743–14744. https://doi.org/10.1073/pnas.1519841112

FAO (2005) Impact of climate change, pests and diseases on food security and poverty reduction. Special event background document for the 31st Session of the Committee on World Food Security. FAO, Rome

Fuller DQ, Stevens CJ (2019) Between domestication and civilization: the role of agriculture and arboriculture in the emergence of the first urban societies. Veg Hist Archaeobot 28:263–282. https://doi.org/10.1007/s00334-019-00727-4

Ghosh S, Watson A, Gonzalez-Navarro OE et al (2018) Speed breeding in growth chambers and glasshouses for crop breeding and model plant research. Nat Protoc 13:2944–2963. https://doi.org/10.1038/s41596-018-0072-z

Guarino L, Lobell DB (2011) A walk on the wild side. Nat Clim Chang 1:374–375. https://doi.org/10.1038/nclimate1272

Gudi S, Kumar P, Singh S et al (2022) Strategies for accelerating genetic gains in crop plants: special focus on speed breeding. Physiol Mol Biol Plants 28:1921–1938. https://doi.org/10.1007/s12298-022-01247-8

Hajjar R, Hodgkin T (2007) The use of wild relatives in crop improvement: a survey of developments over the last 20 years. Euphytica 156:1–13. https://doi.org/10.1007/s10681-007-9363-0

Hernandez-Ochoa IM, Luz Pequeno DN, Reynolds M et al (2019) Adapting irrigated and rainfed wheat to climate change in semi-arid environments: management, breeding options and land use change. Eur J Agron 109:125915. https://doi.org/10.1016/j.eja.2019.125915

Hooghvorst I, Nogués S (2021) Chromosome doubling methods in doubled haploid and haploid inducer-mediated genome-editing systems in major crops. Plant Cell Rep 40:255–270. https://doi.org/10.1007/s00299-020-02605-0

Huang M, Balimponya EG, Mgonja EM et al (2019) Use of genomic selection in breeding rice (Oryza sativa L.) for resistance to rice blast (Magnaportheoryzae). Mol Breed 39:114. https://doi.org/10.1007/s11032-019-1023-2

Intergovernmental Panel on Climate Change (2013) IPCC climate change 2013: the physical science basis: Working Group I Contribution to the Fifth Assessment Report of the Intergovernmental Panel on Climate Change. IPCC, Cambridge

Intergovernmental Panel on Climate Change (2019) IPCC special report on climate change, desertification, land degradation, sustainable land management, food security, and greenhouse gas fluxes in terrestrial ecosystems. IPCC, Cambridge

Kanojia A, Dijkwel PP (2018) Abiotic stress responses are governed by reactive oxygen species and age. In: Annual plant reviews online, 1–32 edn. Wiley, Hoboken, pp 295–326

Kumar N, Kumar N, Shukla A, Shankhdhar SC, Shankhdhar D (2015) Impact of terminal heat stress on pollen viability and yield attributes of rice (Oryza sativa L.). Cereal Res Commun 43(4):616–626

Kumar N, Shankhdhar SC, Shankhdhar D (2016) Impact of elevated temperature on antioxidant activity and membrane stability in different genotypes of rice (Oryza sativa L.). Indian J Plant Physiol 21(1):37–43

Kumar N, Suyal DC, Sharma IP, Verma A, Singh H (2017) Elucidating stress proteins in rice (Oryza sativa L.) genotype under elevated temperature: a proteomic approach to understand heat stress response. 3 Biotech 7:205

Kumar N, Jeena N, Singh H (2019) Elevated temperature modulates rice pollen structure: a study from foothill Himalayan agro-ecosystem in India. 3 Biotech 9:175

Kumar A, Dwivedi GK, Tewari S, Paul J, Sah VK, Singh H, Kumar P, Kumar N, Kaushal R (2020a) Soil organic carbon pools under Terminalia chebula Retz. based agroforestry system in Himalayan foothills, India. Curr Sci 118(7):1098–1103

Kumar A, Dwivedi GK, Tewari S, Paul J, Anand R, Kumar N, Kumar P, Singh H, Kaushal R (2020b) Carbon mineralization and inorganic nitrogen pools under Terminalia chebula Retz.-based agroforestry system in Himalayan foothills, India. For Sci 66(5):634–643

Kumar A, Kumar P, Singh H, Kumar N (2020c) Adaptation and mitigation potential of roadside trees with bio-extraction of heavy metals under vehicular emissions and their impact on physiological traits during seasonal regimes. Urban For Urban Green 58:126900

Kumar A, Kumar P, Singh H, Bisht S, Kumar N (2021a) Relationship of physiological plant functional traits with soil carbon stock in temperate forest of Garhwal Himalaya. Curr Sci 120(8):1368–1373

Kumar A, Kumar P, Singh H, Kumar N (2021b) Modulation of plant functional traits under essential plant nutrients during seasonal regime in natural forests of Garhwal Himalayas. Plant Soil 465:197–212

Kumar A, Tewari S, Singh H, Kumar P, Kumar N, Bisht S, Kushwaha S, Tamta N, Kaushal R (2021c) Biomass accumulation and carbon stocks in different agroforestry system prevalent in Himalayan foothills, India. Curr Sci 120(6):1083–1088

Kumar A, Kumar P, Singh H, Kumar N (2021d) Impact of plant functional traits on near saturated hydraulic conductivity of soil under different forests of Kempty watershed in Garhwal Himalaya. Indian J Soil Conserv 49(1):38–44

Kumar N, Jeena N, Kumar A, Khairakpam R, Singh H (2021e) Comparative response of rice cultivars to elevated air temperature in Bhabar region of Indian Himalaya: status on yield attributes. Heliyon 7:e07474

Kumar A, Singh H, Kumari G, Bisht S, Malik A, Kumar N, Singh M, Raturi A, Barthwal S, Thakur A, Kaushal R (2022) Adaptive resilience of roadside trees to vehicular emissions via leaf enzymatic, physiological, and anatomical trait modulations. Environ Pollut 313:120191

Li H, Yang Y, Hong W et al (2020) Applications of genome editing technology in the targeted therapy of human diseases: mechanisms, advances and prospects. Signal Transduct Target Ther 5:1. https://doi.org/10.1038/s41392-019-0089-y

Liu B, Asseng S, Müller C et al (2016) Similar estimates of temperature impacts on global wheat yield by three independent methods. Nat Clim Chang 6:1130–1136. https://doi.org/10.1038/nclimate3115

Marrou H, Ghanem ME, Amri M et al (2021) Restrictive irrigation improves yield and reduces risk for FABA bean across the Middle East and North Africa: a modeling study. Agric Syst 189:103068. https://doi.org/10.1016/j.agsy.2021.103068

Mendelsohn R, Dinar A (1999) Climate change, agriculture, and developing countries: does adaptation matter? World Bank Res Obs 14:277–293. https://doi.org/10.1093/wbro/14.2.277

Messina CD, van Eeuwijk F, Tang T et al (2022) Crop improvement for circular bioeconomy systems. J ASABE 65:491–504. https://doi.org/10.13031/ja.14912

Miller JF, Seiler GJ (2003) Registration of five oilseed maintainer (HA 429-HA 433) sunflower germplasm lines. Crop Sci 43:2313

Mishra V, Ganguly AR, Nijssen B, Lettenmaier DP (2015) Changes in observed climate extremes in global urban areas. Environ Res Lett 10:024005. https://doi.org/10.1088/1748-9326/10/2/024005

Murthy CS, Sesha Sai MVR (2010) Agricultural drought monitoring and assessment. In: Roy PS, Dwivedi RS, Vijayan D (eds) Remote sensing applications, 1st edn. National Remote Sensing Center, Hyderabad, pp 303–331

Nguyen BD, Brar DS, Bui BC et al (2003) Identification and mapping of the QTL for aluminum tolerance introgressed from the new source, ORYZA RUFIPOGON Griff., into indica rice (Oryza sativa L.). Theor Appl Genet 106:583–593. https://doi.org/10.1007/s00122-002-1072-4

Orlowsky B, Seneviratne SI (2012) Global changes in extreme events: regional and seasonal dimension. Clim Chang 110:669–696. https://doi.org/10.1007/s10584-011-0122-9

Ortigosa A, Gimenez-Ibanez S, Leonhardt N, Solano R (2019) Design of a bacterial speck resistant tomato by CRISPR/Cas9-mediated editing of *Sl*jaz2. Plant Biotechnol J 17:665–673. https://doi.org/10.1111/pbi.13006

Pachauri RK (2014) Climate change 2014: synthesis report. Contribution of Working Groups I, II and III to the Fourth Assessment Report of the Intergovernmental Panel on Climate Change. University of Southern Queensland, Geneva

Pal AK, Singh J, Soni R et al (2020) The role of microorganism in bioremediation for sustainable environment management. In: Bioremediation of pollutants. Elsevier, London, pp 227–249

Parmar S, Deshmukh DB, Kumar R et al (2021) Single seed-based high-throughput genotyping and rapid generation advancement for accelerated groundnut genetics and breeding research. Agronomy 11:1226. https://doi.org/10.3390/agronomy11061226

Paschon DE, Lussier S, Wangzor T et al (2019) Diversifying the structure of zinc finger nucleases for high-precision genome editing. Nat Commun 10:1133. https://doi.org/10.1038/s41467-019-08867-x

Peng B, Guan K, Tang J et al (2020) Towards a multiscale crop modelling framework for climate change adaptation assessment. Nat Plants 6:338–348. https://doi.org/10.1038/s41477-020-0625-3

Pickering R, Ruge-Wehling B, Johnston PA et al (2006) The transfer of a gene conferring resistance to scald (*Rhynchosporiumsecalis*) from *Hordeumbulbosum* into *H. vulgare*chromosome 4HS. Plant Breed 125:576–579. https://doi.org/10.1111/j.1439-0523.2006.01253.x

Price HJ, Hodnett GL, Burson BL et al (2005) A Sorghum bicolor × S. macrospermum hybrid recovered by embryo rescue and culture. Aust J Bot 53:579. https://doi.org/10.1071/BT04213

Pugh TAM, Müller C, Elliott J et al (2016) Climate analogues suggest limited potential for intensification of production on current croplands under climate change. Nat Commun 7:12608. https://doi.org/10.1038/ncomms12608

Razzaq A, Kaur P, Akhter N et al (2021) Next-generation breeding strategies for climate-ready crops. Front Plant Sci 12:620420. https://doi.org/10.3389/fpls.2021.620420

Reilly JM, Schimmelpfennig D (1999) Agricultural impact assessment, vulnerability, and the scope for adaptation. Clim Chang 43:745–788. https://doi.org/10.1023/A:1005553518621

Rogelj J, den Elzen M, Höhne N et al (2016) Paris agreement climate proposals need a boost to keep warming well below 2 °C. Nature 534:631–639. https://doi.org/10.1038/nature18307

Sallam AH, Endelman JB, Jannink J-L, Smith KP (2015) Assessing genomic selection prediction accuracy in a dynamic barley breeding population. Plant Genome 8:eplantgenome2014.05.0020. https://doi.org/10.3835/plantgenome2014.05.0020

Sharma RK, Kumar S, Vatta K et al (2022) Impact of recent climate change on corn, rice, and wheat in southeastern USA. Sci Rep 12:16928. https://doi.org/10.1038/s41598-022-21454-3

Singh H, Yadav M, Kumar N, Kumar A, Kumar M (2020) Assessing adaptation and mitigation potential of roadside trees under the influence of vehicular emissions: a case study of Grevillea robusta and Mangifera indica planted in an urban city of India. PLoS One 15(1):e0227380

Smit B, Wandel J (2006) Adaptation, adaptive capacity and vulnerability. Glob Environ Chang 16:282–292. https://doi.org/10.1016/j.gloenvcha.2006.03.008

Snowdon RJ, Wittkop B, Chen T-W, Stahl A (2021) Crop adaptation to climate change as a consequence of long-term breeding. Theor Appl Genet 134:1613–1623. https://doi.org/10.1007/s00122-020-03729-3

Spinoni J, Barbosa P, De Jager A et al (2019) A new global database of meteorological drought events from 1951 to 2016. J Hydrol Reg Stud 22:100593. https://doi.org/10.1016/j.ejrh.2019.100593

Tadesse W, Sanchez-Garcia M, Assefa SG et al (2019) Genetic gains in wheat breeding and its role in feeding the world. Crop Breed Genet Genom 1:e190005. https://doi.org/10.20900/cbgg20190005

Thornton PK, Ericksen PJ, Herrero M, Challinor AJ (2014) Climate variability and vulnerability to climate change: a review. Glob Chang Biol 20:3313–3328. https://doi.org/10.1111/gcb.12581

Tiwari PC, Joshi B (2012) Natural and socio-economic factors affecting food security in the Himalayas. Food Secur 4:195–207. https://doi.org/10.1007/s12571-012-0178-z

Tong H, Nikoloski Z (2021) Machine learning approaches for crop improvement: leveraging phenotypic and genotypic big data. J Plant Physiol 257:153354. https://doi.org/10.1016/j.jplph.2020.153354

Varshney RK, Singh VK, Kumar A et al (2018) Can genomics deliver climate-change ready crops? Curr Opin Plant Biol 45:205–211. https://doi.org/10.1016/j.pbi.2018.03.007

Vaughan MM, Block A, Christensen SA et al (2018) The effects of climate change associated abiotic stresses on maize phytochemical defenses. Phytochem Rev 17:37–49. https://doi.org/10.1007/s11101-017-9508-2

Vernooy R (2015) Effective implementation of crop diversification strategies for Cambodia, Lao PDR and Vietnam: Insights from past experiences and ideas for new research. Bioversity International, Rome

Voss-Fels KP, Cooper M, Hayes BJ (2019) Accelerating crop genetic gains with genomic selection. Theor Appl Genet 132:669–686. https://doi.org/10.1007/s00122-018-3270-8

Wanga MA, Shimelis H, Mashilo J, Laing MD (2021) Opportunities and challenges of speed breeding: a review. Plant Breed 140:185–194. https://doi.org/10.1111/pbr.12909

Wilson JP, Hess DE, Hanna WW (2000) Resistance to *Striga hermonthica* in wild accessions of the primary Gene Pool of *Pennisetumglaucum*. Phytopathology 90:1169–1172. https://doi.org/10.1094/PHYTO.2000.90.10.1169

Xiao D, Liu DL, Wang B et al (2020) Designing high-yielding maize ideotypes to adapt changing climate in the North China Plain. Agric Syst 181:102805. https://doi.org/10.1016/j.agsy.2020.102805

Ye Z, Wu F, Hennessy DA (2021) Environmental and economic concerns surrounding restrictions on glyphosate use in corn. Proc Natl Acad Sci:118, e2017470118. https://doi.org/10.1073/pnas.2017470118

Zhao C, Liu B, Piao S et al (2017) Temperature increase reduces global yields of major crops in four independent estimates. Proc Natl Acad Sci 114:9326–9331. https://doi.org/10.1073/pnas.1701762114

Zhu X-G, Hasanuzzaman M, Jajoo A et al (2022) Improving photosynthesis through multidisciplinary efforts: the next frontier of photosynthesis research. Front Plant Sci 13:967203. https://doi.org/10.3389/fpls.2022.967203

Plant Functional Traits: A Key Framework for Understanding and Managing Ecosystem Responses to Global Environmental Challenges

15

Amrender Singh Rao, Rahul Chhawri, Ajay Chauhan, Surender Singh Yadav, K. C. Meena, and Pardeep Bansal

Abstract

The greatest threat to present and future biodiversity is climate change. It is estimated that climatic and related environmental changes will threaten global food security and sustainable development goals. Climate change profoundly impacts ecosystem services and disrupts the intricate interactions between biotic and abiotic factors. Plant functional diversity provides a complex nexus of biodiversity, ecosystem services, agroecosystems, and soils. Plant functional traits (PFTs) are heritable morphological, physiological, or phenological properties that influence individual performance and fitness and, in turn, the species' responses to ecological shifts and climatic changes. The PFTs are categorized into morphological, physiological, and phenological traits, providing a comprehensive realization of how plants interact with their environment and influence ecosystem processes. The ecological roles of PFTs, their roles in community assembly, and their

A. S. Rao (✉)
Centre of Excellence on Sustainable Land Management, Indian Council of Forestry Research and Education, Dehradun, Uttarakhand, India

Department of Botany, Maharshi Dayanand University, Rohtak, Haryana, India

R. Chhawri · S. S. Yadav
Department of Botany, Maharshi Dayanand University, Rohtak, Haryana, India

A. Chauhan
Centre of Excellence on Sustainable Land Management, Indian Council of Forestry Research and Education, Dehradun, Uttarakhand, India

K. C. Meena
Department of Botany, Maharshi Dayanand University, Rohtak, Haryana, India

Haryana Forest Department, Panchkula, Haryana, India

P. Bansal
Kishan Lal Public College, Rewari, Haryana, India

N. Kumar, H. Singh (eds.), *Plant Functional Traits for Improving Productivity*,
https://doi.org/10.1007/978-981-97-1510-7_15

impact on ecosystem functions such as nutrient cycling and carbon sequestration are critical components in understanding and managing the dynamics of ecosystems and their resilience in the face of environmental change. PFTs significantly influence the species' responses to environmental changes and guide ecosystem restoration and land-use planning decisions. The chapter highlights the crucial role of plant functional traits in advancing ecological research, offering insights into species' adaptations, community dynamics, and responses to environmental challenges. Further, integrating functional traits into interdisciplinary studies will contribute to a holistic understanding of ecological patterns and processes, informing sustainable management and conservation strategies.

Keywords

Plant functional traits · Agri-horti-silviculture · Climate change · Sustainable future · Productivity · Resilient ecosystems

15.1 Introduction

The growing apprehension surrounding the global issues of biodiversity depletion, human-induced accelerated climate change, food security, and the consequential reduction in ecosystem services has sparked the concerns of potential mass extinction. As temperatures rise, ecosystems face unprecedented challenges, including habitat loss, altered precipitation patterns, more frequent extreme weather events, the acidification of our oceans, and higher sea levels (Kumar et al. 2015, 2016, 2017). These changes disrupt the delicate balance of ecosystems, affecting species' ability to adapt, reproduce, and survive (Kumar et al. 2019, Kumar et al. 2021a, d). These changes directly impact the biodiversity and ecosystem services nexus. This crisis is expected to impact the biodiversity-linked ecosystem services assisting the global economy of 44 trillion USD and constituting nearly half of the global gross domestic product (GDP) (Daily and Matson 2008; WEF 2020). Ecosystem services provided by biodiversity include various tangible and intangible benefits, such as pollination, water purification, soil fertility maintenance, climate regulation, disease control, and recreational opportunities. These services are essential for human well-being, supporting agriculture, fisheries, tourism, pharmaceuticals, and numerous other industries. Moreover, the ecosystem services are not only linked to biodiversity but also extend to agroecosystems just like traditional farming methods, and agricultural practices that focus on enhancing soil quality through ecological methods are being explored (Bender et al. 2016).

The complex nexus of biodiversity, ecosystem services, agroecosystems, and soils has seen significant development in the past decade through the concept of functional diversity. Plant functional diversity pertains to the assortment, variability, and proportional extent of functional traits exhibited by organisms within ecological communities (Faucon et al. 2017). These traits influence how organisms respond to the environment and contribute to the functioning of ecosystems. Functional diversity encompasses several aspects, including the richness of functional types, which denotes a collection of species or organisms that exhibit similar functional characteristics or ecological roles concerning environmental variations or their impacts on ecosystem properties. These studies are directed toward formulating and implementing more sustainable strategies for ecosystems. The concept of functional diversity emphasizes the importance of considering not just species richness but also the functional roles and traits of organisms within a community. Assessing functional diversity provides insights into how organisms interact with their environment, how communities respond to changes, and how these variations influence ecosystem functioning. The quantification and analysis of functional diversity vulnerability can predict the impacts of land-use intensification and forecasting the effects in agroecosystems. Further, it incorporates a broader perspective by evaluating the various functions and roles different species perform in the ecosystem, rather than solely focusing on taxonomic diversity.

The plant functional traits (PFTs) are heritable morphological, physiological, or phenological properties that influence individual performance and fitness and, in turn, the species responses to ecological shifts and climatic changes (Kumar et al. 2020a, b, c). The key traits known to trigger the plant responses to varying climatic conditions include plant growth, reproduction, survival strategies, and responses to environmental conditions. Some common PFTs include leaf area, specific leaf area (leaf area per unit leaf dry mass), leaf nitrogen content, root depth, plant height, seed size, photosynthetic rates, and phenological traits like flowering time or leaf senescence (Singh et al. 2013a, b; Kumar et al. 2021a, b). There is a growing consensus on utilization of the functional types in classifying terrestrial species in order to tackle important ecological questions at the level of ecosystem, landscape, or biome. However, the traditional higher taxonomic classification has limitations at different levels of the ecosystem, landscape, or biome.

Plant functional traits aid ecologists, botanists, and environmental scientists in understanding plant-environment interactions, predicting ecosystem responses to disturbances, and implementing effective conservation and management strategies. By examining these traits across different species, researchers gain insights into the fundamental mechanisms governing plant survival, growth, and ecological dynamics, providing a better grasp of how biodiversity and the works of ecosystems are closely connected.

15.2 Classification and Types of Plant Functional Traits

PFTs have been classified into various groups according to their features and roles. These categories often include morphological, physiological, and phenological traits. Studying a combination of these traits allows ecologists and biologists to better understand how plants interact with their environment, compete for resources, adapt to changing conditions, and influence ecosystem processes (Yang et al. 2015) (Table 15.1). The Table 15.1 provides the categorization of Plant functional trait types and functional traits.

Table 15.1 Plant functional trait (PFT) types and functional traits

Class	PFT type	Functional trait
Morphological traits	Leaf traits	Leaf area, specific leaf area (SLA), leaf size, shape, thickness, and texture
	Root traits	Root depth, root architecture (e.g., taproot vs. fibrous root systems), root biomass allocation, root surface area, and mycorrhizal associations
	Stem height traits	Plant height, stem diameter, branching patterns, and wood density
Physiological trait	Photosynthetic traits	Photosynthetic rates, chlorophyll content, maximum quantum yield of photosystem II (Fv/Fm ratio), and water-use efficiency
	Nutrient trait	Leaf nitrogen content, leaf phosphorus content, nutrient-use efficiency, and nutrient resorption efficiency
	Water relations traits	Water potential, stomatal conductance, transpiration rates, and osmotic adjustment under drought stress
Phenological traits	Flowering traits	Flowering time, duration of flowering, and flower size
	Senescence traits	Leaf senescence timing and duration, leaf shedding patterns, and dormancy periods
	Reproductive traits	Seed size, seed mass, seed number, and seed dispersal mechanisms
Allocation traits	Biomass allocation	Allocation of resources to different plant parts such as leaves, stems, roots, and reproductive structures
	Resource partitioning	Allocation strategies for acquiring resources like light, water, nutrients, and space
Defense traits	Chemical defense	Production of secondary metabolites, such as toxins or compounds that deter herbivores or pathogens
	Physical defense	Traits like thorns, spines, or tough leaf textures that deter herbivory
Interspecies interaction traits	Pollination traits	Floral traits that attract specific pollinators, such as flower color, shape, scent, and nectar production
	Dispersal traits	Traits related to seed dispersal mechanisms, including seed morphology, size, and adaptations for wind, water, or animal dispersal

15.3 PFTs' Role in Ecology, Evolution, and Ecosystem Functioning

Plant functional traits play pivotal roles in ecology through assembly and composition of plant communities. Traits affect competitive interactions among species, leading to the coexistence of different plants in an ecosystem. PFTs influence interactions with other organisms, including herbivores, pollinators, and symbiotic organisms. Specific traits may attract certain pollinators or deter herbivores, shaping plant survival and reproduction. Lavorel and Garnier (2002) came up with a way of thinking about plants by separating traits that show how plants react to their environment (response traits) from traits that influence how ecosystems work (effect traits). Díaz and Cabido (2001) state that by just looking at plant traits helps us see strong links between how ecosystems function and the usual groupings we use for plants, like separating them into grasses, forbs, and legumes. However, these broad groupings can hide differences and might not fully recognize the important role that diversity in plant functions plays in keeping key ecosystem functions going, like making sure that high productivity and nutrient cycling are achieved (Wright et al. 2006). To explain further, not all types of perennial grasses that use C_4 photosynthesis or annual forbs that use C_3 photosynthesis respond the same way to disturbances or changes in resources, as shown by studies (Badgery et al. 2005; Firn et al. 2010, 2012; Han et al. 2012; Kumar et al. 2021c, d). Moreover, there is evidence suggesting that these broad functional groups cannot always predict how species will react to changes in their environment, especially when it comes to invasive species. Even if they belong to the same functional group, native and invasive species might react differently to changes in the environment (Funk 2008; Firn et al. 2010, 2012; Han et al. 2012). Furthermore, relying on simplistic categorical functional groups might be inadequate, especially in ecosystems like grasslands, where the number of such groups is limited. Consequently, observed statistically significant correlations between basic functional groups and alterations in ecosystem function might arise not due to a unified response to disturbances but rather due to reduced variability within these groups.

Variation in response traits is believed to be connected to how organisms respond to different factors such as stress from the environment or competition, and this corresponds to broader effects on how populations of species respond demographically and their overall abundance, as highlighted by Suding et al. (2003). The choice of response traits to study relies on the specific combination of nonliving (abiotic) and living (biotic) factors found within a plant community. One significant trait in distinguishing ecological strategies among communities is plant growth rate (Grime 1977; Reich et al. 2014). Generally, growth rate shows a positive association with shade tolerance but a negative correlation with drought tolerance (Suding et al. 2003). Rapid growth tends to be more prevalent in productive, high-nutrient communities, and this implies that having a high growth rate is beneficial when resources are plentiful. Yet, in some cases, rapid growth can help plants deal with limited resources, especially as seen in invasive species (Funk 2013). Relative growth rate (RGR), which shows how quickly plants add dry mass relative to their existing dry mass, is considered a good indicator of a species' potential to succeed and is the most important measure of plant growth. However, its measurement is complex and time-consuming. However, RGR encompasses many processes (such as photosynthesis, respiration, nutrient allocation, and life strategies) that are connected to other traits that we can measure, like leaf nitrogen concentration, photosynthesis rate, plant tissue density, and the specific leaf area (SLA) (Singh and Verma 2013a, b). Some of these traits, like SLA or wood density, have been shown to explain why RGR varies among different types of plants, whether they are herbaceous (like grasses) or woody (like trees). Additionally, plants on land often balance these growth-related traits. For example, having a high SLA often means having more nitrogen concentration in the leaves and faster photosynthetic rate, but it can also mean the tissue is less dense and does not last long (Singh et al. 2010a, b). So, softer traits like SLA or plant tissue density can also tell us a lot about how plants get and keep nutrients in lots of different environments and types of ecosystems (Singh et al. 2020) (Fig. 15.1). Figure 15.1 represents the nexus of plant functional traits, ecosystems services, and forest ecosystems.

Although considerable advancements have been achieved in comprehending the role of leaf economic spectrum (LES) traits, our understanding of how other traits correlate with plant and community responses remains constrained. Root traits, for instance, are challenging to measure, yet evidence suggests the existence of an economic axis for roots; in situations where plant growth is slow, the roots typically have slower elongation rates, shorter specific root lengths (SRL), larger root diameters, and lower nutrient concentrations. In arid environments, root traits such as depth and elongation rate may be more indicative of how plants respond to changes in water availability compared to leaf traits. Furthermore, the traits most closely associated with plant performance in a specific environment can vary. For instance, in *Polygonum persicaria*, leaf-level water-use efficiency was linked to plant fitness in water-limited habitats, while root biomass allocation was more closely related to fitness in moist environments. New research indicates connections between the characteristics of leaves and stems, like wood density, and traits described in the leaf economic spectrum (LES). These traits of different species are often used in models

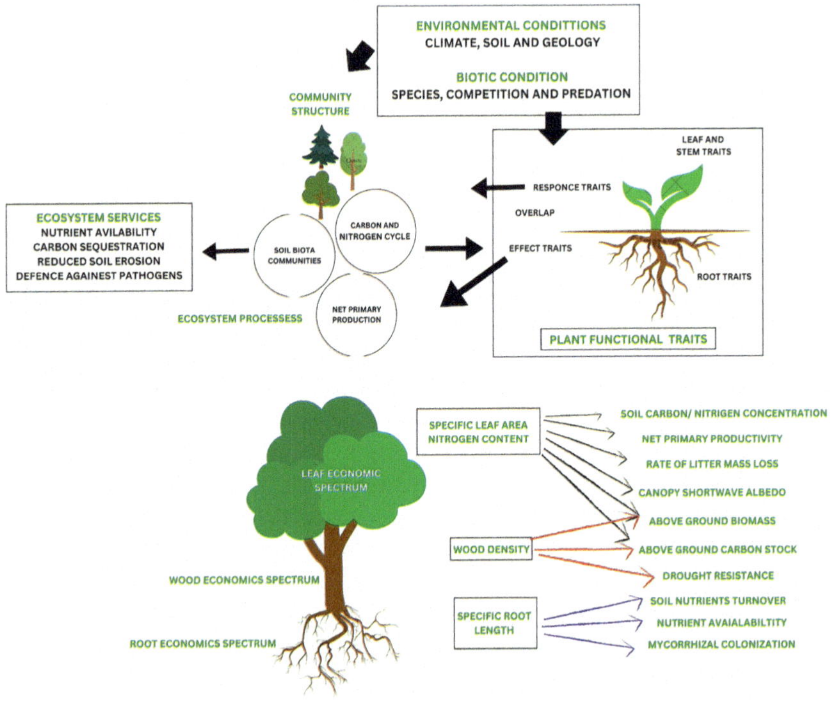

Fig. 15.1 Nexus of plant functional traits, ecosystem services, and forest ecosystems

to make predictions. Yet, there is mounting evidence indicating that different species react differently in terms of their observable traits when they encounter changes in both environmental and living conditions. This ability of species to adjust, known as phenotypic plasticity (explained in Sect. 15.3), plays a big role in creating diversity in how communities function. Several studies (Funk 2008; Firn et al. 2012; Siebenkas et al. 2015; Larson and Funk 2016) have emphasized this aspect. While there are not many concrete links yet between how adaptable species are and how well they fare in various environments (as seen in studies by Firn et al. 2012), if this adaptability turns out to be beneficial, it could become a key measure in understanding how populations, species, and communities react to changes in their surroundings. The importance of this adaptive plasticity in dealing with environmental changes has been explored in several review papers (Berg and Ellers 2010; Nicotra et al. 2010; Valladares et al. 2014).

To establish a thorough set of ideas and concepts in ecology through functional traits, it is imperative to comprehend how the composition and diversity of traits impact the way ecosystems work (Lavorel and Garnier 2002). Characteristic traits play a role in altering the various abiotic and biotic factors involved in a broad range of ecosystem activities, which have been highlighted in recent review articles (Eviner and Chapin III 2003; de Bello et al. 2010; Garnier and Navas 2012). Despite the advancements in understanding effect traits following the framework established by Lavorel and Garnier (2002), predictive models have not progressed as much in incorporating these traits compared to those integrating response traits (Nash Suding et al. 2003). Besides being able to predict how species and communities react to changes in the environment, researchers have thoroughly investigated the connections between leaf characteristics described in the leaf economic spectrum (LES) and how ecosystems function. Significantly, characteristics like relative growth rate (RGR), specific leaf area (SLA), and leaf nitrogen content (N) have been extensively researched (Singh et al. 2014, 2021, 2024). These traits consistently show connections with various aspects of ecosystem health, such as how fast plants grow, how leaves are structured, and the amount of nitrogen present. These studies reveal positive links with primary productivity, how quickly litter decomposes, the availability of nitrogen in the soil for plants, rates at which nitrogen cycles through ecosystems, how appetizing plants are to herbivores, and negative relationships with soil carbon and nitrogen retention. For instance, research by De Deyn et al. (2008), Lavorel and Grigulis (2012), and Loranger et al. (2012) has explored these connections. Similarly, when we look at traits on a community level which resemble those in the leaf economic spectrum (LES), like canopy nitrogen and leaf area index (LAI), we find they scale positively with SLA and leaf nitrogen values. These community-level traits are also associated with aboveground net primary productivity (ANPP), as indicated by Reich (2012).

One extensively researched area regarding traits is how the chemistry and structure of leaves affect how quickly they decompose (de Bello et al. 2010). Many studies have shown that traits related to the leaf economic spectrum (LES) influence decomposition rates. For example, research by Santiago (2007), Cornwell et al. (2008), and Bakker et al. (2011) has explored this connection. Species with traits that match the "fast return" end of the LES, such as fast growth, thin leaves, high nutrient levels, and rapid photosynthesis, typically break down faster than species with traits at the "slow return" end, like slow growth, thicker, tougher leaves with stronger defenses, and lower photosynthesis rates. This suggests that the combination of leaf traits optimized for photosynthesis not only affects nutrient recycling but also plays a crucial role in the global carbon cycle. However, the impact of plant communities on biogeochemical cycles likely involves more than just individual LES traits. While plant traits serve as crucial predictors of decomposition, it is important to note that both biotic and abiotic factors play significant roles as well. Sutton-Grier et al. (2012) found in a restored riparian wetland that environmental factors like soil organic matter and soil nitrogen are almost as important as plant traits in affecting decomposition rates. Additionally, outside factors such as rainfall, grazing, or land use can indirectly influence ecosystem function by changing which plants grow where and altering the average traits of the plant community. These changes then affect decomposition indirectly. This concept has been explored in studies by Santiago et al. (2005), Garibaldi et al. (2007), and Bakker et al. (2011).

Furthermore, the influence of plant traits on soil chemistry is not just about the direct impact from plant litter and exudates. It also includes indirect effects through these inputs on soil organisms. These soil organisms can play a significant role, accounting for over 70% of the differences in nitrogen cycling processes, as indicated by Grigulis et al. (2013). This highlights the intricate relationship between plant traits, their influence on soil chemistry, and the significant role played by soil organisms in mediating these effects. Reported by Orwin et al. (2010), de Vries et al. (2012) have emphasized how the properties of soil biota contribute substantially to the variation in nitrogen cycling processes beyond the direct effects of plant inputs. Understanding how root and wood characteristics affect ecosystem function is not as clear as it is for traits like those in the leaf economic spectrum (LES). However, recent studies suggest that certain root and stem traits related to water may follow strategies similar to those seen in the LES, as noted by Chave et al. (2009). For example, traits like lower sapwood density and higher sap flux, which are positively associated with specific leaf area (SLA) according to O'Grady et al. (2009), might contribute to higher rates of evapotranspiration observed in invasive tree species compared to native ones, as discussed by Swaffer and Holland (2015). Additionally, root characteristics, regardless of LES traits, have been shown to influence factors such as soil moisture (Gross et al. 2008), soil stability, erosion (Stokes et al. 2009), and potentially soil structure, leaching, infiltration (Six et al. 2004; De Deyn et al. 2008), evapotranspiration, and climate cycles (Lee et al. 2005).

Just like leaves, certain root and wood traits, such as their density, lignin, or dry matter content, are expected to slow down decomposition and help store carbon in the soil, as noted by Chambers et al. (2000), De Deyn et al. (2008), Klumpp and Soussana (2009), and Freschet et al. (2012). However, unlike leaf tissue, the amount of nitrogen in roots does not always predict how fast they break down. This is because many factors, like the chemistry of the soil they are in or the presence of mycorrhizae, can influence root decomposition, as Langley et al. (2006) and Freschet et al. (2012) explain. Additionally, the quantity and quality of substances that roots release into the soil can affect how carbon and nitrogen move around in it, possibly boosting microbial activity. But we are still trying to understand exactly how different root substances affect the microbes, as discussed by De Deyn et al. (2008), Dijkstra et al. (2006), Giri et al. (2023), and Kastovska et al. (2015). The connections between plant root, mycorrhizae, and nitrogen fixation bacteria play a big role in how chemicals move though the ecosystem. The symbiotic associations between plants and microbes usually boost availability of nitrogen and phosphorus, which increase primary productivity and carbon stock in the soil. Moreover the symbiotic associations are instrumental in locking carbon and nitrogen with soil particles, according to Langley et al. (2006) and De Deyn et al. (2008). However, the generalizability of these trends remains uncertain as effects may differ across various plant, fungi, or microbial species (Rillig and Mummey 2006). More investigation is required to comprehensively understand the complex interactions and impacts of root and wood traits on ecosystem functions across different species and environmental conditions.

15.4 Conclusion

PFTs are essential characteristics of plant species that influence their responses to the environment, interactions with other species, and contributions to ecosystem functioning. PFTs encompass a wide range of characteristics, including morphological, physiological, and phenological attributes. The variations in these traits reflects species' adaptations to different environmental conditions and ecological niches. Functional traits play an important role in determining species coexistence by influencing resource use, competitive abilities, and niche differentiation. Trait differences among species allow for partitioning of resources, reducing competition, and promoting biodiversity. Incorporating functional traits into ecological models enhances the predictive capacity of these models. Trait-based approaches enable better predictions of species' responses to environmental changes, ecosystem dynamics, and community assembly processes. Functional traits influence ecosystem functions such as nutrient cycling, primary productivity, carbon sequestration, and resilience to disturbances. Understanding trait-environment relationships helps predict and manage ecosystem functions and services. Furthermore, trait-based approaches guide decisions regarding species selection, ecosystem restoration, and land-use planning. Integrating functional traits with other fields, such as genetics, biogeochemistry, and remote sensing, provides interdisciplinary insights into how traits drive ecological patterns and processes at multiple scales. In conclusion, plant functional traits serve as fundamental tools in ecological research, offering insights into species' adaptations, community dynamics, ecosystem functions, and responses to environmental changes. Understanding and applying functional traits contribute significantly to our ability to manage and conserve ecosystems in the face of global environmental challenges. Their integration into ecological studies continues to advance our understanding of the natural world and inform strategies for sustainable management and conservation.

References

Badgery W, Kemp D, Michalk D, King W (2005) Competition for nitrogen between Australian native grass and the introduced weed *Nassella trichotoma*. Ann Bot 96:799–809
Bakker MA, Carreno-Rocabado G, Poorter L (2011) Leaf economics traits predict litter decomposition of tropical plants and differ among land use types. Funct Ecol 25:473–483
Bender SF, Wagg C, van der Heijden MG (2016) An underground revolution: biodiversity and soil ecological engineering for agricultural sustainability. Trends Ecol Evol 31(6):440–452
Berg M, Ellers J (2010) Trait plasticity in species interactions: a driving force of community dynamics. Evol Ecol 24:617–629
Chambers JQ, Higuchi N, Schimel JP, Ferreira LV, Melack JM (2000) Decomposition and carbon cycling of dead trees in tropical forests of the Central Amazon. Oecologia 122:380–388
Chave J, Coomes D, Jansen S, Lewis S, Swenson NG, Zanne AE (2009) Towards a worldwide wood economics spectrum. Ecol Lett 12:351–366
Cornwell WK, Cornelissen JHC, Amatangelo K, Dorrepaal E, Eviner VT, Godoy O, Hobbie SE, Hoorens B, Kurokawa H, P'erez-Harguindeguy N, Quested HM, Santiago LS, Wardle DA, Wright IJ, Aerts R et al (2008) Plant species traits are the predominant control on litter decomposition rates within biomes worldwide. Ecol Lett 11:1065–1071
Daily GC, Matson PA (2008) Ecosystem services: from theory to implementation. Proc Natl Acad Sci 105(28):9455–9456
de Bello F, Lavorel S, Díaz S, Harrington R, Cornelissen JC, Bardgett R, Berg M, Cipriotti P, Feld C, Hering D, Martins da Silva P, Potts S, Sandin L, Sousa J, Storkey J, Wardle D, Harrison P (2010) Towards an assessment of multiple ecosystem processes and services via functional traits. Biodivers Conserv 19:2873–2893
De Deyn GB, Cornelissen JH, Bardgett RD (2008) Plant functional traits and soil carbon sequestration in contrasting biomes. Ecol Lett 11:516–531
de Vries FT, Manning P, Tallowin JRB, Mortimer SR, Pilgrim ES, Harrison KA, Hobbs PJ, Quirk H, Shipley B, Cornelissen JHC, Kattge J, Bardgett RD (2012) Abiotic drivers and plant traits explain landscape-scale patterns in soil microbial communities. Ecol Lett 15:1230–1239
Díaz S, Cabido M (2001) Vive la difference: plant functional diversity matters to ecosystem processes. Trends Ecol Evol 16:646–655
Dijkstra FA, Hobbie SE, Reich PB (2006) Soil processes affected by sixteen grassland species grown under different environmental conditions. Soil Sci Soc Am J 70:770–777
Eviner VT, Chapin FS III (2003) Functional matrix: a conceptual framework for predicting multiple plant effects on ecosystem processes. Annu Rev Ecol Evol Syst 34:455–485
Faucon MP, Houben D, Lambers H (2017) Plant functional traits: soil and ecosystem services. Trends Plant Sci 22(5):385–394
Firn J, Mac Dougall AS, Schmidt S, Buckley YM (2010) Early emergence and resource availability can competitively favour natives over a functionally similar invader. Oecologia 163:775–784
Firn J, Prober SM, Buckley YM (2012) Plastic traits of an exotic grass contribute to its abundance but are not always favorable. PLoS One 7:e35870. https://doi.org/10.1371/journal.pone.0035870
Freschet GT, Aerts R, Cornelissen JHC (2012) A plant economics spectrum of litter decomposability. Funct Ecol 26:56–65
Funk JL (2008) Differences in plasticity between invasive and native plants from a low resource environment. J Ecol 96:1162–1174
Funk JL (2013) The physiology of invasive plants in low-resource environments. Conserv Physiol 1:cot026. https://doi.org/10.1093/conphys/cot1026
Garibaldi LA, Semmartin M, Chaneton EJ (2007) Grazing-induced changes in plant composition affect litter quality and nutrient cycling in flooding Pampa grasslands. Oecologia 151:650–662
Garnier E, Navas M-L (2012) A trait-based approach to comparative functional plant ecology: concepts, methods and applications for agroecology. A review. Agron Sustain Dev 32:365–399

Giri K, Mishra G, Suyal DC, Kumar N, Doley B, Das N, Baruah RC, Bhattacharyya R, Bora N (2023) Performance evaluation of native plant growth-promoting rhizobacteria for paddy yield enhancement in the jhum fields of Mokokchung, Nagaland, North East India. Heliyon 9:e14588

Grigulis K, Lavorel S, Krainer U, Legay N, Baxendale C, Dumont M, Kastl E, Arnoldi C, Bardgett RD, Poly F, Pommier T, Schloter M, Tappeiner U, Bahn M, Clement JC (2013) Relative contributions of plant traits and soil microbial properties to mountain grassland ecosystem services. J Ecol 101:47–57

Grime JP (1977) Evidence for the existence of three primary strategies in plants and its relevance to ecological and evolutionary theory. Am Nat 111:1169–1194

Gross N, Robson TM, Lavorel S, Albert C, Le Bagousse-Pinguet Y, Guillemin R (2008) Plant response traits mediate the effects of subalpine grasslands on soil moisture. New Phytol 180:652–662

Han Y, Buckley YM, Firn J (2012) An invasive grass shows colonization advantages over native grasses under conditions of low resource availability. Plant Ecol 213:1117–1130

Kastovska E, Edwards K, Picek T, Santruckova H (2015) A larger investment into exudation by competitive versus conservative plants is connected to more coupled plant–microbe N cycling. Biogeochemistry 122:47–59

Klumpp K, Soussana JF (2009) Using functional traits to predict grassland ecosystem change: a mathematical test of the response-and-effect trait approach. Glob Chang Biol 15:2921–2934

Kumar N, Kumar N, Shukla A, Shankhdhar SC, Shankhdhar D (2015) Impact of terminal heat stress on pollen viability and yield attributes of rice (Oryza sativa L.). Cereal Res Commun 43(4):616–626

Kumar N, Shankhdhar SC, Shankhdhar D (2016) Impact of elevated temperature on antioxidant activity and membrane stability in different genotypes of rice (Oryza sativa L.). Indian J Plant Physiol 21(1):37–43

Kumar N, Suyal DC, Sharma IP, Verma A, Singh H (2017) Elucidating stress proteins in rice (Oryza sativa L.) genotype under elevated temperature: a proteomic approach to understand heat stress response. 3 Biotech 7:205

Kumar N, Jeena N, Singh H (2019) Elevated temperature modulates rice pollen structure: a study from foothill Himalayan agro-ecosystem in India. 3 Biotech 9:175

Kumar A, Dwivedi GK, Tewari S, Paul J, Anand R, Kumar N, Kumar P, Singh H, Kaushal R (2020a) Carbon mineralization and inorganic nitrogen pools under Terminalia chebula Retz.-based agroforestry system in Himalayan foothills, India. For Sci 66(5):634–643

Kumar A, Kumar P, Singh H, Kumar N (2020b) Adaptation and mitigation potential of roadside trees with bio-extraction of heavy metals under vehicular emissions and their impact on physiological traits during seasonal regimes. Urban For Urban Green 58:126900

Kumar A, Dwivedi GK, Tewari S, Paul J, Sah VK, Singh H, Kumar P, Kumar N, Kaushal R (2020c) Soil organic carbon pools under Terminalia chebula Retz. based agroforestry system in Himalayan foothills, India. Curr Sci 118(7):1098–1103

Kumar A, Kumar P, Singh H, Bisht S, Kumar N (2021a) Relationship of physiological plant functional traits with soil carbon stock in temperate forest of Garhwal Himalaya. Curr Sci 120(8):1368–1373

Kumar A, Kumar P, Singh H, Kumar N (2021b) Modulation of plant functional traits under essential plant nutrients during seasonal regime in natural forests of Garhwal Himalayas. Plant Soil 465:197–212

Kumar A, Kumar P, Singh H, Kumar N (2021c) Impact of plant functional traits on near saturated hydraulic conductivity of soil under different forests of Kempty watershed in Garhwal Himalaya. Indian J Soil Conserv 49(1):38–44

Kumar A, Tewari S, Singh H, Kumar P, Kumar N, Bisht S, Kushwaha S, Tamta N, Kaushal R (2021d) Biomass accumulation and carbon stocks in different agroforestry system prevalent in Himalayan foothills, India. Curr Sci 120(6):1083–1088

Langley JA, Chapman SK, Hungate BA (2006) Ectomycorrhizal colonization slows root decomposition: the post-mortem fungal legacy. Ecol Lett 9:955–959

Larson JE, Funk JL (2016) Seedling root responses to soil moisture and the identification of a belowground trait spectrum across three growth forms. New Phytol 210:827–838. https://doi.org/10.1111/nph.13829

Lavorel S, Garnier E (2002) Predicting changes in community composition and ecosystem functioning from plant traits: revisiting the Holy Grail. Funct Ecol 16:545–556

Lavorel S, Grigulis K (2012) How fundamental plant functional trait relationships scale-up to trade-offs and synergies in ecosystem services. J Ecol 100:128–140

Lee JE, Oliveira RS, Dawson TE, Fung I (2005) Root functioning modifies seasonal climate. Proc Natl Acad Sci 102:17576–17581

Loranger J, Meyer ST, Shipley B, Kattge J, Loranger H, Roscher C, Weisser WW (2012) Predicting invertebrate herbivory from plant traits: evidence from 51 grassland species in experimental monocultures. Ecology 93:2674–2682

Nicotra AB, Atkin OK, Bonser SP, Davidson AM, Finnegan EJ, Mathesius U, Poot P, Purugganan MD, Richards CL, Valladares F, van Kleunen M (2010) Plant phenotypic plasticity in a changing climate. Trends Plant Sci 15:684–692

O'Grady AP, Cook PG, Eamus D, Duguid A, Wischusen JDH, Fass T, Worldege D (2009) Convergence of tree water use within an arid-zone woodland. Oecologia 160:643–655

Orwin KH, Buckland SM, Johnson D, Turner BL, Smart S, Oakley S, Bardgett RD (2010) Linkages of plant traits to soil properties and the functioning of temperate grassland. J Ecol 98:1074–1083

Reich PB (2012) Key canopy traits drive forest productivity. Proc R Soc B Biol Sci 279:2128–2134

Reich PB, Rich RL, Lu X, Wang YP, Oleksyn J (2014) Biogeographic variation in evergreen conifer needle longevity and impacts on boreal forest carbon cycle projections. Proc Natl Acad Sci 111:13703–13708

Rillig MC, Mummey DL (2006) Mycorrhizas and soil structure. New Phytol 171:41–53

Santiago LS (2007) Extending the leaf economics spectrum to decomposition: evidence from a tropical forest. Ecology 88:1126–1131

Santiago LS, Schuur EAG, Silvera K (2005) Nutrient cycling and plant-soil feedbacks along a precipitation gradient in lowland Panama. J Trop Ecol 21:461–470

Siebenkas A, Schumacher J, Roscher C (2015) Phenotypic plasticity to light and nutrient availability alters functional trait ranking across eight perennial grassland species. AoB Plants 7:plv029. https://doi.org/10.1093/aobpla/plv029

Singh H, Verma A (2013a) Physiological responses of rice cultivars to various nitrogen levels. Int J Agric Environ Biotechnol 6(3):383–388

Singh H, Verma A (2013b) Characterization and screening of high nitrogen efficient rice genotype to curtail environmental pollution. Int J Agric Environ Biotechnol 6(1):23–30

Singh H, Verma A, Shukla A (2010a) Comparative study of yield and yield components of hybrid and inbred genotypes of rice (*Oryza Sativa L.*). Int J Agric Environ Biotechnol 3:355–360

Singh H, Verma A, Krishnamoorthy Mand Shukla A (2010b) Consequence of diverse nitrogen levels on leaf pigments in five rice genotypes under field emergent circumstance. Int J Bioresour Stress Manage 1:189–193

Singh H, Verma A, Rai SK (2013a) Biochemical evaluation of different rice genotypes grown at various nitrogen levels using SDS-PAGE. Curr Adv Agric Sci 5(1):144–146

Singh H, Verma A, Shukla A (2013b) Guttation fluid as a physiological marker for selection of nitrogen efficient rice (*Oryza sativa* L.) genotypes. Afr J Biotechnol 12(44):6276–6281

Singh H, Verma A, Ansari MW, Shukla A (2014) Physiological response of rice (*Oryza sativa* L.) genotypes to elevated nitrogen applied under field conditions. Plant Signal Behav 9:e29015. https://doi.org/10.4161/psb.29015

Singh H, Yadav M, Kumar N, Kumar A, Kumar M (2020) Assessing adaptation and mitigation potential of roadside trees under the influence of vehicular emissions: a case study of Grevillea robusta and Mangifera indica planted in an urban city of India. PLoS One 15(1):e0227380

Singh S, Singh H, Sharma SK, Nautiyal R (2021) Seasonal variation in biochemical responses of bamboo clones in the sub-tropical climate of Indian Himalayan foothills. Heliyon 7(4):e06859. https://doi.org/10.1016/j.heliyon.2021.e06859

Singh M, Singh H, Kumar A, Kumar M, Barthwal S, Thakur A (2024) Soil nitrogen availability determines the CO_2 fertilization effect on tree species (*Neolamarckia cadamba*): growth and physiological evidence. Environ Sustain. https://doi.org/10.1007/s42398-023-00300-w

Six J, Bossuyt H, Degryze S, Denef K (2004) A history of research on the link between (micro) aggregates, soil biota, and soil organic matter dynamics. Soil Tillage Res 79:7–31

Stokes A, Atger C, Bengough A, Fourcaud T, Sidle R (2009) Desirable plant root traits for protecting natural and engineered slopes against landslides. Plant Soil 324:1–30

Nash Suding K, Goldberg DE, Hartman KM (2003) Relationships among species traits: separating levels of response and identifying linkages to abundance. Ecology 84:1–16

Sutton-Grier AE, Wright J, Richardson C (2012) Different plant traits affect two pathways of riparian nitrogen removal in a restored freshwater wetland. Plant Soil 365:41–57

Swaffer BA, Holland KL (2015) Comparing ecophysiological traits and evapotranspiration of an invasive exotic, *Pinus halepensisin* native woodland overlying a karst aquifer. Ecohydrology 8:230–242

Valladares F, Matesanz S, Guilhaumon F, Ara'ujo MB, Balaguer L, Benito-Garz'on M, Cornwell W, Gianoli E, van Kleunen M, Naya DE, Nicotra AB, Poorter H, Zavala MA (2014) The effects of phenotypic plasticity and local adaptation on forecasts of species range shifts under climate change. Ecol Lett 17:1351–1364

WEF (2020) New nature economy report. https://www.weforum.org/reports/new-nature-economy-report-series/. Accessed 24 Nov 2023

Wright JP, Naeem S, Hector A, Lehman C, Reich PB, Schmid B, Tilman D (2006) Conventional functional classification schemes underestimate the relationship with ecosystem functioning. Ecol Lett 9:111–120

Yang Y, Zhu Q, Peng C, Wang H, Chen H (2015) From plant functional types to plant functional traits: a new paradigm in modelling global vegetation dynamics. Prog Phys Geogr 39(4):514–535

16 Challenges and Opportunities in Scaling Up Plant Functional Trait-Based Approaches

Hem C. Joshi, Prabha Pant, and Harsh K. Chauhan

Abstract

Plant functional traits, such as leaf size, root depth, and flowering time, have become pivotal in understanding plant ecology and ecosystem functioning. These traits provide critical insights into how plants respond to environmental changes, facilitating predictions of community dynamics and ecosystem services. As the application of plant functional traits gains momentum, many challenges and exciting opportunities emerge in the effort to scale up these approaches. Challenges include the pressing need for standardizing trait measurements across diverse ecosystems. Variability in traits within and between species poses a significant challenge, necessitating the development of robust statistical tools and the integration of genetic and ecological data to understand the mechanisms underlying this variability. Additionally, scaling up these approaches from individual plants to ecosystems requires bridging the gap between field observations and large-scale ecological processes. Data gaps are evident for non-model species, and there is an urgent need to expand trait databases to include a broader spectrum of organisms. Improving the accessibility and interoperability of trait databases and analytical tools is paramount to enhancing the usability of functional trait data. Overcoming the data collection and sharing challenges can be facilitated by establishing global trait networks and encouraging cross-disciplinary collaboration model. Citizen science can also play a pivotal role in data collection efforts, further broadening the scope and depth of trait-based research. Despite these challenges, promising opportunities exist to advance plant functional trait-based approaches. Leveraging recent technological advancements in remote sensing, genomics, and computational methods can enable more efficient collection and analysis of trait data at larger scales. This

H. C. Joshi (✉) · P. Pant · H. K. Chauhan
Department of Botany, D.S.B. Campus, Kumaun University, Nainital, Uttarakhand, India

N. Kumar, H. Singh (eds.), *Plant Functional Traits for Improving Productivity*,
https://doi.org/10.1007/978-981-97-1510-7_16

may help the development of more accurate predictive models for ecological responses to global change, enhancing conservation and sustainable land management efforts. Interdisciplinary collaboration between ecologists, biologists, and data scientists is vital in expanding our knowledge of plant functional traits. This approach can unravel the ecological significance of these traits and elucidate their potential in guiding ecosystem management decisions.

Keywords

Plant functional traits · Ecosystem functioning · Ecosystem services · Trait-based research · Sustainable land management

16.1 Introduction

Plant functional traits (PFTs) encompass a range of morphological, physiological, and phenological characteristics that exert indirect influences on individual plant fitness by affecting growth, reproduction, and survival (Violle et al. 2007; Kumar et al. 2020a, b, c). These traits serve as pivotal attributes of plants, offering valuable insights into their ecological roles, interactions with the environment, and potential applications across various domains (Lavorel and Garnier 2002; Migliavacca et al. 2021). These traits also play a significant role in determining plant productivity and stress resilience, influencing the competitive dynamics of plants in diverse environmental settings (Funk et al. 2017; Kumar et al. 2021a, b), for example, leaf mass per area (LMA), which exhibits a positive correlation with photosynthetic productivity and a negative correlation with structural robustness. LMA is contingent upon resource availability and environmental conditions (Díaz et al. 2015). Leaf pigments like chlorophyll and carotenoids dictate photosynthetic capabilities, with variations indicating changes in plant health due to stress. Additionally, constituents such as nitrogen and carbon in leaves are directly intertwined with biosphere-atmosphere cycles and hold great significance in the parameterization of vegetation within Earth system models (Singh et al. 2013a, b; Yang et al. 2015). Therefore, to comprehend ecosystem functioning comprehensively, it is imperative to compile a comprehensive suite of quantitative trait measurements. Functional traits, which usually develop over evolutionary time scales to maximize plant survivorship and functional performances in changing environment, are important indices to explore how ecosystems respond and adapt to a changing environment (Singh and Verma 2013a, b). Many ecologists have argued that identifying regional- to global-scale patterns in functional traits, at organism to ecosystem scales, in combination with responses to environmental changes, is necessary to increase our ability to predict how ecosystems will function in the future (Wright et al. 2017; Kumar et al. 2021c, d). Traditionally, plant functional types have served as a means to depict shifts in community composition within dynamic global vegetation models (DGVMs). However, this approach has exhibited limitations in accurately predicting ecosystem functions and their responses to climatic changes.

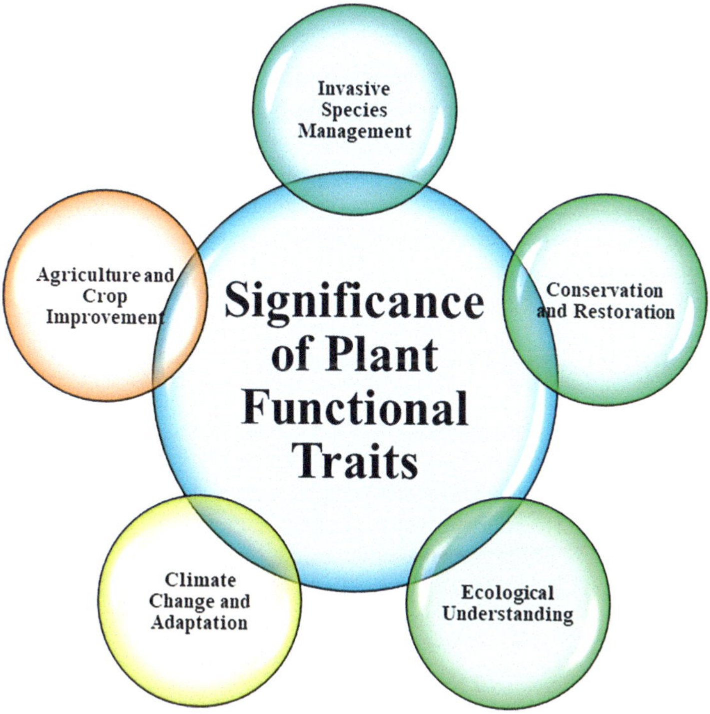

Fig. 16.1 Overview of the significance of plant functional traits in various contexts

Recently, trait-based modeling has emerged as one of the most promising methodologies for simulating community dynamics in the face of global alterations (Violle et al. 2012). To bolster this trait-based modeling approach, there is an urgent need to gather empirical evidence, create comprehensive datasets, and uncover large-scale patterns and the factors influencing variations in functional traits across environmental gradients (Fig. 16.1).

16.1.1 Ecological Understanding

Plant traits aid in characterizing and quantifying the diversity of plant species within ecosystems. They enable researchers to assess how different species coexist and compete for resources based on their unique functional traits. Plant functional traits play a crucial role in determining ecosystem processes such as nutrient cycling, carbon sequestration, and energy flow (Singh et al. 2010a, b; Westerband et al. 2021). Traits influence how plants interact with other organisms, impacting ecosystem stability and resilience (Biggs et al. 2020; Singh et al. 2020).

16.1.2 Climate Change and Adaptation

Changes in plant functional traits can serve as early indicators of the impacts of climate change on plant communities (Griffin-Nolan et al. 2019). For instance, alterations in leaf phenology or photosynthetic rates can reveal shifts in response to changing temperatures or precipitation patterns (Green et al. 2020). Understanding plant traits related to stress tolerance (e.g., drought resistance) can guide the development of climate-resilient crop varieties, enhancing agricultural sustainability in the face of climate uncertainties.

16.1.3 Agriculture and Crop Improvement

Identifying and selecting plant traits that contribute to increased crop yields are essential for addressing global food security challenges (Lenaerts et al. 2019; Kumar et al. 2015, 2016). Traits related to growth rate, nutrient-use efficiency, and pest resistance can be optimized to improve crop performance. Plant traits can also help design more sustainable agricultural practices by minimizing the need for synthetic inputs like fertilizers and pesticides (Singh et al. 2014, 2021, 2024; Bacha and Iqbal 2023).

16.1.4 Conservation and Restoration

Restoration efforts benefit from knowledge of plant functional traits to select appropriate native species for restoring degraded habitats (Carlucci et al. 2020). Traits related to nitrogen fixation, mycorrhizal associations, and competitive abilities can influence restoration success. Understanding the functional diversity of plant communities aids in conserving biodiversity. Identifying key traits that contribute to ecosystem resilience and stability is vital for effective conservation management.

16.1.5 Invasive Species Management

Plant functional traits are valuable for identifying invasive plant species and predicting their impacts on native ecosystems (van Kuijk et al. 2021). Traits related to rapid growth, high reproductive capacity, and resource acquisition are the common characteristic features of invasive species (Fig. 16.2).

16.2 Challenges in Scaling Up Plant Functional Trait-Based Approaches

Scaling up plant functional trait-based approaches from the laboratory to the field and from small-scale studies to larger ecosystems presents several significant challenges. These challenges can hinder the practical application of trait-based research

Fig. 16.2 Purpose of the chapter: explore scaling up challenges and opportunities

in agriculture, ecology, and environmental management. Some key challenges are as follows (Fig. 16.3):

Spatial and Temporal Variability Plant functional traits can exhibit substantial spatial and temporal variability (Kattenborn et al. 2019). Scaling up requires data collection over larger geographic areas and longer timeframes, which can be resource-intensive and logistically challenging.

Trait Plasticity Plants often exhibit trait plasticity, meaning that their traits can change in response to environmental conditions (Gratani 2014). Scaling up requires accounting for this plasticity and developing models that can predict trait responses under different stressors accurately.

Interactions and Trade-offs Plant traits can interact with each other and exhibit trade-offs (Laughlin et al. 2021). Identifying and quantifying these interactions and

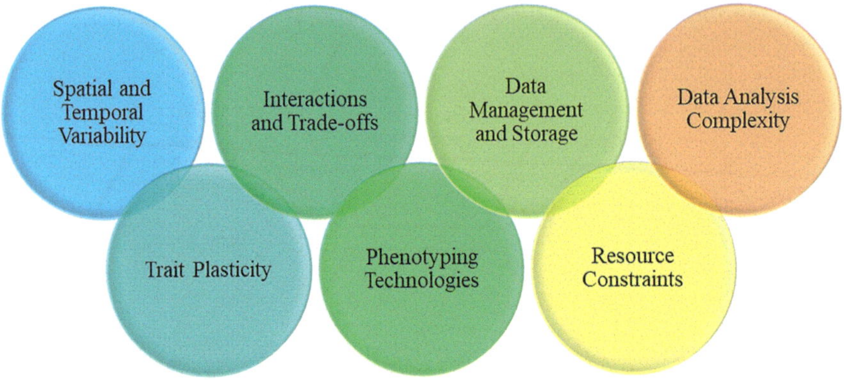

Fig. 16.3 Challenges in scaling up plant functional trait-based approaches

trade-offs at larger scales can be complex and may require advanced statistical and modeling approaches.

Phenotyping Technologies Existing phenotyping technologies are often limited in terms of throughput and scalability. Scaling up plant functional traits, research requires the development of cost-effective, high-throughput phenotyping tools that can handle large sample sizes (Smith et al. 2021).

Data Management and Storage Managing and storing the vast amounts of data generated in large-scale trait-based studies can be a logistical challenge (Derocles et al. 2018). Effective data management systems are needed to ensure data integrity and accessibility.

Resource Constraints Conducting large-scale field experiments or surveys can be resource-intensive in terms of funding, manpower, and equipment. Securing the necessary resources for scaling up research can be a hurdle.

Data Analysis Complexity Analyzing large-scale trait data can be computationally intensive and require advanced statistical and modeling techniques. Researchers must be well-equipped with the necessary computational tools and expertise.

Translation to Practical Applications While trait-based research generates valuable insights, translating these findings into practical applications, such as crop breeding or ecosystem management, can be challenging (Parsons et al. 2011). Bridging the gap between research and implementation is essential for realizing the potential benefits of trait-based approaches.

Ethical and Social Considerations As trait-based research expands, ethical and social considerations, such as intellectual property rights, data sharing, and the impact on local communities, become increasingly important and require careful attention.

16.2.1 Spatial and Temporal Variability

Spatial and temporal variability is an inherent characteristic of all-natural phenomena, especially when considering dynamic and complex systems, such as soil-plant. To ensure agricultural success, it's crucial to precisely assess the extent and fluctuations of variation over space and time caused by all factors influencing crop yield. The study of variability involves the spatial analysis techniques particularly geostatistics and must be performed respecting the basic assumptions of the methods, from data collection planning to the choice of the investigation models. Comprehending variability within a holistic system perspective facilitates decision-making and improves the efficiency of management practices.

Spatial and temporal variability are fundamental concepts in various scientific fields, including geology, ecology, meteorology, and economics. They refer to the fluctuations, differences, or changes that occur in space (across locations) and time (over periods) within a given system or dataset. Understanding spatial and temporal variability is crucial for making informed decisions, predicting patterns, and managing resources effectively. It can be seen in various contexts (Fig. 16.4):

Environmental Studies In ecology, researchers study spatial variability to understand biodiversity patterns across landscapes, for instance, examining how species diversity changes from one habitat to another within a forest ecosystem.

Geology Geologists analyze the spatial variability of rock formations, minerals, and soil properties across different geographical areas. This information can be critical for resource exploration and land-use planning.

Urban Planning In city planning, spatial variability can relate to population density, land use, traffic patterns, and infrastructure distribution. Urban planners use this data to make informed decisions about zoning, transportation, and public services.

Temporal Variability Temporal variability, on the other hand, focuses on changes that occur over time. It is crucial in understanding how systems evolve or fluctuate across different time scales.

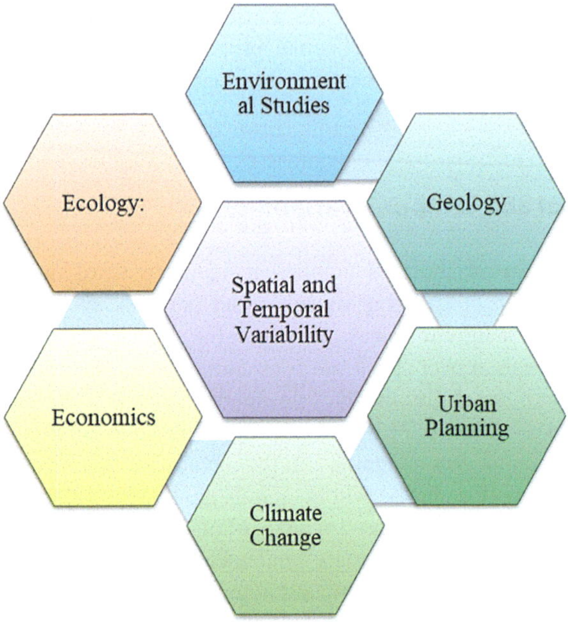

Fig. 16.4 Spatial and temporal variability can be seen in various contexts

Climate Change Temporal variability plays a central role in climate science. Researchers study long-term climate data to identify trends and variations in temperature, precipitation, and other climate parameters.

Economics Economists analyze temporal variability in economic indicators like GDP, inflation rates, and unemployment rates. This information helps governments and businesses make economic policies and investment decisions.

Ecology Temporal variability is essential in ecology to study population dynamics, phenology (seasonal patterns), and the impact of disturbances like wildfires or disease outbreaks on ecosystems.

Spatial and temporal variability is crucial in many scientific disciplines. For example, in climate science, researchers not only study how temperature varies from one location to another (spatial variability) but also how it changes over decades or centuries (temporal variability). Similarly, in ecological studies, scientists examine how species distribution varies across different habitats (spatial) and how these distributions change seasonally or over longer time frames (temporal).

16.2.2 Geographic and Temporal Variations in Plant Traits

Geographic and temporal variations in plant traits are essential aspects of ecological and botanical research. These variations refer to how the characteristics or traits of plants change across different geographic locations and over time. Understanding these variations can provide insights into the adaptation of plants to diverse environments, the effects of climate change, and the functioning of ecosystems. Here's an overview of geographic and temporal variations in plant traits (Kumar et al. 2022).

16.2.2.1 Geographic Variations in Plant Traits

S.N.	Characteristics	Geographic variations
1.	Climate and elevation gradients	One of the most prominent geographic factors influencing plant traits is climate. As we move from equatorial regions to polar regions or from lowlands to high-altitude areas, we'll observe variations in traits such as leaf size, shape, and texture, as well as the types of plants that can thrive in these conditions.
2.	Soil composition	Different soil types and nutrient levels in various geographic regions can lead to variations in plant traits. For example, plants in nutrient-poor soils may develop traits that enhance nutrient uptake and conservation.
3.	Water availability	In areas with varying levels of precipitation, plants adapt by exhibiting differences in traits related to water use efficiency, such as leaf size, stomatal density, and root architecture.
4.	Biotic interactions	Geographic variations in plant traits can also be influenced by interactions with other organisms, including herbivores, pollinators, and pathogens. Coevolution with these organisms can lead to the development of specific defensive or reproductive traits.
5.	Species diversity	Geographic regions with high species diversity may exhibit greater trait diversity, as different species adapt to niche-specific conditions.

16.2.2.2 Temporal Variations in Plant Traits

S.N.	Characteristics	Temporal variations
1.	Seasonal changes	Many plant traits exhibit temporal variations throughout the year, driven by seasonal changes in temperature, light, and precipitation. For example, deciduous trees shed their leaves in the fall, and plants adjust their photosynthetic rates and flowering times in response to changing seasons.
2.	Long-term trends	Over longer time scales, such as decades or centuries, plant traits can change in response to climate change, land use, and other environmental factors. For instance, plants may shift their flowering times or migrate to higher altitudes in response to warming temperatures.
3.	Successional changes	In ecosystems undergoing ecological succession, plant traits can change as the composition of species shifts over time. Pioneer species may have traits optimized for colonizing disturbed areas, while late-successional species may exhibit traits related to competition and shade tolerance.

(continued)

S.N.	Characteristics	Temporal variations
4.	Human-induced changes	Human activities, such as selective breeding, agriculture, and urbanization, can lead to rapid changes in plant traits. Cultivated crops, for instance, have been modified over generations to enhance traits like yield and disease resistance.

Geographic and temporal variations in plant traits are critical for understanding how plants respond to their environment and how ecosystems adapt to changing conditions. This knowledge is essential for conservation efforts, land management, and predicting the impacts of climate change on vegetation patterns and biodiversity. Researchers use various methods, including field surveys, remote sensing, and trait databases, to gather data and analyze these variations across different spatial and temporal scales.

16.2.3 Trait Plasticity

Trait plasticity in plants refers to the ability of plants to adjust their morphological, physiological, and phenological traits in response to changes in their environment (Pérez-Ramos et al. 2019). Plants exhibit a wide range of trait plasticity, allowing them to adapt to various environmental conditions and ecological niches. Some key aspects of trait plasticity in plants are (Fig. 16.5) as follows.

16.2.3.1 Morphological Traits

Plants can alter the size and shape of their leaves based on light availability (Westerband et al. 2021). In low light conditions, they may produce larger, thinner leaves to capture more light, while in high light conditions, they may have smaller,

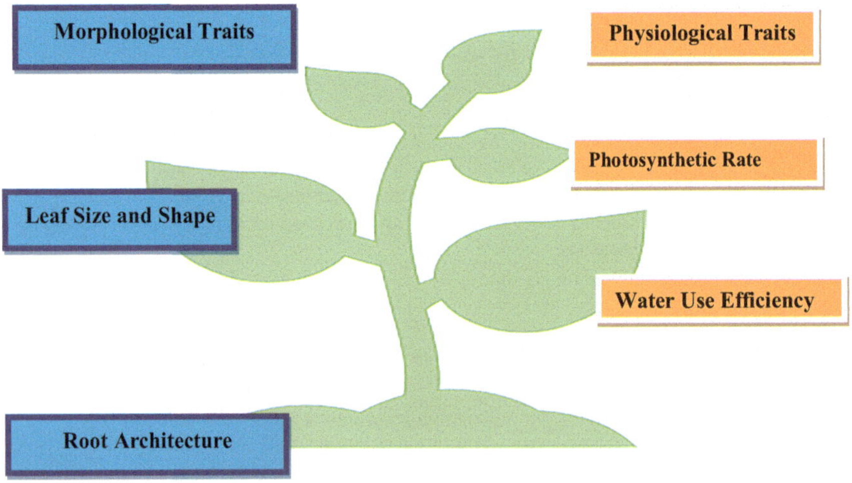

Fig. 16.5 Overview of morphological and physiological traits

thicker leaves to reduce water loss and excessive exposure to light. Root systems can adapt to soil conditions. In nutrient-poor soils, plants may develop more extensive root systems to explore a larger soil volume, while in nutrient-rich soils, they may invest less in root growth (Zhou et al. 2020).

16.2.3.2 Physiological Traits

Plants can adjust their photosynthetic rates in response to changing light levels, temperature, and carbon dioxide concentrations (Dusenge et al. 2019). In response to water availability, plants can modify their water use efficiency by adjusting stomatal conductance and transpiration rates (Condon 2020).

16.2.3.3 Phenological Traits

Plants can alter their flowering time in response to temperature and photoperiod changes (Osnato et al. 2022). This adaptation ensures that they flower when conditions are favorable for pollination and seed production. The timing of fruit production can also be plastic, allowing plants to synchronize fruit maturation with optimal conditions for seed dispersal. Plants can allocate resources differently depending on environmental conditions. For example, in nutrient-poor environments, they may allocate more resources to root growth for nutrient uptake, while in nutrient-rich environments, they may invest more in aboveground biomass.

16.2.3.4 Response to Biotic and Abiotic Stress

Plants can also modify traits in response to biotic interactions. For example, they may produce defensive chemicals or structures in response to herbivore attacks or attract pollinators through changes in floral traits (Kumar et al. 2019, 2021a, b, c, d, e; Kessler and Chautá 2020). Plants exhibit plasticity in response to abiotic stress factors such as drought, salinity, and extreme temperatures (Shaar-Moshe et al. 2019). They may activate stress-responsive genes, accumulate osmolytes, or adjust their leaf morphology to minimize stress effects.

16.2.4 Ecological Implications and Climate Change

Trait plasticity in plants can have significant ecological implications. It can affect competitive interactions with other plant species, herbivore-plant interactions, and the composition of plant communities. Plants with high trait plasticity may be better able to colonize a range of habitats (Gioria et al. 2023). Trait plasticity is of particular importance in the context of climate change. As environmental conditions shift, plants with the ability to exhibit plastic responses may have a higher likelihood of survival and reproduction in new habitats. Trait plasticity in plants is essential for predicting how plant communities will respond to changing environmental conditions, including climate change. Researchers use experiments, field studies, and genetic analyses to investigate the extent and mechanisms of plasticity in different plant species, contributing to our understanding of plant ecology and adaptation.

16.2.5 Experiments and Modeling to Account for Plastic Responses

To account for plastic responses in plants, researchers employ a combination of experiments and modeling approaches. These methods help in understanding how plants adjust their traits in response to changing environmental conditions and how these responses influence plant performance, population dynamics, and ecological interactions. Here's an overview of how experiments and modeling are used to study plasticity in plants.

16.2.5.1 Experiments to Account for Plastic Responses

Common garden experiments	These experiments involve growing plants from different populations or ecotypes in a common environment. By exposing genetically distinct plants to the same conditions, researchers can identify differences in trait expression that are due to plastic responses rather than genetic variation.
Manipulative field experiments	Researchers alter specific environmental factors in natural or controlled field settings to study how plants respond. For example, they may manipulate water availability, light levels, or nutrient content to observe trait changes in response to these variables.
Growth chamber experiments	Controlled growth chambers allow researchers to precisely control environmental conditions such as temperature, humidity, and light. This enables them to study how individual plant traits change under different controlled conditions.
Phenotyping platforms	High-throughput phenotyping platforms use advanced imaging and sensor technologies to monitor plant traits continuously over time. This can provide valuable data on how traits change dynamically in response to environmental cues.
Common garden and field transplant experiments	Researchers transplant individuals from one environment or population to another to assess how plasticity influences performance and fitness. This helps in understanding local adaptation and the importance of plastic responses in plant survival and reproduction.

16.2.5.2 Modeling to Account for Plastic Responses

Phenotypic plasticity models	These mathematical models incorporate the concept of plasticity to predict how plant traits change in response to environmental factors. Models like reaction norms describe the relationship between genotype, environment, and phenotype.
Ecological models	Ecological models integrate plastic responses into larger ecosystem models. For example, they can predict how plasticity in plant traits affects species interactions, community composition, and ecosystem functioning.
Climate change models	Climate change models often include plastic responses to predict how plant species distributions and phenology may shift in response to changing climate conditions.
Population genetics models	These models assess how plasticity and genetic variation interact to influence plant population dynamics and adaptation over time.

(continued)

Trait-based models	Trait-based ecological models use plant trait data to understand ecosystem processes. By incorporating plasticity, researchers can better predict how changes in plant traits affect ecosystem functioning and services.
Agent-based models	These models simulate individual plants within a population and allow researchers to explore how plasticity influences plant competition, dispersal, and evolution.
Machine learning and data-driven models	Machine learning algorithms can analyze large datasets of plant traits and environmental variables to uncover complex relationships and patterns related to plastic responses.

Combining experiments and modeling is a powerful approach to study plasticity in plants (Arnold et al. 2019). Experiments provide empirical data, while models help generalize and make predictions based on this data. This integrated approach is particularly valuable for understanding how plasticity contributes to plant adaptation, species interactions, and ecosystem responses to environmental change. It also aids in developing strategies for conservation and land management in the face of ongoing environmental challenges such as climate change and habitat degradation.

16.2.6 Phenotyping Technologies

Phenotyping, the process of measuring an organism's physical and biochemical traits, is crucial in various scientific fields, including biology, agriculture, and ecology. Traditional phenotyping methods can indeed be labor-intensive and time-consuming. However, advancements in technology have led to the development of high-throughput phenotyping technologies that can significantly streamline the process. Here's an overview of both traditional and modern phenotyping methods.

Traditional phenotyping methods	
Manual measurements	Many phenotypic traits, such as plant height, leaf size, and fruit yield, were traditionally measured manually. Researchers use rulers, calipers, and other tools to record these traits individually for each sample.
Visual scoring	Qualitative traits, like flower color or disease symptoms, often required visual assessment and scoring by human observers. This can be subjective and time-consuming.
Harvest and destructive sampling	To measure traits like root morphology or biomass, researchers harvest destructively samples of the plants, which not only required time and effort but also prevented further measurements on the same individuals.
Labor-intensive data collection	Collecting phenotypic data for a large number of samples or over time could be extremely labor-intensive, leading to limitations in the scale and scope of research projects.

(continued)

Modern high-throughput phenotyping technologies	
Automated imaging	High-resolution cameras and imaging systems can capture phenotypic data rapidly and non-destructively. For example, plant phenotyping platforms can capture images of thousands of plants, allowing researchers to measure traits like leaf area and color.
Remote sensing	Satellite and drone-based remote sensing technologies enable the monitoring of large agricultural fields or ecosystems, providing data on traits like crop health, vegetation cover, and more.
Sensors and probes	Various sensors and probes can measure environmental parameters (e.g., soil moisture, temperature, and nutrient levels) and plant physiological traits (e.g., chlorophyll fluorescence) continuously and automatically.
Robotics	Robots equipped with sensors and cameras can perform tasks such as planting, watering, and harvesting while also collecting data on plant traits. These robots can work around the clock, increasing data collection efficiency.
Machine learning and AI	Advanced data analysis techniques, such as machine learning and artificial intelligence, can process large datasets generated by high-throughput phenotyping technologies to extract meaningful patterns and insights.
Phenomics platforms	Dedicated phenotyping platforms and facilities have been established, offering a controlled environment for high-throughput data collection. These platforms often integrate multiple technologies for comprehensive phenotyping.
Noninvasive techniques	Technologies like nuclear magnetic resonance (NMR) and magnetic resonance imaging (MRI) can provide noninvasive insights into plant physiology and root system architecture.

While modern phenotyping technologies have revolutionized the field, challenges remain. These include the high cost of equipment, data management and analysis, and ensuring the accuracy and reproducibility of measurements. Nevertheless, the transition from labor-intensive and time-consuming methods to high-throughput technologies has greatly enhanced the efficiency and scope of phenotypic research, enabling scientists to better understand complex biological systems, improve crop breeding programs, and address environmental and agricultural challenges.

16.2.7 The Necessity of High-Throughput and Automated Phenotyping

High-throughput and automated phenotyping are essential in contemporary scientific research and various fields for several compelling reasons (Fig. 16.6).

High-throughput and automated phenotyping significantly accelerate data collection processes (Jangra et al. 2021). They can analyze a large number of samples or subjects rapidly, saving valuable time and resources. This efficiency is particularly critical in research projects with tight timelines or when monitoring rapidly changing conditions. For studies involving a large number of samples, such as genetic screens, crop breeding programs, or ecological surveys, manual phenotyping is often impractical due to its labor-intensive nature. High-throughput methods

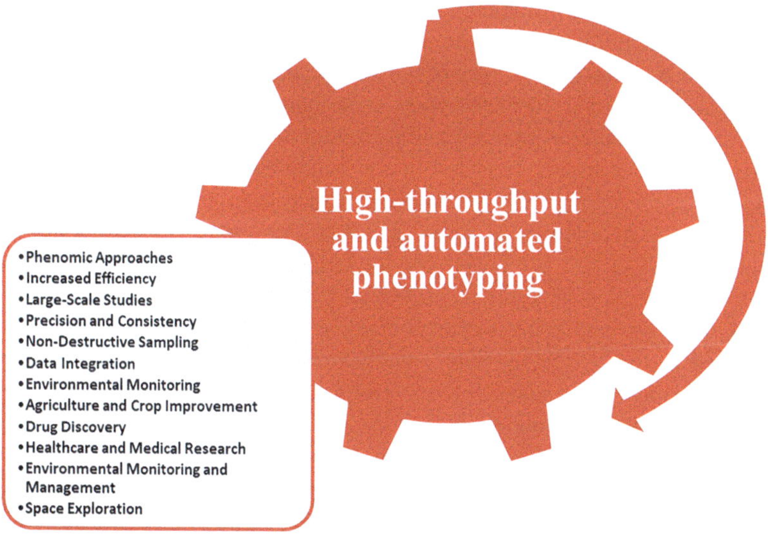

Fig. 16.6 Overviews of essentials in contemporary scientific research

enable researchers to handle the scale required for robust statistical analyses and meaningful results. Automated systems can consistently and precisely measure traits without human error or subjectivity. This ensures data reliability and reproducibility, which are essential for scientific rigor and comparison between studies. Many high-throughput phenotyping methods, such as imaging and remote sensing, are non-destructive. This allows for continuous monitoring of the same samples over time, providing valuable insights into dynamic processes and reducing the need for constant replanting or resampling. High-throughput phenotyping technologies often incorporate various sensors and imaging modalities. This integration allows researchers to collect multiple types of data simultaneously, providing a more comprehensive understanding of complex biological systems or environmental interactions. High-throughput phenotyping supports a phenomics approach, which seeks to understand how the entire phenotype of an organism (all observable traits) responds to genetic, environmental, and physiological factors (Mir et al. 2019). This holistic perspective is crucial for deciphering complex biological processes. In fields like ecology and environmental science, automated phenotyping helps researchers track changes in ecosystems and species responses to environmental shifts, such as climate change. This information is vital for conservation efforts and understanding ecosystem dynamics. High-throughput phenotyping is fundamental in crop breeding programs (Jangra et al. 2021). It allows breeders to screen thousands of plant varieties for desired traits like drought resistance, disease resistance, and yield potential, accelerating the development of new crop varieties that can address food security challenges. In pharmaceutical research, automated phenotyping is used to screen thousands of compounds for their effects on cell cultures, tissues, or model organisms. This accelerates drug discovery and development processes. Automated phenotyping is increasingly used in healthcare for high-throughput diagnostics, disease monitoring, and personalized medicine. It allows for the rapid analysis of patient samples, leading to faster diagnosis and treatment decisions. Automated phenotyping contributes to monitoring air and water quality, tracking pollution sources, and assessing the impact of pollutants on ecosystems and human health. Essential in contemporary scientific research in space research, high-throughput phenotyping helps assess the viability of plants and ecosystems for long-duration space missions and extraterrestrial colonization.

High-throughput and automated phenotyping are indispensable tools that not only enhance the efficiency of data collection but also enable researchers to tackle complex questions and address critical challenges in fields ranging from biology and agriculture to ecology, medicine, and environmental science. These technologies play a pivotal role in advancing scientific knowledge, innovation, and decision-making in a rapidly changing world.

16.3 Opportunities in Scaling Up Plant Functional Trait-Based Approaches

16.3.1 Crop Improvement

Crop improvement is a critical endeavor in agriculture aimed at enhancing the quality, productivity, and sustainability of crop plants (Schröder et al. 2019). It involves various strategies and techniques to achieve specific goals, such as breeding for desirable traits and increasing crop yields while reducing inputs. Here's an overview of these two key aspects of crop improvement.

Breeding for desirable traits involves selecting and developing crop varieties with specific characteristics that meet the needs of farmers, consumers, and the agricultural industry. This process often combines traditional and modern breeding techniques to create improved crop varieties. Some key aspects of breeding for desirable traits include:

Breeding for desirable traits	
Trait selection	Identifying the traits that are most desirable for a particular crop, such as disease resistance, drought tolerance, increased yield, improved nutritional content, or better taste.
Genetic diversity	Maintaining and harnessing genetic diversity within a crop species to introduce new traits or improve existing ones. This diversity can be found in wild relatives of cultivated crops.
Hybridization	Crossbreeding different varieties or even species to combine desirable traits, often leading to hybrid vigor or heterosis, which results in superior crop performance.
Marker-assisted breeding	Using molecular markers and genetic information to accelerate the breeding process by identifying and selecting plants with desired traits more efficiently.

(continued)

Breeding for desirable traits	
Genome editing	Techniques like CRISPR-Cas9 enable precise editing of the crop's genome to introduce or enhance specific traits.
Phenotypic screening	Evaluating plant phenotypes in the field or controlled environments to select individuals with the desired traits.
Field testing	Conducting extensive field trials to assess the performance and adaptability of new crop varieties in different environments.

Increasing crop yields while reducing inputs is a fundamental goal of sustainable agriculture. It involves optimizing resource use, minimizing environmental impacts, and ensuring economic viability for farmers. Several strategies are employed to achieve this:

Enhancing crop yields and reducing inputs	
Improved agronomy	Implementing best practices in crop management, such as proper planting density, timing, and spacing, to maximize yield potential.
Irrigation management	Efficient water use through techniques like drip irrigation and soil moisture monitoring to reduce water wastage.
Nutrient management	Precision application of fertilizers based on soil testing to ensure optimal nutrient levels and reduce overuse of chemical inputs.
Pest and disease management	Integrated pest management (IPM) strategies that combine biological control, crop rotation, and reduced pesticide use to control pests and diseases.
Conservation agriculture	Implementing no-till or reduced tillage practices to improve soil health, reduce erosion, and conserve moisture.
Genetic improvement	Developing crop varieties that are inherently more resistant to pests and diseases, tolerate adverse environmental conditions, and have higher yield potential.
Remote sensing and precision agriculture	Using technologies like satellite imagery, drones, and GPS-guided equipment to optimize resource use and monitor crop health.
Crop rotation and diversification	Rotating crops and diversifying agricultural landscapes to reduce pest pressure and improve soil fertility.
Post-harvest management	Reducing post-harvest losses through improved storage, transportation, and processing techniques.
Economic and policy incentives	Implementing policies and incentives that encourage sustainable agricultural practices and reduce the financial burden on farmers.

Crop improvement is an ongoing process that requires collaboration among plant breeders, agronomists, geneticists, and farmers. By combining advances in breeding techniques with sustainable agricultural practices, the goal of enhancing crop yields and reducing inputs can be achieved, contributing to global food security and environmental sustainability (Wright et al. 2004, Kumar et al. 2017).

16.3.2 Ecosystem Restoration

Ecosystem restoration is a crucial strategy for rehabilitating degraded ecosystems, conserving biodiversity, and mitigating the impacts of climate change (Simonson et al. 2021). In the context of ecosystem restoration, identifying plant traits and promoting biodiversity are essential components of successful restoration efforts. Ecosystem restoration often involves replanting native or desirable plant species to restore ecological functions and habitat quality. Identifying the right plant traits is crucial for successful restoration. An overview of how these aspects contribute to ecosystem restoration is as follows:

Identifying plant traits for restoration	
Adaptation to local conditions	Selecting plant species and genotypes adapted to local environmental conditions (e.g., climate, soil, and hydrology) increases their chances of survival and establishment.
Resilience	Choosing plant species with traits that confer resilience to environmental stressors, such as drought tolerance or resistance to pests and diseases, enhances their long-term survival and performance.
Nutrient cycling	Plants that contribute to nutrient cycling through traits like nitrogen fixation or mycorrhizal associations can improve soil fertility and ecosystem function.
Habitat restoration	Selecting plants that provide habitat and food resources for native wildlife, such as pollinators or herbivores, can aid in restoring ecological interactions and enhancing biodiversity.
Invasive species management	Some plant traits can help in outcompeting invasive species, contributing to the restoration of native vegetation.
Erosion control	Plants with deep root systems or ground-covering traits can help stabilize soil, reduce erosion, and improve water quality in restoration areas.
Diversity of plant traits	Promoting a diversity of plant traits within a restored ecosystem can enhance its resilience to changing environmental conditions and provide a wider range of ecosystem services.

16.3.3 Biodiversity Enhancement and Climate Change Mitigation

Effective ecosystem restoration efforts integrate the identification of suitable plant traits with a focus on enhancing biodiversity and addressing climate change challenges. Collaboration among scientists, conservationists, policymakers, and local communities is crucial for planning, implementing, and monitoring restoration projects that achieve these objectives. By restoring ecosystems with the right plant traits and fostering biodiversity, we can contribute to more resilient and sustainable landscapes (Fig. 16.7).

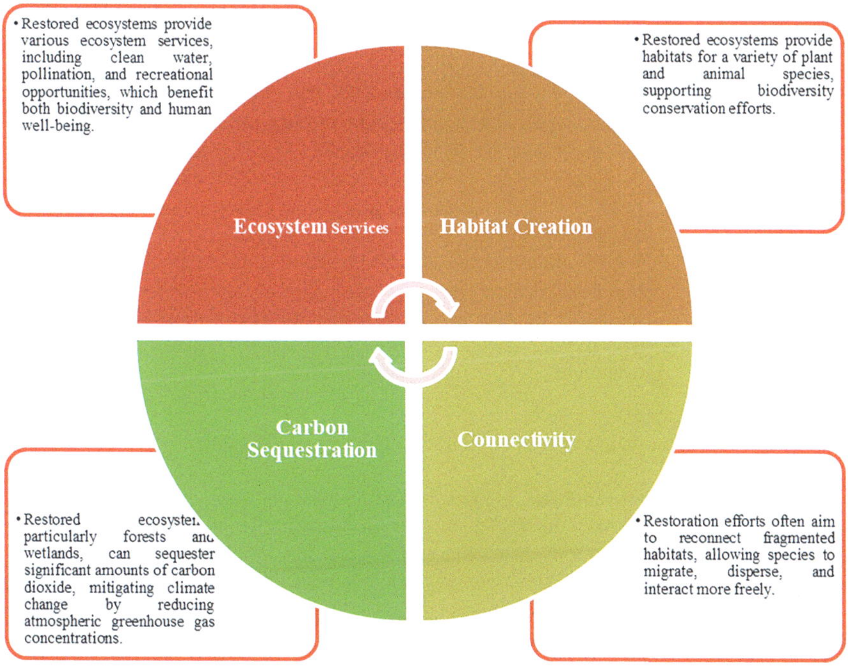

Fig. 16.7 Ecosystem restoration plays a significant role in enhancing biodiversity and mitigating the impacts of climate change

16.3.4 Biodiversity Conservation

Biodiversity conservation is a critical endeavor aimed at protecting and preserving the variety of life forms on Earth. It involves understanding and safeguarding different aspects of biodiversity, including species diversity, genetic diversity, and functional diversity (Corlett 2020). Insights into functional diversity play a vital role in guiding and enhancing conservation efforts. Functional diversity contributes and guides biodiversity conservation in the following ways:

Functional diversity refers to the variety of functional traits and roles that species and individuals within a community or ecosystem perform (Mammola et al. 2021). These traits can include feeding habits, reproductive strategies, niche occupation, and more. Understanding functional diversity provides valuable insights:

Ecosystem functioning	Functional diversity influences the functioning of ecosystems. Different species with unique functional traits contribute to processes such as nutrient cycling, pollination, seed dispersal, and decomposition.
Resilience and stability	Diverse functional traits within a community or ecosystem enhance its resilience and stability. Ecosystems with a range of functional traits are often better able to adapt to environmental changes and disturbances.

(continued)

Redundancy and complementarity	Functional diversity can reveal redundancy (multiple species with similar traits) and complementarity (species with different traits that complement each other) in ecological communities. This understanding aids in predicting how communities may respond to species loss or gain.
Invasion resistance	Functional diversity can influence a community's resistance to invasive species. A diverse community may be less susceptible to invasion because of niche saturation or competition with resident species.
Ecosystem services	Many ecosystem services, such as carbon sequestration, pest control, and water purification, depend on functional diversity. Conserving functional diversity ensures the provision of these services.
Response to environmental change	Functional diversity provides information on how ecosystems may respond to environmental changes, including climate change. Species with specific functional traits may be more or less vulnerable to altered conditions.

Insights into functional diversity guide conservation efforts in several ways:

Priority setting	Conservationists can prioritize the protection of species with unique functional traits that are crucial for ecosystem stability and functioning.
Restoration planning	When restoring degraded ecosystems, understanding functional diversity helps select appropriate species to reintroduce and guide ecosystem restoration efforts.
Habitat management	Functional diversity considerations influence habitat management practices, such as controlled burns, grazing, and invasive species control, to maintain a range of ecological functions.
Reserve design	In designing protected areas and reserves, knowledge of functional diversity helps ensure that a variety of functional roles are represented, contributing to the long-term viability of ecosystems.
Climate adaptation	Functional diversity can inform strategies for conserving species that are resilient to changing climate conditions or identifying potential corridors for species movement in response to climate change.
Conservation monitoring	Monitoring changes in functional diversity can serve as an early warning system for ecosystem health and potential threats.

By incorporating insights into functional diversity into conservation planning and management, conservation efforts become more effective and tailored to the specific needs of ecosystems and species. This approach contributes to the preservation of biodiversity and the services and functions it provides to ecosystems and society.

16.4 Climate Change Adaptation

Climate change adaptation is the process of adjusting to the changing climate conditions to reduce vulnerability, enhance resilience, and maintain or improve social, economic, and environmental well-being. In the context of climate change adaptation, predicting and managing plant responses and promoting climate-resilient ecosystems and agriculture are crucial strategies. Here's how these strategies contribute to climate change adaptation in the following ways:

Climate change affects plant species and ecosystems in various ways, including shifts in temperature, precipitation patterns, and the frequency of extreme weather events (Menezes-Silva et al. 2019). Predicting and managing plant responses are essential components of climate change adaptation:

Phenology and timing	Understanding how climate change alters the timing of plant life cycle events (phenology) helps in predicting when plants will bud, flower, and produce fruit. This information assists in planning agricultural activities, such as planting and harvesting, to align with changing conditions.
Species distribution	Climate models can predict changes in the geographic distribution of plant species. Conservation efforts can then focus on protecting and managing these species in their new or expanding ranges.
Invasive species control	As the climate changes, some invasive species may become more problematic. Monitoring and managing these species are crucial to prevent them from outcompeting native vegetation.
Crop selection and breeding	Farmers can adapt to climate change by selecting crop varieties or species that are better suited to changing conditions, such as heat-resistant or drought-tolerant cultivars. Plant breeding programs can develop climate-resilient crop varieties.
Ecosystem restoration	Restoring ecosystems with native or climate-adapted plant species helps ecosystems adapt to changing conditions, maintain biodiversity, and provide ecosystem services.
Preserving genetic diversity	Preserving genetic diversity within plant populations ensures that they have the raw material for adaptation to changing climates. Seed banks and conservation efforts are essential for safeguarding genetic diversity.

Climate-resilient ecosystems and agriculture involve practices and strategies that enhance the ability of ecosystems and agricultural systems to withstand and adapt to climate change:

Diversified cropping systems	Crop diversification, intercropping, and agroforestry can increase resilience by reducing the risk associated with climate-related disturbances or pests.
Water management	Efficient water management practices, such as drip irrigation and rainwater harvesting, help conserve water resources and provide stability in water availability for crops.
Soil health	Building healthy soils through practices like cover cropping and reduced tillage improves soil water retention, nutrient cycling, and resistance to erosion, benefiting agriculture and ecosystems.
Ecosystem-based adaptation	Incorporating ecosystem-based adaptation approaches, such as restoring wetlands and forests, can provide natural solutions for climate change mitigation and adaptation.
Protected areas and corridors	Establishing protected areas and wildlife corridors allows species to move and adapt to changing habitats, reducing the risk of extinction due to climate change.
Climate-smart agriculture	Climate-smart agricultural practices integrate climate adaptation and mitigation strategies. These practices include precision agriculture, early warning systems for pests and diseases, and climate-informed decision-making.
Policy and governance	Governments and organizations play a critical role in developing policies and incentives that support climate-resilient agriculture and ecosystems.

By predicting and managing plant responses and promoting climate-resilient ecosystems and agriculture, societies can better prepare for and adapt to the challenges posed by climate change. These strategies help safeguard food security, conserve biodiversity, and maintain the health and functioning of ecosystems in a changing climate.

16.5 Data-Driven Insights

Data-driven insights play a crucial role in various fields, from business and healthcare to science and government (Sarker 2021). They enable evidence-based decision-making, uncover patterns and trends, and provide valuable information for improving processes and outcomes. Data science encompasses the entire data life cycle, from data collection and cleaning to storage and retrieval. It ensures that data is accurate, reliable, and well-organized for analysis. Data scientists explore data to understand its characteristics and identify potential patterns or outliers. Data visualization techniques help in presenting complex information in a comprehensible manner. Machine learning algorithms enable predictive modeling, where historical data is used to make future predictions or classifications. Examples include predicting customer behavior, disease outbreaks, or stock prices. Machine learning can identify patterns in data that may not be apparent through traditional statistical analysis. This is valuable for tasks like image recognition, natural language processing, and anomaly detection. Machine learning algorithms power recommendation engines, which suggest products, content, or services based on user preferences and behavior. These systems are widely used in e-commerce and content platforms. Data-driven insights help optimize processes, such as supply chain management, transportation routes, and energy consumption, by analyzing data to identify areas for improvement. Machine learning models can assess risk in various domains, such as credit scoring, insurance, and cybersecurity, by analyzing historical data and identifying factors associated with risk. Data-driven personalization tailors user experiences, content, and services to individual preferences, increasing user engagement and satisfaction. Data science and machine learning are critical in genomics, drug discovery, disease diagnosis, and patient care, helping to improve medical outcomes and treatments.

16.6 Integration and Analysis of Large Datasets

The integration and analysis of large datasets often require big data technologies like Apache Hadoop and Spark, which can efficiently process and analyze massive amounts of data distributed across clusters of computers (Rao et al. 2019). Data warehouses are used to consolidate and manage large datasets from various sources, making them accessible for analysis. This is particularly important in business intelligence and analytics. Integration tools and platforms enable the merging of data from diverse sources, including databases, cloud services, IoT devices, and external

APIs, to create a unified dataset for analysis. Scalable analytics platforms allow organizations to perform complex analyses on large datasets using parallel processing and distributed computing, ensuring timely insights. Protecting large datasets is crucial. Data encryption, access controls, and compliance with data privacy regulations are essential considerations when handling sensitive information. Tools for data visualization and dashboard creation help users explore and interact with large datasets, enabling them to extract insights and make data-driven decisions. Data exploration platforms facilitate the discovery of trends, patterns, and anomalies in large datasets through interactive exploration and visualization. Establishing data governance practices ensures that large datasets are managed, maintained, and used in a way that aligns with organizational goals and complies with regulations. Data-driven insights, powered by data science and machine learning, have the potential to transform industries, drive innovation, and inform decision-making across diverse sectors. The ability to integrate and analyze large datasets efficiently is critical in harnessing the full potential of data for actionable insights and informed strategies (Sofi et al. 2022).

16.7 Conclusion

In conclusion, scaling up plant functional trait-based approaches presents both significant challenges and exciting opportunities in the field of ecology and environmental science. The study of plant functional traits has revolutionized our understanding of plant-environment interactions, ecosystem functioning, and biodiversity conservation. However, as we embark on the journey of expanding these approaches to broader scales, several challenges emerge. One of the primary challenges is the need for standardized data collection and integration across various ecosystems and regions. Achieving consistency in trait measurements and data sharing is essential to build comprehensive trait databases and facilitate cross-site comparisons.

Another challenge is accounting for trait plasticity and variability across different spatial and temporal scales. Understanding how plant traits respond to environmental changes and how these responses influence ecosystem dynamics requires advanced modeling and analytical techniques. Furthermore, the scalability of trait-based approaches necessitates collaboration among multidisciplinary teams, including ecologists, data scientists, and remote sensing experts. Bridging the gap between field data collection and cutting-edge technology is vital for successful scaling. Despite these challenges, there are remarkable opportunities ahead. Scaling up plant functional trait-based approaches can provide insights into global patterns of biodiversity, ecosystem functioning, and responses to environmental changes. These approaches have the potential to inform conservation strategies, sustainable land management practices, and climate change mitigation efforts.

Moreover, advancements in remote sensing technologies, machine learning, and big data analytics offer unprecedented opportunities to collect and analyze trait data at larger scales. Integrating these technological innovations with ecological expertise can unlock new frontiers in our understanding of plant-environment interactions.

References

Arnold PA, Kruuk LE, Nicotra AB (2019) How to analyse plant phenotypic plasticity in response to a changing climate. New Phytol 222(3):1235–1241

Bacha SAS, Iqbal B (2023) Advancing agro-ecological sustainability through emerging genetic approaches in crop improvement for plants. Funct Integr Genomics 23(2):145

Biggs CR, Yeager LA, Bolser DG, Bonsell C, Dichiera AM, Hou Z, Keyser SR, Khursigara AJ, Lu K, Muth AF, Negrete B Jr (2020) Does functional redundancy affect ecological stability and resilience? A review and meta-analysis. Ecosphere 11(7):e03184

Carlucci MB, Brancalion PH, Rodrigues RR, Loyola R, Cianciaruso MV (2020) Functional traits and ecosystem services in ecological restoration. Restor Ecol 28(6):1372–1383

Condon AG (2020) Drying times: plant traits to improve crop water use efficiency and yield. J Exp Bot 71(7):2239–2252

Corlett RT (2020) Safeguarding our future by protecting biodiversity. Plant Divers 42(4):221–228

Derocles SA, Bohan DA, Dumbrell AJ, Kitson JJ, Massol F, Pauvert C, Plantegenest M, Vacher C, Evans DM (2018) Biomonitoring for the 21st century: integrating next-generation sequencing into ecological network analysis. Adv Ecol Res 58:1–62

Díaz S, Kattge J, Cornelissen JHC, Wright IJ, Lavorel S, Dray S, Reu B, Kleyer M, Wirth C, Prentice IC, Garnier E, Bönisch G, Westoby M, Poorter H, Reich PB, Moles AT, Dickie J, Gillison AN, Zanne AE, Chave J, Wright SJ, Sheremet'ev SN, Jactel H, Christopher B, Cerabolini B, Pierce S, Shipley B, Kirkup D, Casanoves F, Joswig JS, Günther A, Falczuk V, Rüger N, Mahecha MD, Gorné LD (2015) The global spectrum of plant form and function. Nature 529:1–17. https://doi.org/10.1038/nature16489

Dusenge ME, Duarte AG, Way DA (2019) Plant carbon metabolism and climate change: elevated CO_2 and temperature impacts on photosynthesis, photorespiration and respiration. New Phytol 221(1):32–49

Funk JL, Larson JE, Ames GM, Butterfield BJ, Cavender-Bares J, Firn J, Laughlin DC, Sutton-Grier AE, Williams L, Wright J (2017) Revisiting the Holy Grail: using plant functional traits to understand ecological processes. Biol Rev 92:1156–1173. https://doi.org/10.1111/brv.12275

Gioria M, Hulme PE, Richardson DM, Pyšek P (2023) Why are invasive plants successful? Annu Rev Plant Biol 74:635–670

Gratani L (2014) Plant phenotypic plasticity in response to environmental factors. Adv Bot 2014:208747

Green JK, Berry J, Ciais P, Zhang Y, Gentine P (2020) Amazon rainforest photosynthesis increases in response to atmospheric dryness. Sci Adv 6(47):eabb7232

Griffin-Nolan RJ, Blumenthal DM, Collins SL, Farkas TE, Hoffman AM, Mueller KE, Ocheltree TW, Smith MD, Whitney KD, Knapp AK (2019) Shifts in plant functional composition following long-term drought in grasslands. J Ecol 107(5):2133–2148

Jangra S, Chaudhary V, Yadav RC, Yadav NR (2021) High-throughput phenotyping: a platform to accelerate crop improvement. Phenomics 1(2):31–53

Kattenborn T, Fassnacht FE, Schmidtlein S (2019) Differentiating plant functional types using reflectance: which traits make the difference? Remote Sensing Ecol Conserv 5(1):5–19

Kessler A, Chautá A (2020) The ecological consequences of herbivore-induced plant responses on plant–pollinator interactions. Emerg Topics Life Sciences 4(1):33–43

Kumar N, Kumar N, Shukla A, Shankhdhar SC, Shankhdhar D (2015) Impact of terminal heat stress on pollen viability and yield attributes of rice (Oryza sativa L.). Cereal Res Commun 43(4):616–626

Kumar N, Shankhdhar SC, Shankhdhar D (2016) Impact of elevated temperature on antioxidant activity and membrane stability in different genotypes of rice (Oryza sativa L.). Indian J Plant Physiol 21(1):37–43

Kumar N, Suyal DC, Sharma IP, Verma A, Singh H (2017) Elucidating stress proteins in rice (Oryza sativa L.) genotype under elevated temperature: a proteomic approach to understand heat stress response. 3 Biotech 7:205

Kumar N, Jeena N, Singh H (2019) Elevated temperature modulates rice pollen structure: a study from foothill Himalayan agro-ecosystem in India. 3 Biotech 9:175

Kumar A, Dwivedi GK, Tewari S, Paul J, Anand R, Kumar N, Kumar P, Singh H, Kaushal R (2020a) Carbon mineralization and inorganic nitrogen pools under Terminalia chebula Retz.-based agroforestry system in Himalayan foothills, India. For Sci 66(5):634–643

Kumar A, Kumar P, Singh H, Kumar N (2020b) Adaptation and mitigation potential of roadside trees with bio-extraction of heavy metals under vehicular emissions and their impact on physiological traits during seasonal regimes. Urban For Urban Green 58:126900

Kumar A, Dwivedi GK, Tewari S, Paul J, Sah VK, Singh H, Kumar P, Kumar N, Kaushal R (2020c) Soil organic carbon pools under Terminalia chebula Retz. based agroforestry system in Himalayan foothills, India. Curr Sci 118(7):1098–1103

Kumar N, Jeena N, Kumar A, Khairakpam R, Singh H (2021a) Comparative response of rice cultivars to elevated air temperature in Bhabar region of Indian Himalaya: status on yield attributes. Heliyon 7:e07474

Kumar A, Kumar P, Singh H, Bisht S, Kumar N (2021b) Relationship of physiological plant functional traits with soil carbon stock in temperate forest of Garhwal Himalaya. Curr Sci 120(8):1368–1373

Kumar A, Kumar P, Singh H, Kumar N (2021c) Modulation of plant functional traits under essential plant nutrients during seasonal regime in natural forests of Garhwal Himalayas. Plant Soil 465:197–212

Kumar A, Tewari S, Singh H, Kumar P, Kumar N, Bisht S, Kushwaha S, Tamta N, Kaushal R (2021d) Biomass accumulation and carbon stocks in different agroforestry system prevalent in Himalayan foothills, India. Curr Sci 120(6):1083–1088

Kumar A, Kumar P, Singh H, Kumar N (2021e) Impact of plant functional traits on near saturated hydraulic conductivity of soil under different forests of Kempty watershed in Garhwal Himalaya. Indian J Soil Conserv 49(1):38–44

Kumar A, Singh H, Kumari G, Bisht S, Malik A, Kumar N, Singh M, Raturi A, Barthwal S, Thakur A, Kaushal R (2022) Adaptive resilience of roadside trees to vehicular emissions via leaf enzymatic, physiological, and anatomical trait modulations. Environ Pollut 313:120191

Laughlin DC, Mommer L, Sabatini FM, Bruelheide H, Kuyper TW, McCormack ML, Bergmann J, Freschet GT, Guerrero-Ramírez NR, Iversen CM, Kattge J (2021) Root traits explain plant species distributions along climatic gradients yet challenge the nature of ecological trade-offs. Nat Ecol Evolut 5(8):1123–1134

Lavorel S, Garnier E (2002) Predicting changes in community composition and ecosystem functioning from plant traits: revisiting the Holy Grail. Funct Ecol 16:545–556. https://doi.org/10.1046/j.1365-2435.2002.00664.x

Lenaerts B, Collard BC, Demont M (2019) Improving global food security through accelerated plant breeding. Plant Sci 287:110207

Mammola S, Carmona CP, Guillerme T, Cardoso P (2021) Concepts and applications in functional diversity. Funct Ecol 35(9):1869–1885

Menezes-Silva PE, Loram-Lourenço L, Alves RDFB, Sousa LF, Almeida SEDS, Farnese FS (2019) Different ways to die in a changing world: consequences of climate change for tree species performance and survival through an ecophysiological perspective. Ecol Evol 9(20):11979–11999

Migliavacca M, Musavi T, Mahecha MD, Nelson JA, Knauer J, Baldocchi DD et al (2021) The three major axes of terrestrial ecosystem function. Nature 598(7881):468–472

Mir RR, Reynolds M, Pinto F, Khan MA, Bhat MA (2019) High-throughput phenotyping for crop improvement in the genomics era. Plant Sci 282:60–72

Osnato M, Cota I, Nebhnani P, Cereijo U, Pelaz S (2022) Photoperiod control of plant growth: flowering time genes beyond flowering. Front Plant Sci 12:805635

Parsons AJ, Edwards GR, Newton PCD, Chapman DF, Caradus JR, Rasmussen S, Rowarth JS (2011) Past lessons and future prospects: plant breeding for yield and persistence in cool-temperate pastures. Grass Forage Sci 66(2):153–172

Pérez-Ramos IM, Matías L, Gómez-Aparicio L, Godoy Ó (2019) Functional traits and phenotypic plasticity modulate species coexistence across contrasting climatic conditions. Nat Commun 10(1):2555

Rao TR, Mitra P, Bhatt R, Goswami A (2019) The big data system, components, tools, and technologies: a survey. Knowl Inf Syst 60:1165–1245

Sarker IH (2021) Data science and analytics: an overview from data-driven smart computing, decision-making and applications perspective. SN Comput Sci 2(5):377

Schröder P, Sauvêtre A, Gnädinger F, Pesaresi P, Chmeliková L, Doğan N, Gerl G, Gökçe A, Hamel C, Millan R, Persson T (2019) Discussion paper: sustainable increase of crop production through improved technical strategies, breeding and adapted management–a European perspective. Sci Total Environ 678:146–161

Shaar-Moshe L, Hayouka R, Roessner U, Peleg Z (2019) Phenotypic and metabolic plasticity shapes life-history strategies under combinations of abiotic stresses. Plant Direct 3(1):e00113

Simonson WD, Miller E, Jones A, García-Rangel S, Thornton H, McOwen C (2021) Enhancing climate change resilience of ecological restoration—a framework for action. Perspect Ecol Conserv 19(3):300–310

Singh H, Verma A (2013a) Physiological responses of rice cultivars to various nitrogen levels. Int J Agric Environ Biotechnol 6(3):383–388

Singh H, Verma A (2013b) Characterization and screening of high nitrogen efficient rice genotype to curtail environmental pollution. Int J Agric Environ Biotechnol 6(1):23–30

Singh H, Verma A, Shukla A (2010a) Comparative study of yield and yield components of hybrid and inbred genotypes of rice (Oryza Sativa L.). Int J Agric Environ Biotechnol 3:355–360

Singh H, Verma A, Krishnamoorthy Mand Shukla A (2010b) Consequence of diverse nitrogen levels on leaf pigments in five rice genotypes under field emergent circumstance. Int J Bioresour Stress Manage 1:189–193

Singh H, Verma A, Rai SK (2013a) Biochemical evaluation of different rice genotypes grown at various nitrogen levels using SDS-PAGE. Curr Adv Agric Sci 5(1):144–146

Singh H, Verma A, Shukla A (2013b) Guttation fluid as a physiological marker for selection of nitrogen efficient rice (Oryza sativa L.) genotypes. Afr J Biotechnol 12(44):6276–6281

Singh H, Verma A, Ansari MW, Shukla A (2014) Physiological response of rice (Oryza sativa L.) genotypes to elevated nitrogen applied under field conditions. Plant Signal Behav 9:e29015. https://doi.org/10.4161/psb.29015

Singh H, Yadav M, Kumar N, Kumar A, Kumar M (2020) Assessing adaptation and mitigation potential of roadside trees under the influence of vehicular emissions: a case study of *Grevillea robusta* and *Mangifera indica* planted in an urban city of India. PLoS One 15(1):e0227380

Singh S, Singh H, Sharma SK, Nautiyal R (2021) Seasonal variation in biochemical responses of bamboo clones in the sub-tropical climate of Indian Himalayan foothills. Heliyon 7(4):e06859. https://doi.org/10.1016/j.heliyon.2021.e06859

Singh M, Singh H, Kumar A, Kumar M, Barthwal S, Thakur A (2024) Soil nitrogen availability determines the CO_2 fertilization effect on tree species (*Neolamarckia cadamba*): growth and physiological evidence. Environ Sustain. https://doi.org/10.1007/s42398-023-00300-w

Smith DT, Potgieter AB, Chapman SC (2021) Scaling up high-throughput phenotyping for abiotic stress selection in the field. Theor Appl Genet 134(6):1845–1866

Sofi II, Verma S, Charles B, Ganie AH, Sharma N, Shah MA (2022) Predicting distribution and range dynamics of *Trillium govanianum* under climate change and growing human footprint for targeted conservation. Plant Ecol 223:53–69

van Kuijk T, Biesmeijer JC, van der Hoorn BB, Verdonschot PF (2021) Functional traits explain crayfish invasive success in The Netherlands. Sci Rep 11(1):2772

Violle C, Navas ML, Vile D, Kazakou E, Fortunel C, Hummel I, Garnier E (2007) Let the concept of trait be functional! Oikos 116:882–892. https://doi.org/10.1111/j.2007.0030-1299.15559.x

Violle C, Enquist BJ, McGill BJ, Jiang L, Albert CH, Hulshof C, Jung V, Messier J (2012) The return of the variance: intraspecific variability in community ecology. Trends Ecol Evol 27:244–252. https://doi.org/10.1016/j.tree.2011.11.014

Westerband AC, Funk JL, Barton KE (2021) Intraspecific trait variation in plants: a renewed focus on its role in ecological processes. Ann Bot 127(4):397–410

Wright M, Siegel DS, Mustar P (2017) An emerging ecosystem for student start-ups. The J TechTrans 42:909–922

Wright IJ, Reich PB, Westoby M, Ackerly DD, Baruch Z, Bongers F, Cavender-Bares J, Chapin T, Cornelissen JHC, Diemer M, Flexas J, Garnier E, Groom PK, Gulias J, Hikosaka K, Lamont BB, Lee T, Lee W, Lusk C, Midgley JJ, Navas M-L, Niinemets U, Oleksyn J, Osada N, Poorter H, Poot P, Prior L, Pyankov VI, Roumet C, Thomas SC, Tjoelker MG, Veneklaas EJ, Villar R (2004) The worldwide leaf economics spectrum. Nature 428:821–827. https://doi.org/10.1038/nature02403

Yang Y, Wang G, Xiong Y (2015) The contribution of intraspecific trait variability to plant community assembly patterns on the niche-neutral continuum. Pak J Bot 47:1039–1050

Zhou Y, Wigley BJ, Case MF, Coetsee C, Staver AC (2020) Rooting depth as a key woody functional trait in savannas. New Phytol 227(5):1350–1361

Integration of Plant Functional Traits in Sustainable Agriculture

17

Satish Chandra

Abstract

Trait is defined as a characteristic measured at an individual organism's morphological, phenological, physiological, behavioral, or cultural level or another relevant level of organization. Functional trait is any morphological, phenological, and physiological trait that impacts an individual's fitness via its effect on vegetative biomass, reproductive output, and survival. Plant morphological traits include growth form, plant height, number of branches, length and width of the leaf (leaf surface), leaf number, inflorescence pattern, flower size, flower shape, fruit size, fruit shape, fruit color, number of seeds, seed color, and seed mass. Root traits include root type, rooting depth, root hairs, root branching, association with mycorrhiza, and other microbes. Physiological traits are procedures and functions that help plants acquire nutrients and other resources.

Physiological traits include rate of photosynthesis, chlorophyll contents, light saturation point, rate of transpiration, stomatal conductance, leaf temperature, water-use efficiency, and respiration. Phenological traits include different events recorded throughout the year in the plants. Functional traits at the community, population, and individual levels play pivotal roles in ecosystem functioning and ecosystem services. For example, leguminous species are essential in fixing atmospheric nitrogen in the soil and affecting the agricultural ecosystems' functioning and productivity. In sustainable agriculture, crop productivity is increased by utilizing organic farming and other nature-friendly agricultural practices. A trait-based approach in agriculture can lead us to understand better the interactions between species, crop-environment interaction, the impact of agricultural practices and crop domestication on the environment, and trends in crop produc-

S. Chandra (✉)
Department of Botany, Pandit Shivram Government Degree College Tyuni, Dehradun, Uttarakhand, India

N. Kumar, H. Singh (eds.), *Plant Functional Traits for Improving Productivity*, https://doi.org/10.1007/978-981-97-1510-7_17

tivity. In the present chapter, the author tried to reconcile the functional traits in the sustainable agriculture system.

Keywords

Agroecology · Climate change · Productivity · Rice · Sustainable agriculture · Traits · Agricultural crops

17.1 Introduction

In ecology, trait is defined as a measurable characteristic that is measured at morphological, physiological, phonological, behavioral, or cultural level of an individual organism or other relevant levels of organization (Dawson et al. 2021). Trait responses vary as per the environmental conditions and have an effect on the environmental conditions. Functional trait is any morphological, phenological, or physiological trait which impact fitness of an individual via its effect on vegetative biomass, reproductive output, and survival (Violle et al. 2007; Kumar et al. 2015, 2016; Singh et al. 2020).

Plant morphological traits include above- and belowground features. Aboveground traits include plant growth form, plant height, number of branches, length and width of the leaf (leaf surface), leaf number, inflorescence pattern, flower size, flower shape, fruit size, fruit shape, fruit color, number of seeds, seed color, and seed mass. Root traits are considered as belowground features. Root traits include root type, rooting depth, root hairs, root branching, and association with mycorrhiza and other microbes (Portis et al. 2015; Burylo et al. 2009; Paez-Garcia et al. 2015; Pierce et al. 2014; Giri et al. 2023). Physiological traits are procedures and functions that help plants in the acquisition of nutrients and other resources. Physiological traits include rate of photosynthesis, chlorophyll contents, light saturation point, rate of transpiration, stomatal conductance, leaf temperature, water-use efficiency, and respiration. (Singh et al. 2010a, b, 2013a, b; Singh and Verma 2013a, b). Physiological traits define growth condition and environmental condition of the plant and their effect imparted on the morphological traits (Foulkes et al. 2007; Laliberté 2017; Cossani and Reynolds 2012). Phenological traits includes different events recorded throughout the year in the plants. These are initial growth, onset of leaf unfolding, onset of flowering, end of flowering, flowering duration, peak of flowering, onset of fruiting, end of fruiting, fruiting duration, onset of senescence, end of senescence, growing season length, etc. Phenological traits have impact on environmental conditions and are influenced by changes in climatic conditions (Sporbert et al. 2022; Kumar et al. 2017, 2019).

Agroecology is a branch of science that deals with the ecological approach to agriculture. It has great importance today. One of the goals of agroecology is sustainable agriculture (Wezel et al. 2009). Sustainable agriculture is envisioned around the central theme that states "meet the needs of the present without compromising the ability of future generations to meet their own needs" (Brodt et al. 2011). Sustainability in agriculture can be achieved by utilizing environmental resources in way that is economically profitable and brings social and economic equality without polluting and overexploiting the environment. In sustainable agriculture, crop productivity is increased by utilizing organic farming and other nature-friendly agricultural practices (Singh et al. 2014, 2021, 2024). It also focused on lesser use of chemical fertilizers and pesticides. The negative impact of agriculture on the broader ecosystem is minimized to certain extent that should not beyond the field's edge. In this system of agriculture, biologically integrated practices and internal cycling of nutrients are promoted (Brodt et al. 2011).

In the present scenario, agricultural sustainability is important for human existence. Various functional traits of different crop species are exploited commercially. Harnessing sustainably the functional traits of different species is the need of the hour.

17.2 Functional Traits and Agriculture

Functional traits at community, population, and individual levels play a pivotal role in the ecosystem functioning and ecosystem services. For example, atmospheric nitrogen fixation in the soil by a group of leguminous species is important for functioning of the agricultural ecosystem and their productivity. However, functional traits may be accessed at different levels. As in the aforesaid example, the group of the leguminous species (community) have the ability to fix atmospheric nitrogen, though in the community at the species level (population), their average nitrogen fixation rate varies, and at the individual level, their average nitrogen fixation rate may also vary depending considerably on the morphological and environmental factors. Trait-based approach in the agriculture can lead us to better understanding of the species interaction (Wood et al. 2015). As in the intercropping of the maize with cowpeas, traits such as nitrogen fixation and biomass production are defined. Maize in the rhizosphere releases phytosiderophores that chelate iron and other minerals. On the contrary, legumes release protons in the rhizosphere which acidify the environment and increase the uptake of minerals through the activity of ferric reductase (Namatsheve et al. 2021). Increased nitrogen fixation by the leguminous plants results into increased biomass production or protein content of the maize (Wood et al. 2015). Similarly, pea-barley and faba bean-wheat mixed culture have some trait-specific outcomes. In the mixture, cereals benefit more than legumes by increasing their nitrogen uptake efficiency. Traits such as grain yield, tiller formation, and shoot biomass production in cereals positively correlated to nitrogen (Ajal et al. 2021). Trait-based approach helps us to understand the relationship between two and more traits. Traditionally, it is believed that higher leaf level resource acquisition resulted into better plant productivity. However, a study on the *Coffea arabica* variety shows that plants expressing increased leaf-level resource acquisition have lower yield (Gagliardi et al. 2015). Thus, trait-based research can be useful for understanding the interaction between different traits of different species.

Trait-based study reveals that during the domestication of the herbaceous crops increased size (biomass) and efficiently compete with weeds for light which is an evolutionary trajectory and expected in earlier resource-rich environments. Plants aggressively compete with their neighbors for light and divert their resources to stems which resulted in the elongation of crops. It diminished the community productivity and increased the risk of lodging in the agricultural crop. To produce more productive crops, during green revolution, breeding programs were performed to shorten the plant height. In the wild, relatives of the crop species traits such as total plant dry mass, growth rate of seedlings, canopy height, leaf area, root length, and root density are highly coordinated. However, in the recent breeding, coordinated evolution of the traits is decreased and consequently new trait combinations are formed which are harnessed as new varieties. These varieties are highly productive in the controlled environment but vulnerable to stress or climate change. Moreover, disruption of coordination between traits at the genetic and biochemical level provides opportunity to develop new combinations of the desired traits through breeding and genetic engineering (Milla et al. 2014; Kumar et al. 2021a, 2021b, 2021c, 2021d, 2021e). In the domestication of the tetraploid wheat *Triticum turgidum*, Roucou et al. (2018) observed decreased biomass in roots and increased shoot parts over the course of domestication. With domestication, traits such as longevity of leaves and structural defence reduce over a period of time, while metabolic rate and nutrient content of the leaves increased. In the ancestral form of the species, aboveground traits and belowground traits are coordinated. However, human selection has broken this coordination; thus, the modern form of the species invests more on the aboveground traits than the belowground traits. It is postulated that domestication has a direct effect on the seed enlargement of the domesticated species. Seeds of the domesticated vegetable crops are about 20–25% larger than their wild ancestors. Though seeds of these crops are not harvested for consumption, the edible parts such as stems, leaves, and roots are consumed. However, it is also proposed that seed size enlargement is associated with burying of seeds deeper in the farms that would germinate more effectively and produce larger vigorous seedlings with greater fitness and more competitive plants (Purugganan 2019).

During the process of domestication, co-evolutionary relationship among species has changed. Domestication benefited all the parties (associated species). In the altered ecology of the domestication process, dramatic niche modification has been increasing. Novel artificial ecological environment induces plastic response in the crops compared to naturally occurring mutations. Natural mutations are slow and their transmission is reliant on the individuals and their reproductive fitness. In the artificially formed environment (agriculture), niche differentiation directly affects a number of individuals simultaneously, and it is comparatively a speedy process. During the domestication process, certain traits such as water use, tilling, and grain shape and number of a particular crop are favored over other traits. Artificial engineering of ecology and artificial selection has an impact on the structure and productivity of the ecosystem (Bates 2022; Zeder 2017).

In the agricultural ecosystem, farmers favor certain functional traits over other traits that may or may not be related to the crop yield. For instance, crop yield in the rice cultivation was given second preference by farmers in West Africa over plant height and phenology (Dalton 2004). Traditionally, cocoa farmers in West Africa favor shade canopy trait that positively affects cocoa biomass by improved light regulation and nutrient cycling. However, it is believed that excessive shade contributes to the reduction in the productivity of the cocoa (Isaac et al. 2008).

17.3 Trait-Based Approach and Evolution of the Crop Varieties

17.3.1 Evolution of the Rice Varieties

Rice is an important crop and over 3.5 billion people around the world have been dependent on it as a staple food. To meet the demand of increasing population, by 2050, it is estimated that rice cultivation needs to be increased around 25%. Ecologically, rice cultivation is a highly water demanding system. Due to the present environmental conditions, water availability is shrinking. Thus, highly water consuming rice cultivation is not sustainable. Therefore, rice cultivation needs to be shifted from more water consuming system to less water consuming system (Bates 2022). Rice belongs to the genus *Oryza* and comprises more than 20 species. Cytogenetically, *Oryza* has 11 genome types, 6 are diploid (AA, BB, CC, EE, FF, and GG), and the rest are tetraploid (BBCC, CCDD, HHJJ, HHKK, and KKLL). There are two species of cultivated rice; Asian (*O. sativa*) and African (O. *glaberrima*) rice are AA-genome species. Both are domesticated separately from the wild species *O. rufipogon* and *O. barthii*, respectively (Xu and Sun 2021). Rice domestication began around 8000 to 9000 years ago (Callaway 2014). During the course of the domestication of the rice, certain traits are favored over other traits by farmers. For instance, in the Indian subcontinent, people favor of rice flavor, long grain, and non-glutinous traits. Consequently, most of the Indian rice varieties have these traits. However, in the East Asian countries, glutinous trait of rice is predominant (Jiang et al. 2022; Sharma et al. 2021a, b). In grasses, seeds after maturation usually shatter from the spikelet (ear) and necessary for the guaranteed propagation in the wild. Shattering is related to the reduced plant yield and therefore artificial selection by humans does not favor it. Loss of grain-shattering habit was one of the key steps of rice domestication. Guaranteeing propagation of wild rice was assured through natural grain shattering after maturity. Grain shattering is not beneficial for harvesting. In cultivated rice, loss of the grain-shattering habit increased its harvest and reduces yield loss (Xu and Sun 2021). During domestication, sickle harvesting leads to the selection of the non-shattering ear traits (seeds do not fall off), and consequently human interaction with the rice triggers coevolution of the species as per human desirable traits (Bates 2022). The progenitors of the rice had prostrate growth habit that was necessary for its survival and propagation. Wild rice has loose panicles (ear) with lesser and smaller grains, while cultivated rice has compact panicle with more and larger grains. In the wild rice, grains are with black hull and the pericarp is red. But in the cultivated rice, hull is golden colored or yellowish and the pericarp is generally white (only a small fraction has a red pericarp). In the wild form of rice, grains exhibit dormancy that leads to the desynchronized germination of species, and it is essential for the propagation and survival of the species in the unplanned wild environment. Cultivated rice exhibit weak to moderate dormancy and that is required for the synchronized germination and harvest of the crop. After domestication, farmers selected the erect form that was required for the increased plant density in the limited farmed area and therefore yield of the crop. Some of the other traits such as erect growth form, shortened awns or awnlessless, and compact panicles are favored by humans during early steps of rice domestication. However, traits like number, color size and dormancy of grain, and ecological adaptability of the cultivars were selected during the late domestication or the improvement phase of the crop (Kumar et al. 2020a, b, Xu and Sun 2021).

Oryza nivara, which is a descendant of *O. rufipogon* and a progenitor of the Asian rice, was grown in the fringes of pools of the water in the wild. It is an annual species and exposed to the frequent grassing and trampling. Thus, over the course of evolution, the species produces more seeds for the survival and compacted soil in which seed can survive and germinate. These traits were crucial for the development of the modern rice. Anthropogenic selection favored high-seed producing populations during cultivation of rice that ultimately impacted productivity of the ecosystem (Vaughan et al. 2008). Scientists presume that the earlier rice ((proto-) *indica*) of the Indian subcontinent were grown in the dry conditions and had low productivity. However, the East Asian species (*japonica*) were grown in the wet conditions and had comparatively high productivity. Because of hybridization, the wet condition is more favored by the Indian rice (Fig. 17.1). Thus, domestication of the rice has a great impact on the availability and management of water and water availability has shaped various traits of the rice. However, this generalization is still debated and need more scientific data to support (Bates 2022).

With the advent of the Green Revolution in the mid-twentieth century, scientists favored some morphological, physiological, and phenological traits of the rice over other traits for the improvement of the varieties. Some of the favored traits were earliness of harvest, deep water and flood resistant, drought resistant, lodging resistant, dormancy of seed, non-shredding of grains, control of wild rice, disease resistant, and higher response to heavy manuring (https://drdpat.bih.nic.in/Rice%20Varieties%20in%20India.htm). Consequently, surplus productivity was achieved but it also enhanced use of some unsustainable practices. Monoculture cultivation, loss of genetic diversity, intensive use of fertilizers, diminishing water resources, loss of soil fertility, soil erosion, underground water pollution, various incidence of human and livestock diseases, and global warming are some disadvantages of the innovations used during the Green Revolution (Rahman 2015). High-yielding, dwarf varieties of rice were developed during the Green Revolution. Intensive use of fertilizers in the traditional varieties made them prone to lodging (Hirano et al. 2017), and intensive use of underground water made the soil saline (Rahman 2015).

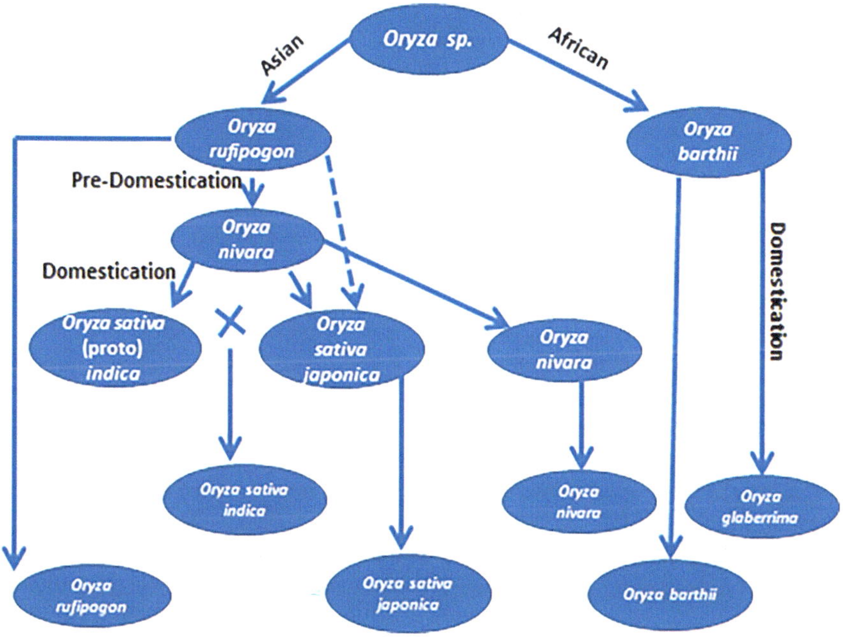

Fig. 17.1 Evolution of the modern rice from the progenitors (dotted lines for japonica rice indicates origination of species. Some literature trace their ancestry to *O. nivara*, while others to *O. rufipogon*)

Some of the wild relatives of the cultivated rice have physiological traits that can be exploited for the improvement of the rice. *Oryza alta* wild species of rice are grown in the river basins, in canals, and in open and sunny areas of South American regions. It is tolerant to salinity and resistant to striped stem borer and bacterial grain rot disease. Gene editing of the species forms shorter plants, larger grains, non-shattering, and optimized flowering time. It is also an important gene pool for improving *O. sativa* (Gireesh 2018; Fornasiero et al. 2022). *Oryza coarctata* is also a wild rice grown in the coastal regions of South Asia, from Pakistan to Myanmar. The species is adapted to grow naturally under saline conditions. Salinity stress is a global threat to the productivity of the rice and *O. coarctata* can be a solution to this problem (Fornasiero et al. 2022). For the sustainability of the rice cultivation, traits such as insect pests, fungal and bacterial disease resistance, and tolerance to drought, high temperature, salinity, and flooding are important. Various landraces that are adapted to specific environmental conditions are important sources of such genetic diversity (Kumar et al. 2022; Fornasiero et al. 2022).

17.3.2 Evolution of the Wheat Varieties

Wheat is an annual grass species and one of the staple foods of the human civilization. Wheat is cultivated in around 89 countries and part of the diet of about 2.5 billion people. There are two cultivated species of wheat; one is the bread wheat (*Triticum aestivum*) that is grown in about 90–95% of the total wheat cropped area and the other is the durum wheat (*Triticum durum*) that forms the remaining 5–10% area. However, there are several other species of wheat grown at local levels by farmers in different parts of the globe (Gohar et al. 2022; Haas et al. 2019; Levy and Feldman 2022).

Evolution of wheat species and domestication of bread wheat have always fascinated researchers. Domestication played a pioneering role in the evolution of bread wheat. It is believed that domestication of wheat started in the Fertile Crescent area. Evolution of bread wheat was traced to the hybridization between species of the two closely related genera *Aegilops* and *Triticum*. Bread wheat (*Triticum aestivum*) is a hexaploid ($2n = 6x = 42$) and contains BBAADD genome derived from the three diploid ($2n = 2x = 14$) species. The AA genome is provided by *Triticum urartu*, the BB genome by *Aegilops speltoides*, and the DD genome by *Aegilops tauschii* (Fig. 17.2). However, the BB genome is a bit controversial, and it is also believed

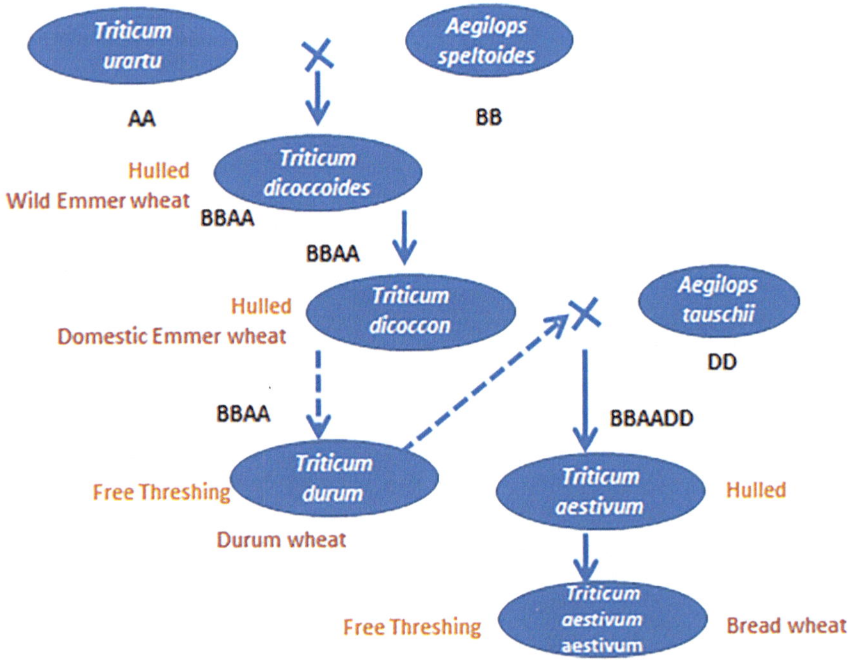

Fig. 17.2 Evolution of bread wheat. (Dotted lines from *T. dicoccon* to *T. durum* indicate evolution of the second form the first after some mutation. Similarly, it is debated which one, *T. dicoccon* or *T. durum*, provided the BBAA genome for bread wheat)

that *Aegilops speltoides* is not a direct contributor but rather contributed by the ancestor of *Aegilops speltoides* (that is yet to be discovered or extinct). Initially, *Triticum urartu* and the ancestor of *Aegilops speltoides* hybridized and formed the tetraploid *T. dicoccoides* (*T. turgidum* ssp. *dicoccoides*) ($2n = 4x = 28$, BBAA). The first cereal was domesticated in the Fertile Crescent and it is believed that wild emmer wheat was one of them. The domesticated form (emmer wheat) is known as *T. dicoccon* (*T. turgidum* ssp. *dicoccon*) ($2n = 4x = 28$, BBAA). The origin of tetraploid durum wheat, *T. durum* ((*T. turgidum* ssp. *durum*) ($2n = 4x = 28$, BBAA)), is debated and believed that hulled emmer form transformed into the naked durum. *T. aestivum* (hexaploid ($2n = 6x = 42$) BBAADD) emerged about 9000 years ago as a result of hybridization between domesticated tetraploid emmer wheat and the wild diploid *Aegilops tauschii* (($2n = 2x = 14$) DD) in the Fertile Crescent. *T. aestivum* has both hulled and naked species. *T. aestivum* ssp. *spelta* and *T. aestivum* ssp. *macha* are hulled species and *T. aestivum* ssp. a*estivum*, *T. aestivum* ssp. *compactum*, and *T. aestivum* ssp. *sphaerococcum* are free threshing or naked species. Among these, *T. aestivum* ssp. *aestivum* is the globally cultivated species and related to the food security of human civilization. Other subspecies are cultivated in the smaller areas (Gohar et al. 2022; Haas et al. 2019; Levy and Feldman 2022; Sharma et al. 2021a, b).

Earliest domesticated wheat had hulled seeds. *Triticum monococcum*, which is a diploid wheat and also known as einkorn wheat, is considered the first domesticated hulled wheat. Being a member of the grasses, wheat domestication also went through same trait transformation as discussed earlier in the case of the rice. For example, seeds after maturation usually shatter from the spikelet (ear) in the wild species, and loss of grain-shattering habit was the key step of domestication. Naked or free threshing trait was favored by earlier farmers and consequently the present day cultivated wheat (either durum or bread) has naked seeds. Some other traits such as high yielding, resistance to diseases, and resistance to changing climatic conditions were favored by early farmers. In the early1900s, some traits such as rust resistance, early flowering, early maturation, and resistance to biotic stress were the focus during wheat hybridization programs. In the 1950s–1960s, the International Maize and Wheat Improvement Center (CIMMYT), Mexico, initiated a wheat improvement program under the supervision of Norman Borlaug through the introduction of the short height gene from the semi-dwarf "Norin-10" wheat. This hybridization produces spring wheat with some new traits such as short and stiff strawed, tillered profusely, produce more grain per head, less likely to lodge, and resistant to rust. Two famous varieties from the hybridization were Sonora 64 and LermaRojo 64. These two varieties paved the way for the Green Revolution and saved millions of people from hunger (Lumpkin 2015; Gohar et al. 2022).

However, it is important to note here that the lineage of the semi-dwarf "Norin-10" wheat is traced back to the Japanese short-straw variety Daruma crossed with the two American varieties Fultz and Turkey Red. Both the Fultz and Turkey Red varieties of wheat were introduced in Japan. Here, Norin is an acronym (first letter of each word) of the Japanese Agricultural Experiment Station as spelled out using Latin letters. The Japanese wheat breeder Mr. Gonjiro Inazuka was the chief of the

Wheat Breeding Program during the hybridization process. The number 10 in the Norin 10 indicates the tenth variety produced. Genes Rht1 and Rht2 are responsible for the semi-dwarf stature of the Norin 10 (Lumpkin 2015).

Green Revolution has increased wheat yield many folds, but the nutritional quality was compromised and many old varieties disappeared. It is estimated that in the twenty-first century, almost 63% of wheat varieties contain "Green Revolution" alleles. Varieties having these alleles do not perform satisfactory in the dry and warm regions. Different races of the rusts have been evolved post Green Revolution era and threatened the growth of the high-yielding varieties. Cloning of several male sterility genes provided new tools for producing hybrid seeds in the recent years and in wheat as dominant male-sterility gene *Ms2* facilitated the evolution of several lines and cultivars. DNA markers in marker-assisted breeding for traits such as lodging, disease resistance, stress resistance kernel traits, herbicide tolerant, and vernalization response are utilized for various hybridization programs. Genome editing and genetically modified wheat are some of the recent paradigms utilized for the improvement of wheat. Glyphosate-resistant wheat, which was round-up ready (MON-71800) and developed by Monsanto in 2004, was the first genetically modified wheat. There are several transgenic wheat lines expressing DREB1A, HDG11, WRKY2, TaSHN1, NAC, and bZIP2 which have high tolerance to drought and increased yield potential. Researchers also developed many other transgenic varieties to cope up with disease epidemic and the changing environmental conditions (Gohar et al. 2022).

17.4 Sustainable Agriculture and Functional Traits

The research on how a functional trait-based approach leads to the agroecosystem functioning is in the initial stage. Sustainable agriculture is based on the sustainable use of the natural resources such as water, soil carbon, genetic diversity, and other available resources (Kumar et al. 2021a, b, c, d). As discussed in the earlier sections, that trait-based approach of crops is focused on low water consumption and utilization of genetic diversity for the development of the resistant varieties for biotic and abiotic stresses, diseases, and increased yields. Trait-based approach can help us to understand the causes and consequences of a particular ecological shift on the ecosystem functioning and services. It also helps us to understand the relationship between the process of domestication of a particular crop and the productivity of the ecosystem. It also enhances our understanding about the interaction among the species and their cumulative effect on the ecosystem productivity. Trait-based approach also deepens our understanding about evolution and coevolution of species under consideration and their impact on the surrounding environmental conditions. Further, more research need to be done in this particular field for better understanding of the trait-based approach in the agroecosystem.

17.5 Conclusion

Functional trait is any morphological, phenological, and physiological trait that impacts the fitness of an individual via its effect on vegetative biomass, reproductive output, and survival. Functional traits at community, population, and individual levels play a pivotal role in the ecosystem functioning and ecosystem services. For example, leguminous species play an important role in fixing the atmospheric nitrogen into the soil and therefore affect the functioning of the agricultural ecosystems and their productivity. In the sustainable agriculture, productivity of the crop is increased by utilizing organic farming and other nature-friendly agriculture practices. Trait-based approach in the agriculture can help us to better understand species interaction, crop-environment interaction, impact of agricultural practices, and crop domestication on environment and trend of crop productivity.

References

Ajal J, Jäck O, Vico G, Weih M (2021) Functional trait space in cereals and legumes grown in pure and mixed cultures is influenced more by cultivar identity than crop mixing. Perspect Plant Ecol Evol Syst 50:125612. https://doi.org/10.1016/j.ppees.2021.125612

Bates J (2022) The fits and starts of Indian rice domestication: how the movement of rice across Northwest India impacted domestication pathways and agricultural stories. Front Ecol Evol 10:924977. https://doi.org/10.3389/fevo.2022.924977

Brodt S, Six J, Feenstra G, Ingels C, Campbell D (2011) Sustainable agriculture. Nat Educ Knowl 3(10):1

Burylo M, Rey F, Roumet C, Buisson E, Dutoit T (2009) Linking plant morphological traits to uprooting resistance in eroded marly lands (Southern Alps, France). Plant Soil 324:31–42. https://doi.org/10.1007/s11104-009-9920-5

Callaway E (2014) Domestication: the birth of rice. Nature 514(7524):S58–S59. https://doi.org/10.1038/514S58a

Cossani CM, Reynolds MP (2012) Physiological traits for improving heat tolerance in wheat. Plant Physiol 160(4):1710–1718. https://doi.org/10.1104/pp.112.207753

Dalton TJ (2004) A household hedonic model of rice traits: economic values from farmers in West Africa. Agric Econ 31:149–159. https://doi.org/10.1016/j.agecon.2004.09.003

Dawson SK, Carmona CP, González-Suárez M, Jönsson M, Chichorro F, Mallen-Cooper M, Duthie AB (2021) The traits of "trait ecologists": an analysis of the use of trait and functional trait terminology. Ecol Evol 11(23):16434–16445. https://doi.org/10.1002/ece3.8321

Fornasiero A, Wing RA, Ronald P (2022) Rice domestication. Curr Biol 32(1):R20–R24. https://doi.org/10.1016/j.cub.2021.11.025

Foulkes MJ, Sylvester-Bradley R, Weightman R, Snape JW (2007) Identifying physiological traits associated with improved drought resistance in winter wheat. Field Crop Res 103(1):11–24. https://doi.org/10.1016/j.fcr.2007.04.007

Gagliardi S, Martin AR, VirginioFilho EDM, Rapidel B, Isaac ME (2015) Intraspecific leaf economic trait variation partially explains coffee performance across agroforestry management regimes. Agric Ecosyst Environ 200:151–160. https://doi.org/10.1016/j.agee.2014.11.014

Gireesh C (2018) *Oryza alta* Swollen. In: Mondal T, Henry R (eds) The wild Oryza genomes. Springer, Cham, pp 55–59. https://doi.org/10.1007/978-3-319-71997-9_4

Giri K, Mishra G, Suyal DC, Kumar N, Doley B, Das N, Baruah RC, Bhattacharyya R, Bora N (2023) Performance evaluation of native plant growth-promoting rhizobacteria for paddy yield enhancement in the jhum fields of Mokokchung, Nagaland, North East India. Heliyon 9:e14588

Gohar S, Sajjad M, Zulfiqar S, Liu J, Wu J, Rahman MU (2022) Domestication of newly evolved hexaploid wheat—a journey of wild grass to cultivated wheat. Front Genet 13:1022931. https://doi.org/10.3389/fgene.2022.1022931

Haas M, Schreiber M, Mascher M (2019) Domestication and crop evolution of wheat and barley: genes, genomics, and future directions. J Integr Plant Biol 61(3):204–225. https://doi.org/10.1111/jipb.12737

Hirano K, Ordonio RL, Matsuoka M (2017) Engineering the lodging resistance mechanism of post-Green Revolution rice to meet future demands. Proc Jpn Acad Ser B Phys Biol Sci 93(4):220–233. https://doi.org/10.2183/pjab.93.014

Isaac ME, Dawoe E, Sieciechowicz K (2008) Assessing local knowledge use in agroforestry management with cognitive maps. Environ Manag 43(6):1321–1329. https://doi.org/10.1007/s00267-008-9201-8

Jiang C, Rashid MAR, Zhang Y, Zhao Y, Pan Y (2022) Genome wide association study on development and evolution of glutinous rice. BMC Genomic Data 23(1):33. https://doi.org/10.1186/s12863-022-01033-1

Kumar N, Kumar N, Shukla A, Shankhdhar SC, Shankhdhar D (2015) Impact of terminal heat stress on pollen viability and yield attributes of rice (Oryza sativa L.). Cereal Res Commun 43(4):616–626

Kumar N, Shankhdhar SC, Shankhdhar D (2016) Impact of elevated temperature on antioxidant activity and membrane stability in different genotypes of rice (Oryza sativa L.). Indian J Plant Physiol 21(1):37–43

Kumar N, Suyal DC, Sharma IP, Verma A, Singh H (2017) Elucidating stress proteins in rice (Oryza sativa L.) genotype under elevated temperature: a proteomic approach to understand heat stress response. 3 Biotech 7:205

Kumar N, Jeena N, Singh H (2019) Elevated temperature modulates rice pollen structure: a study from foothill Himalayan agro-ecosystem in India. 3 Biotech 9:175

Kumar A, Dwivedi GK, Tewari S, Paul J, Anand R, Kumar N, Kumar P, Singh H, Kaushal R (2020a) Carbon mineralization and inorganic nitrogen pools under Terminalia chebula Retz.-based agroforestry system in Himalayan foothills, India. For Sci 66(5):634–643

Kumar A, Kumar P, Singh H, Kumar N (2020b) Adaptation and mitigation potential of roadside trees with bio-extraction of heavy metals under vehicular emissions and their impact on physiological traits during seasonal regimes. Urban For Urban Green 58:126900

Kumar A, Kumar P, Singh H, Bisht S, Kumar N (2021a) Relationship of physiological plant functional traits with soil carbon stock in temperate forest of Garhwal Himalaya. Curr Sci 120(8):1368–1373

Kumar A, Kumar P, Singh H, Kumar N (2021b) Modulation of plant functional traits under essential plant nutrients during seasonal regime in natural forests of Garhwal Himalayas. Plant Soil 465:197–212

Kumar A, Kumar P, Singh H, Kumar N (2021c) Impact of plant functional traits on near saturated hydraulic conductivity of soil under different forests of Kempty watershed in Garhwal Himalaya. Indian J Soil Conserv 49(1):38–44

Kumar A, Tewari S, Singh H, Kumar P, Kumar N, Bisht S, Kushwaha S, Tamta N, Kaushal R (2021d) Biomass accumulation and carbon stocks in different agroforestry system prevalent in Himalayan foothills, India. Curr Sci 120(6):1083–1088

Kumar N, Jeena N, Kumar A, Khairakpam R, Singh H (2021e) Comparative response of rice cultivars to elevated air temperature in Bhabar region of Indian Himalaya: status on yield attributes. Heliyon 7:e07474

Kumar A, Singh H, Kumari G, Bisht S, Malik A, Kumar N, Singh M, Raturi A, Barthwal S, Thakur A, Kaushal R (2022) Adaptive resilience of roadside trees to vehicular emissions via leaf enzymatic, physiological, and anatomical trait modulations. Environ Pollut 313:120191

Laliberté E (2017) Below-ground frontiers in trait-based plant ecology. New Phytol 213(4):1597–1603. https://doi.org/10.1111/nph.14247

Levy AA, Feldman M (2022) Evolution and origin of bread wheat. Plant Cell 34(7):2549–2567. https://doi.org/10.1093/plcell/koac130

Lumpkin TA (2015) How a gene from Japan revolutionized the world of wheat: CIMMYT's quest for combining genes to mitigate threats to global food security. In: Advances in wheat genetics: from genome to field: Proceedings of the 12th International Wheat Genetics Symposium. Springer, Tokyo, pp 13–20. https://doi.org/10.1007/978-4-431-55675-6_2

Milla R, Morente-López J, Miguel Alonso-Rodrigo J, Martín-Robles N, Chapin FS (2014) Shifts and disruptions in resource-use trait syndromes during the evolution of herbaceous crops. Proc Biol Sci 281:20141429. https://doi.org/10.1098/rspb.2014.1429

Namatsheve T, Chikowo R, Corbeels M, Mouquet-Rivier C, Icard-Vernière C, Cardinael R (2021) Maize-cowpea intercropping as an ecological intensification option for low input systems in sub-humid Zimbabwe: productivity, biological N2-fixation and grain mineral content. Field Crop Res 263:108052. https://doi.org/10.1016/j.fcr.2020.108052

Paez-Garcia A, Motes CM, Scheible WR, Chen R, Blancaflor EB, Monteros MJ (2015) Root traits and phenotyping strategies for plant improvement. Plan Theory 4(2):334–355. https://doi.org/10.3390/plants4020334

Pierce S, Bottinelli A, Bassani I, Ceriani RM, Cerabolini BE (2014) How well do seed production traits correlate with leaf traits, whole-plant traits and plant ecological strategies? Plant Ecol 215:1351–1359. https://doi.org/10.1007/s11258-014-0392-1

Portis E, Cericola F, Barchi L, Toppino L, Acciarri N, Pulcini L et al (2015) Association mapping for fruit, plant and leaf morphology traits in eggplant. PLoS One 10(8):e0135200. https://doi.org/10.1371/journal.pone.0135200

Purugganan MD (2019) Evolutionary insights into the nature of plant domestication. Curr Biol 29(14):R705–R714. https://doi.org/10.1016/j.cub.2019.05.053

Rahman S (2015) Green revolution in India: environmental degradation and impact on livestock. Asian J Water Environ Pollut 12(1):75–80

Roucou A, Violle C, Fort F, Roumet P, Ecarnot M, Vile D (2018) Shifts in plant functional strategies over the course of wheat domestication. J Appl Ecol 55(1):25–37. https://doi.org/10.1111/1365-2664.13029

Sharma A, Srivastava A, Singh S, Mishra S, Mohan S, Singh A, Jaiswal HK (2021a) Aromatic rice of India: it's types and breeding strategies. IntechOpen, London. https://doi.org/10.5772/intechopen.99232

Sharma S, Schulthess AW, Bassi FM, Badaeva ED, Neumann K, Graner A, Kilian B (2021b) Introducing beneficial alleles from plant genetic resources into the wheat germplasm. Biology 10(10):982. https://doi.org/10.3390/biology10100982

Singh H, Verma A (2013a) Physiological responses of rice cultivars to various nitrogen levels. Int J Agric Environ Biotechnol 6(3):383–388

Singh H, Verma A (2013b) Characterization and screening of high nitrogen efficient rice genotype to curtail environmental pollution. Int J Agric Environ Biotechnol 6(1):23–30

Singh H, Verma A, Shukla A (2010a) Comparative study of yield and yield components of hybrid and inbred genotypes of rice (Oryza Sativa L.). Int J Agric Environ Biotechnol 3:355–360

Singh H, Verma A, Krishnamoorthy Mand Shukla A (2010b) Consequence of diverse nitrogen levels on leaf pigments in five rice genotypes under field emergent circumstance. Int J Bioresour Stress Manage 1:189–193

Singh H, Verma A, Rai SK (2013a) Biochemical evaluation of different rice genotypes grown at various nitrogen levels using SDS-PAGE. Curr Adv Agric Sci 5(1):144–146

Singh H, Verma A, Shukla A (2013b) Guttation fluid as a physiological marker for selection of nitrogen efficient rice (*Oryza sativa* L.) genotypes. Afr J Biotechnol 12(44):6276–6281

Singh H, Verma A, Ansari MW, Shukla A (2014) Physiological response of rice (Oryza sativa L.) genotypes to elevated nitrogen applied under field conditions. Plant Signal Behav 9:e29015. https://doi.org/10.4161/psb.29015

Singh H, Yadav M, Kumar N, Kumar A, Kumar M (2020) Assessing adaptation and mitigation potential of roadside trees under the influence of vehicular emissions: a case study of Grevillea robusta and Mangifera indica planted in an urban city of India. PLoS One 15(1):e0227380

Singh S, Singh H, Sharma SK, Nautiyal R (2021) Seasonal variation in biochemical responses of bamboo clones in the sub-tropical climate of Indian Himalayan foothills. Heliyon 7(4):e06859. https://doi.org/10.1016/j.heliyon.2021.e06859

Singh M, Singh H, Kumar A, Kumar M, Barthwal S, Thakur A (2024) Soil nitrogen availability determines the CO2 fertilization effect on tree species (Neolamarckia cadamba): growth and physiological evidence. Environ Sustain. https://doi.org/10.1007/s42398-023-00300-w

Sporbert M, Jakubka D, Bucher SF, Hensen I, Freiberg M, Heubach K et al (2022) Functional traits influence patterns in vegetative and reproductive plant phenology—a multi-botanical garden study. New Phytol 235(6):2199–2210. https://doi.org/10.1111/nph.18345

Vaughan DA, Lu BR, Tomooka N (2008) The evolving story of rice evolution. Plant Sci 174(4):394–408. https://doi.org/10.1016/j.plantsci.2008.01.016

Violle C, Navas ML, Vile D, Kazakou E, Fortunel C, Hummel I, Garnier E (2007) Let the concept of trait be functional! Oikos 116(5):882–892. https://doi.org/10.1111/j.0030-1299.2007.15559.x

Wezel A, Bellon S, Doré T, Francis C, Vallod D, David C (2009) Agroecology as a science, a movement and a practice. A review. Agron Sustain Dev 29:503–515. https://doi.org/10.1051/agro/2009004

Wood SA, Karp DS, DeClerck F, Kremen C, Naeem S, Palm CA (2015) Functional traits in agriculture: agrobiodiversity and ecosystem services. Trends Ecol Evol 30(9):531–539. https://doi.org/10.1016/j.tree.2015.06.013

Xu R, Sun C (2021) What happened during domestication of wild to cultivated rice. Crop J 9(3):564–576. https://doi.org/10.1016/j.cj.2021.02.005

Zeder MA (2017) Domestication as a model system for the extended evolutionary synthesis. Interface Focus 7:20160133. https://doi.org/10.1098/rsfs.2016.013

Future Challenges and Opportunities in Utilizing Plant Functional Traits for Enhanced Crop Productivity

18

Sonali Gaur

Abstract

The future of agriculture faces multifaceted challenges and opportunities in harnessing plant functional traits to optimize crop productivity. As global population growth accelerates and climate change impacts agricultural ecosystems, there is an urgent need to develop resilient and high-yielding crop varieties. Plant functional traits encompass characteristics such as leaf morphology, photosynthetic efficiency, and nutrient uptake to enhance crop performance. The challenges lie in decoding the intricate interplay of these traits within diverse environmental contexts, necessitating a holistic understanding of plant biology. Additionally, advancements in technology, such as high-throughput phenotyping and precision breeding techniques, present unprecedented opportunities to accelerate trait-based crop improvement. Moreover, the integration of big data analytics and machine learning offers a promising avenue for deciphering complex trait interactions and predicting crop responses under changing environmental conditions. However, ethical considerations, regulatory frameworks, and the socioeconomic implications of adopting trait-based approaches must be carefully addressed. The pivotal role of plant functional traits in shaping the future of agriculture emphasizes the need for interdisciplinary collaborations and innovative strategies to overcome challenges and unlock the full potential of these traits for sustainable and enhanced crop productivity.

Keywords

Agricultural resilience · Functional traits · Crop productivity · Climate change · Precision breeding

S. Gaur (✉)
Department of Genetics and Plant Breeding, G.B. Pant University of Agriculture and Technology, Pantnagar, Uttarakhand, India

N. Kumar, H. Singh (eds.), *Plant Functional Traits for Improving Productivity*, https://doi.org/10.1007/978-981-97-1510-7_18

18.1 Introduction

Plant functional traits are now recognized as essential markers of plant performance and ecological strategies. Plant functional traits encompass morphological, phenological, and physiological characteristics that play a pivotal role in shaping a plant's interactions with its environment (Singh et al. 2013a, b). These traits are fundamental to an individual plant's fitness and are essential components in the development of mechanistic vegetation models (Legendre et al. 1997). The exploration of how these attributes influence the presence and abundance of species has long-standing roots in the field of ecology. Gaining insights into these traits is critical for advancing agricultural productivity, promoting biodiversity conservation, and bolstering ecosystem resilience (Singh et al. 2014, 2021, 2024).

The terms "trait" and "functional trait" were historically ambiguous, leading to occasional confusion in their usage. For instance, "trait" was employed to describe measurements at both the individual level (such as leaf traits or characteristics related to plant stature and seed mass) and at a plot (Weiher et al. 1999; Petchey 2004; Kumar et al. 2017, 2019). Another source of confusion was the application of the term "trait" to soil conditions or microbial phosphorus (Eviner 2004). To establish a unified and unambiguous definition of functional traits (Violle et al. 2007), a clear perspective is introduced: "any morphological, physiological, and phenological traits which impact fitness indirectly via their effects on growth, reproduction, and survival, the three components of individual performance."

The significant decline in biodiversity has spurred extensive research aimed at elucidating the impact of biodiversity changes on ecosystem functioning. Hence, the exploration of forthcoming challenges and prospects within the domain of plant functional traits, with a particular emphasis on their utilization to enhance plant productivity, is imperative (Singh and Verma 2013a, b). Research in ecological dynamics based on functional traits has played a crucial role in advancing our comprehension of natural plant communities. While experimental research on biodiversity and ecosystem functioning has highlighted the importance of biodiversity in various ecosystem processes, such as plant productivity, pinpointing the underlying mechanisms remains a central challenge (Singh et al. 2010a, b). Despite this, the principles of functional trait ecology have not been extensively integrated into agricultural research and management practices thus far.

To enhance this comprehension, we advocate employing a functional trait approach in agroecology that incorporates recent advancements in trait research designed for ecosystems characterized by multitrophic interactions and spatial heterogeneity for enhancing productivity. It is crucial to measure trait values under diverse environmental conditions and various agricultural management regimes to anticipate the variability of ecosystem services in response to farm practices and environmental factors. Utilizing this information is imperative for formulating management strategies that are readily applicable for farmers, enabling them to effectively cultivate agriculture that offers a spectrum of ecosystem services. Adopting a trait-based approach in agriculture can enhance our comprehension of species interactions, the interplay between crops and the environment, the effects of agricultural practices, and the influence of crop domestication on the environment, ultimately shedding light on trends in crop productivity.

Plant species exhibit diversity in morphological, anatomical, and physiological traits, even though they share a common fundamental function of gas exchange, encompassing processes such as photosynthesis, respiration, and transpiration. The particular species manifests distinct phenotypes in different environments which play a crucial role in the evolution, development, and functioning of plants within their surroundings. Consequently, it is now widely acknowledged as a significant mechanism contributing to the diversity of plant functional traits. Variations in leaf traits are observable not only across different species but also among populations of the same species and even within organs produced by a single plant. Comparable adjustments in leaf structure and form in response to the environment are evident at each of these levels, whether across species, populations, or individual plants. The utilization of plant functional traits allows researchers to establish connections between processes observed at the individual plant level and broader aspects such as plant population dynamics, species distribution patterns, coexistence, community assembly, and ultimately, ecosystem functioning.

18.2 Scope

This chapter delves into the future challenges and opportunities within the realm of plant functional traits, emphasizing their role in improving productivity. From the intricacies of climate change adaptation to the promises of synthetic biology, we navigate through the uncharted territories of plant trait research. The utilization of plant functional traits is on the rise for constructing mechanistic models capable of predicting the responses of ecological communities to both abiotic and biotic perturbations. These models also aim to forecast the impact of species on ecosystem function and services in the face of a swiftly changing world. Nevertheless, considerable challenges persist in this endeavor.

18.3 Challenges in Plant Functional Traits

Functional traits hold immense promise for advancing productivity in plants, but challenges persist. Overcoming these hurdles requires interdisciplinary collaboration, innovation in measurement techniques, ethical considerations in genetic manipulation, and a concerted effort to bridge the gap between research and practical application. As we navigate these challenges, the potential benefits of harnessing functional traits for enhanced plant productivity remain a beacon for sustainable agriculture.

18.3.1 Genetic Diversity

The genetic complexity of plant populations poses a challenge in deciphering the contribution of individual functional traits to productivity. Understanding the intricacies of genetic diversity and trait interactions remains a hurdle. Despite strides in trait research, unlocking the genetic basis of functional traits and integrating them into breeding programs remains a formidable challenge. Functional traits encompass morphological, (eco)physiological, and reproductive characteristics that have a direct impact on an individual's growth, reproduction, and survival, thereby influencing plant fitness both indirectly and directly (Nock et al. 2016; Violle et al. 2007). Acquired through direct measurements or databases, these traits serve as valuable tools for examining the responses of populations, species, communities, and ecosystems to environmental factors, such as land use and climate, by evaluating the delicate balance required to optimize traits for productivity while preserving genetic diversity.

The domestication of crops involved an evolutionary trajectory, leading to increased size (biomass) and enhanced competitiveness with weeds for light, particularly in resource-rich environments. During this process, plants exhibited aggressive competition for light with neighboring plants, allocating resources to elongated stems. However, this led to decreased community productivity and an increased risk of lodging in crops. To address this, breeding programs during the Green Revolution focused on reducing plant height, aiming to produce more productive crops. In the wild relatives of crop species, traits such as total plant dry mass, seedling growth rate, canopy height, leaf area, root length, and root density are highly synchronized. Contrastingly, recent breeding efforts have resulted in a decreased harmonized evolution of these traits, leading to the formation of new trait combinations utilized in the development of new varieties as per the requirement of the present scenario. While these varieties exhibit high productivity in controlled environments, they are susceptible to stress and climate change. Additionally, the disruption of trait coordination at the genetic and biochemical levels presents an opportunity to develop novel combinations of desired traits through breeding and genetic engineering (Milla et al. 2014). Balancing yield with other essential traits such as resilience and resource-use efficiency poses a complex dilemma.

18.3.2 Trait Trade-Offs

Many functional traits are linked, and alterations in one trait may trigger trade-offs with others. Balancing desirable traits for productivity without compromising other essential characteristics presents a significant challenge. The co-inheritance of such linked alleles makes it challenging to break undesirable linkages between traits; it poses a substantial challenge in plant breeding programs, affecting the accuracy of selection and genomic predictions. This reduces the efficacy of selection, as the breeder may inadvertently select for linked, undesirable traits along with the target traits. This poses a substantial challenge in plant breeding programs, affecting the accuracy of selection and genomic predictions. Mitigating its impact requires a multifaceted approach, including targeted recombination, diverse genetic panels, and dynamic population management.

18.3.3 Trait Quantification and Measurement

18.3.3.1 Standardized Measurements

Variability in methodologies for quantifying functional traits hinders cross-study comparisons. Developing standardized measurement protocols is imperative for creating a cohesive database for trait-based research.

18.3.3.2 High-Throughput Technologies

While high-throughput technologies offer immense potential, integrating these tools for large-scale trait quantification poses challenges. Optimizing methods for efficient trait measurement in diverse plant populations is a priority. Advancements in omics technologies provide unprecedented opportunities for trait research. However, integrating these vast datasets into a cohesive understanding of plant biology and productivity presents computational and analytical challenges. This section explores the potential and hurdles of incorporating these technologies. Advancements in high-throughput technologies offer unparalleled opportunities to explore the genomic, transcriptomic, and metabolomic basis of functional traits. However, integrating omics data into a holistic understanding of plant biology and productivity is a computational and analytical challenge that demands interdisciplinary collaboration.

Elevating plant productivity through the exploration of functional traits is significantly propelled by omics technologies, encompassing genomics, transcriptomics, proteomics, and metabolomics. Genomic insights enable the identification of trait-associated genes and unravel the intricate linkages among different traits. Transcriptomics elucidates dynamic gene expression patterns, while proteomics delves into protein functionality and posttranslational modifications. Metabolomics profiles biochemical pathways, revealing metabolic signatures linked to specific traits. Integrating omics data through systems biology approaches informs strategies for trait optimization, allowing precision breeding programs to target multiple traits simultaneously. Despite its transformative potential, this synergy encounters challenges such as data integration complexities and ethical considerations. For instance, ensuring responsible genetic manipulation and data privacy is paramount. Examples of successful integration include the development of stress-resistant crops and crops with enhanced nutritional content, demonstrating the transformative impact of this approach on sustainable agriculture. Notably, advances in gene editing technologies, such as CRISPR-Cas9, have enabled precise modifications in crop genomes, exemplified by the creation of disease-resistant plants and crops with improved yield potential.

18.3.3.3 Ethical Considerations

Genetic manipulation to enhance specific traits for productivity raises ethical concerns. Balancing the potential benefits with ethical considerations and public acceptance is a critical challenge in the implementation of genetic engineering. Balancing the potential benefits with environmental and socioeconomic impacts requires a careful and transparent approach. By implementing robust regulatory frameworks and engaging stakeholders in participatory decision-making, ethical considerations can be integrated into the fabric of agricultural innovation.

18.4 Environmental Sensitivity

18.4.1 Climate Change

Dynamic environmental conditions, especially in the context of climate change, pose challenges in predicting how functional traits will respond. Unraveling the intricacies of trait-environment interactions is crucial for adapting plants to changing climates. Understanding how plant functional traits can be harnessed to mitigate the impacts of climate change on crop yields is a pressing challenge. The intricate interactions between traits and changing environmental conditions led to diverse responses by species such as adaptation, migration, extinction, or persistence (i.e., tolerance) (Aitken et al. 2008; Dawson et al. 2011; Willis and MacDonald 2011; IPCC 2014). These various reactions are believed to be closely associated with the presence (or absence) of specific functional traits or, more precisely, trait attributes (defined as the value or state of a trait in a given location at a specific time) (Garnier et al. 2016) (e.g., Estrada et al. 2016).

Consequently, possessing "favorable" traits or trait attributes enables a species to navigate a changing climate, while the absence of these attributes can lead to population decline or local extinctions. The consequences extend to determining which ecosystem services will exhibit greater resilience to climate change, given that many of these services are interconnected with specific traits. For instance, the size and structure of root and shoot systems play a regulatory role in the climate, water, and soil dynamics of ecosystems (Garnier et al. 2020). In alpine ecosystems, traits like leaf size and pubescence influence the heat retention capacity of plants, affecting their ability to survive in extreme temperature conditions (Kumar et al. 2015, 2016, 2021a, 2021b, 2021c, 2021d, 2021e). These traits, in turn, shape interactions among alpine plant species. Quantifying heritability serves as a method to gauge the extent of trait variation attributed to genetics and assess the relative influence of natural selection on trait divergence (Geber and Griffen 2003). However, the genetic basis for the variability in functional traits remains predominantly indefinable, encompassing uncertainties about the connections among correlated traits and their adaptive capability to address climate change.

18.4.2 Multifactorial Stressors

Plants often face multiple stressors simultaneously, such as nutrient deficiencies, water scarcity, and pest infestations. Understanding how functional traits respond to and mitigate these multifactorial stressors is a complex challenge. Multifactorial stressors often result in compounded effects, impacting both the quantity and quality of crops. Functional traits such as biochemical composition, disease resistance mechanisms, and stress tolerance mechanisms become pivotal in maintaining and enhancing the quality of agricultural produce. Functional traits, including root architecture and nutrient acquisition mechanisms, play a crucial role in maintaining soil health. By choosing and breeding crops with traits that enhance nutrient cycling and soil structure, we can mitigate the negative impacts of multifactorial stressors on agricultural soils.

18.4.3 Interactions in Multispecies Communities

In natural and agricultural ecosystems, plants seldom exist in isolation. Understanding how functional traits mediate interactions in multispecies communities is a challenge. The dynamics of plant-plant, plant-microbe, and plant-pollinator interactions require further elucidation for optimizing productivity in diverse agroecosystems. Alterations in individual traits or phenotypes have repercussions beyond the performance of a plant. These shifts also influence interactions with other species within the community, providing feedback on the functioning of ecological processes. By examining trait-mediated relationships, valuable insights into the mechanisms driving community dynamics and ecological processes are gained which delves into the intricate relationships between these traits and the dynamics of plant communities.

(a) Competition

Certain functional traits, such as root architecture and nutrient acquisition strategies, can dictate the competitive interactions between plant species. For instance, a deep-rooted species may outcompete a shallow-rooted counterpart for nutrient resources in the soil.

(b) Mutualism

Mutualistic interactions between plants and other organisms, such as mycorrhizal fungi, are often influenced by specific functional traits. Mycorrhizal associations can enhance nutrient uptake, promoting mutual benefits for interacting species.

Facilitation and Allelopathy

(c) Facilitation

Plant functional traits contribute to facilitative interactions, where certain species create favorable microenvironments for others. Traits like canopy architecture can provide shade, benefiting understory plants by mitigating harsh environmental conditions.

(d) Allelopathy

Chemical traits, including the release of allelopathic compounds, can influence plant-plant interactions. For instance, the secretion of allelochemicals from some species may inhibit the growth of neighboring plants, shaping community composition.

(e) Phenological Synchrony

The timing of key life events, such as flowering or leaf senescence, represents another dimension of plant functional traits that influence interactions within multispecies communities. Synchronized phenological patterns can impact pollination dynamics and resource availability.

18.5 Implementation at Scale

18.5.1 Scaling from Individuals to Ecosystems

Translating insights from individual plant traits to ecosystem-level productivity is challenging. Bridging the gap between trait-based research and large-scale agricultural practices requires a comprehensive understanding of trait scaling.

18.5.2 Farmer Adoption

Encouraging farmers to adopt practices based on functional traits demands effective communication and education. Bridging the gap between scientific knowledge and on-the-ground application poses a challenge in enhancing productivity on a broader scale. Traditional methods and long-established routines may create resistance to adopting new approaches. Organizing on-field demonstrations and pilot projects showcasing the positive outcomes of functional trait adoption can serve as powerful persuaders. Seeing tangible benefits firsthand can overcome skepticism and encourage wider adoption.

18.6 Opportunities

18.6.1 Precision Agriculture

The era of precision agriculture beckons, leveraging plant functional traits to tailor management practices at the individual plant level. High-resolution data, combined with machine learning algorithms, can optimize resource-use efficiency, reduce environmental impacts, and enhance overall productivity.

18.6.2 Synthetic Biology and Trait Engineering

Synthetic biology offers unprecedented opportunities for trait engineering. Modifying and optimizing plant functional traits through genetic manipulation holds promise for tailoring crops to specific environmental conditions, improving resistance to pests and diseases, and boosting overall yield.

18.6.3 Ecosystem Services and Agroecology

Harnessing the ecosystem services provided by diverse functional traits contributes not only to increased productivity but also to sustainable agriculture. Promoting agroecological practices that capitalize on plant diversity and trait complementarity can foster resilient and productive agroecosystems (Kumar et al. 2020a, b, c).

18.6.4 Global Collaborations and Data Sharing

The digital age facilitates global collaborations and data sharing, providing a platform for researchers to pool resources and accelerate discoveries. International initiatives focused on building trait databases and sharing research findings can expedite our understanding of plant functional traits and their implications for productivity.

18.7 Conclusion

The future of plant functional trait research for improving productivity lies at the intersection of cutting-edge science, technological innovation, and global collaboration. Addressing the challenges and capitalizing on the opportunities outlined in this chapter will propel us toward sustainable and resilient agricultural systems, ensuring food security in an era of unprecedented environmental change. As we navigate these uncharted territories, interdisciplinary collaboration and a commitment to ethical and sustainable practices will be our guiding beacons.

References

Aitken SN, Yeaman S, Holliday JA, Wang T, Curtis-McLane S (2008) Adaptation, migration or extirpation: climate change outcomes for tree populations. Evol Appl 1(1):95–111

Dawson TP, Jackson ST, House JI, Prentice IC, Mace GM (2011) Beyond predictions: biodiversity conservation in a changing climate. Science 332(6025):53–58

Estrada M, Burnett M, Campbell AG, Campbell PB, Denetclaw WF, Gutiérrez CG et al (2016) Improving underrepresented minority student persistence in STEM. CBE Life Sci Educ 15(3):es5

Eviner VT (2004) Plant traits that influence ecosystem processes vary independently among species. Ecology 85(8):2215–2229

Garnier E, Navas ML, Grigulis K (2016) Plant functional diversity: organism traits, community structure, and ecosystem properties. Oxford University Press, Oxford

Garnier A, Hulot FD, Petchey OL (2020) Manipulating the strength of organism–environment feedback increases nonlinearity and apparent hysteresis of ecosystem response to environmental change. Ecol Evol 10(12):5527–5543

Geber MA, Griffen LR (2003) Inheritance and natural selection on functional traits. Int J Plant Sci 164(S3):S21–S42

IPCC (2014) Fifth Assessment Report (AR5). Cambridge University Press, Cambridge

Kumar N, Kumar N, Shukla A, Shankhdhar SC, Shankhdhar D (2015) Impact of terminal heat stress on pollen viability and yield attributes of rice (Oryza sativa L.). Cereal Res Commun 43(4):616–626

Kumar N, Shankhdhar SC, Shankhdhar D (2016) Impact of elevated temperature on antioxidant activity and membrane stability in different genotypes of rice (Oryza sativa L.). Indian J Plant Physiol 21(1):37–43

Kumar N, Suyal DC, Sharma IP, Verma A, Singh H (2017) Elucidating stress proteins in rice (Oryza sativa L.) genotype under elevated temperature: a proteomic approach to understand heat stress response. 3 Biotech 7:205

Kumar N, Jeena N, Singh H (2019) Elevated temperature modulates rice pollen structure: a study from foothill Himalayan agro-ecosystem in India. 3 Biotech 9:175

Kumar A, Dwivedi GK, Tewari S, Paul J, Anand R, Kumar N, Kumar P, Singh H, Kaushal R (2020a) Carbon mineralization and inorganic nitrogen pools under Terminalia chebula Retz.-based agroforestry system in Himalayan foothills, India. For Sci 66(5):634–643

Kumar A, Dwivedi GK, Tewari S, Paul J, Sah VK, Singh H, Kumar P, Kumar N, Kaushal R (2020b) Soil organic carbon pools under Terminalia chebula Retz. based agroforestry system in Himalayan foothills, India. Curr Sci 118(7):1098–1103

Kumar A, Kumar P, Singh H, Kumar N (2020c) Adaptation and mitigation potential of roadside trees with bio-extraction of heavy metals under vehicular emissions and their impact on physiological traits during seasonal regimes. Urban For Urban Green 58:126900

Kumar A, Kumar P, Singh H, Bisht S, Kumar N (2021a) Relationship of physiological plant functional traits with soil carbon stock in temperate forest of Garhwal Himalaya. Curr Sci 120(8):1368–1373

Kumar A, Kumar P, Singh H, Kumar N (2021b) Modulation of plant functional traits under essential plant nutrients during seasonal regime in natural forests of Garhwal Himalayas. Plant Soil 465:197–212

Kumar A, Kumar P, Singh H, Kumar N (2021c) Impact of plant functional traits on near saturated hydraulic conductivity of soil under different forests of Kempty watershed in Garhwal Himalaya. Indian J Soil Conserv 49(1):38–44

Kumar A, Tewari S, Singh H, Kumar P, Kumar N, Bisht S, Kushwaha S, Tamta N, Kaushal R (2021d) Biomass accumulation and carbon stocks in different agroforestry system prevalent in Himalayan foothills, India. Curr Sci 120(6):1083–1088

Kumar N, Jeena N, Kumar A, Khairakpam R, Singh H (2021e) Comparative response of rice cultivars to elevated air temperature in Bhabar region of Indian Himalaya: status on yield attributes. Heliyon 7:e07474

Legendre P, Galzin R, Harmelin-Vivien ML (1997) Relating behavior to habitat: solutions to the fourth-corner problem. Ecology 78(2):547–562

Milla R, Morente-López J, Alonso-Rodrigo JM, Martín-Robles N, Stuart Chapin F III (2014) Shifts and disruptions in resource-use trait syndromes during the evolution of herbaceous crops. Proc R Soc B 281:20141429

Nock CA, Vogt RJ, Beisner BE (2016) Functional traits. In: eLS. Wiley, Hoboken, pp 1–8

Petchey O (2004) On the statistical significance of functional diversity effects. Funct Ecol 18:297–303

Singh H, Verma A (2013a) Physiological responses of rice cultivars to various nitrogen levels. Int JAgric Environ Biotechnol 6(3):383–388

Singh H, Verma A (2013b) Characterization and screening of high nitrogen efficient rice genotype to curtail environmental pollution. Int J Agric Environ Biotechnol 6(1):23–30

Singh H, Verma A, Shukla A (2010a) Comparative study of yield and yield components of hybrid and inbred genotypes of rice (Oryza Sativa L.). Int J Agric Environ Biotechnol 3:355–360

Singh H, Verma A, Krishnamoorthy Mand Shukla A (2010b) Consequence of diverse nitrogen levels on leaf pigments in five rice genotypes under field emergent circumstance. Int J Bioresour Stress Manage 1:189–193

Singh H, Verma A, Rai SK (2013a) Biochemical evaluation of different rice genotypes grown at various nitrogen levels using SDS-PAGE. Curr Adv Agric Sci 5(1):144–146

Singh H, Verma A, Shukla A (2013b) Guttation fluid as a physiological marker for selection of nitrogen efficient rice (Oryza sativa L.) genotypes. Afr J Biotechnol 12(44):6276–6281

Singh H, Verma A, Ansari MW, Shukla A (2014) Physiological response of rice (Oryza sativa L.) genotypes to elevated nitrogen applied under field conditions. Plant Signal Behav 9:e29015. https://doi.org/10.4161/psb.29015

Singh S, Singh H, Sharma SK, Nautiyal R (2021) Seasonal variation in biochemical responses of bamboo clones in the sub-tropical climate of Indian Himalayan foothills. Heliyon 7(4):e06859. https://doi.org/10.1016/j.heliyon.2021.e06859

Singh M, Singh H, Kumar A, Kumar M, Barthwal S, Thakur A (2024) Soil nitrogen availability determines the CO2 fertilization effect on tree species (Neolamarckia cadamba): growth and physiological evidence. Environ Sustain. https://doi.org/10.1007/s42398-023-00300-w

Violle C, Navas ML, Vile D, Kazakou E, Fortunel C, Hummel I, Garnier E (2007) Let the concept of trait be functional! Oikos 116(5):882–892

Weiher E, Van Der Werf A, Thompson K, Roderick M, Garnier E, Eriksson O (1999) Challenging Theophrastus: a common core list of plant traits for functional ecology. J Veg Sci 10(5):609–620

Willis KJ, MacDonald GM (2011) Long-term ecological records and their relevance to climate change predictions for a warmer world. Annu Rev Ecol Evol Syst 42:267–287

SPRINGER NATURE

GPSR Compliance

The European Union's (EU) General Product Safety Regulation (GPSR) is a set of rules that requires consumer products to be safe and our obligations to ensure this.

If you have any concerns about our products, you can contact us on ProductSafety@springernature.com

In case Publisher is established outside the EU, the EU authorized representative is:

Springer Nature Customer Service Center GmbH
Europaplatz 3
69115 Heidelberg, Germany

The manufacturer's authorised representative in the EU is Springer Nature Customer Service Centre GmbH, Europaplatz 3, 69115 Heidelberg, Germany. If you have any concerns regarding our products, please contact ProductSafety@springernature.com

Printed and bound by CPI Group (UK) Ltd, Croydon, CR0 4YY
01/10/2026
02225935-0007